T0399882

Distributed Artificial Intelligence

Internet of Everything (IoE): Security and Privacy Paradigm

Series Editors: Vijender Kumar Solanki, Raghvendra Kumar, and Le Hoang Son

IOT
Security and Privacy Paradigm
Edited by Souvik Pal, Vicente Garcia Diaz and Dac-Nhuong Le

Smart Innovation of Web of Things
Edited by Vijender Kumar Solanki, Raghvendra Kumar and Le Hoang Son

Big Data, IoT, and Machine Learning
Tools and Applications
Rashmi Agrawal, Marcin Paprzycki, and Neha Gupta

Internet of Everything and Big Data
Major Challenges in Smart Cities
Edited by Salah-ddine Krit, Mohamed Elhoseny, Valentina Emilia Balas, Rachid Benlamri, and Marius M. Balas

Bitcoin and Blockchain
History and Current Applications
Edited by Sandeep Kumar Panda, Ahmed A. Elngar, Valentina Emilia Balas, and Mohammed Kayed

Privacy Vulnerabilities and Data Security Challenges in the IoT
Edited by Shivani Agarwal, Sandhya Makkar, and Tran Duc Tan

Handbook of IoT and Blockchain
Methods, Solutions, and Recent Advancements
Edited by Brojo Kishore Mishra, Sanjay Kumar Kuanar, Sheng-Lung Peng, and Daniel D. Dasig, Jr.

Blockchain Technology
Fundamentals, Applications, and Case Studies
Edited by E Golden Julie, J. Jesu Vedha Nayahi, and Noor Zaman Jhanjhi

Data Security in Internet of Things Based RFID and WSN Systems Applications
Edited by Rohit Sharma, Rajendra Prasad Mahapatra, and Korhan Cengiz

Securing IoT and Big Data
Next Generation Intelligence
Edited by Vijayalakshmi Saravanan, Anpalagan Alagan, T. Poongodi, and Firoz Khan

Distributed Artificial Intelligence
A Modern Approach
Edited by Satya Prakash Yadav, Dharmendra Prasad Mahato, and Nguyen Thi Dieu Linh

Security and Trust Issues in Internet of Things
Blockchain to the Rescue
Edited by Sudhir Kumar Sharma, Bharat Bhushan, and Bhuvan Unhelkar

For more information about this series, please visit: HYPERLINK "https://www.crcpress.com/Internet-of-Everything-IoE-Security-and-Privacy-Paradigm/book-series/CRCIOESPP" https://www.crcpress.com/Internet-of-Everything-IoE-Security-and-Privacy-Paradigm/book-series/CRCIOESPP

Distributed Artificial Intelligence

A Modern Approach

Edited by
Satya Prakash Yadav, Dharmendra Prasad Mahato,
and Nguyen Thi Dieu Linh

CRC Press is an imprint of the
Taylor & Francis Group, an **informa** business

First edition published 2021
by CRC Press
6000 Broken Sound Parkway NW, Suite 300, Boca Raton, FL 33487-2742

and by CRC Press
2 Park Square, Milton Park, Abingdon, Oxon, OX14 4RN

Library of Congress Cataloging-in-Publication Data

Names: Yadav, Satya Prakash, editor.
Title: Distributed artificial intelligence : a modern approach / edited by
Satya Prakash Yadav, Dharmendra Prasad Mahato and Nguyen Thi Dieu Linh.
Description: First edition. | Boca Raton : CRC Press, 2020. | Series:
Internet of everything (ioe): security and privacy paradigm | Includes
bibliographical references and index.
Identifiers: LCCN 2020028557 (print) | LCCN 2020028558 (ebook) | ISBN
9780367466657 (hbk) | ISBN 9781003038467 (ebk)
Subjects: LCSH: Distributed artificial intelligence.
Classification: LCC Q337 .D574 2020 (print) | LCC Q337 (ebook) | DDC
006.30285/436--dc23
LC record available at https://lccn.loc.gov/2020028557
LC ebook record available at https://lccn.loc.gov/2020028558

ISBN: 978-0-367-46665-7 (hbk)
ISBN: 978-1-003-03846-7 (ebk)

Typeset in Times
by Deanta Global Publishing Services, Chennai, India

Contents

Preface

Distributed artificial intelligence (DAI) has emerged as a powerful paradigm for representing and solving complex problems. DAI is the branch of AI concerned with how to coordinate behavior among a collection of semi-autonomous problem-solving agents: how they can coordinate their knowledge, goals, and plans to act together, to solve joint problems, or to make individually or globally rational decisions in the face of uncertainty and multiple, conflicting perspectives. Growth of this field has been spurred by the advances in distributed computing environments and widespread information connectivity. Although DAI started as a branch of artificial intelligence, it has emerged as an independent research discipline in its own right, representing a confluence of ideas from several disciplines. Distributed, coordinated systems of problem-solvers are rapidly becoming practical partners in critical human problem-solving environments, and DAI is a rapidly developing field of both application and research, experiencing explosive growth around the world.

This book presents a collection of articles surveying several major recent developments in DAI. The book focuses on issues and challenges that arise in building practical DAI systems in real-world settings and covers some solutions of the issues. It provides a synthesis of recent thinking, both theoretical and applied, on major problems of DAI.

Editors

Satya Prakash Yadav is currently on the faculty of the Information Technology Department, ABES Institute of Technology (ABESIT), Ghaziabad, India. He submitted his Ph.D. thesis entitled "Fusion of Medical Images in Wavelet Domain" to Dr. A.P.J. Abdul Kalam Technical University (AKTU) (formerly UPTU). A seasoned academician with more than 12 years of experience, he has published three books (*Programming in C*, *Programming in C++*, and *Blockchain and Cryptocurrency*). He has undergone industrial training programs during which he was involved in live projects with companies in the areas of SAP, Railway Traffic Management Systems, and Visual Vehicles Counter and Classification (used in the Metro rail network design). He is an alumnus of Netaji Subhas Institute of Technology (NSIT), Delhi University, India. A prolific writer, Mr. Satya Prakash Yadav has published two patents and authored many research papers. He also has presented research papers at many conferences in the areas of Image Processing and Programming. He is a lead editor in the Science Publishing Group and Eureka Journals.

Dharmendra Prasad Mahato is currently an assistant professor in the Department of Computer Science and Engineering at National Institute of Technology Hamirpur, Himachal Pradesh, India. He earned his AMIETE degree in Computer Science and Engineering with distinction from the Institute of Electronics and Telecommunication Engineers (IETE), India, in 2011. He earned his Master of Technology in Computer Science and Engineering from Atal Bihari Vajpayee-Indian Institute of Information Technology and Management Gwalior in 2013 and his Ph.D. in Computer Science and Engineering from the Indian Institute of Technology (Banaras Hindu University), Varanasi, India, in January 2018. His research interests include distributed computing, artificial intelligence, operating systems, databases, and modeling and simulation. He has published in journals such as *Applied Soft Computing*, *Swarm and Evolutionary Computation*, *ISA Transactions*, *Cluster Computing*, *Concurrency and Computation: Practice and Experience*, and presented at conferences such as AINA, ICPP, ICDCN, and E-Science.

Nguyen Thi Dieu Linh, Ph.D., works in the Division of Electronics and Telecommunication Engineering in Electronics and Information, Faculty of Electronics Engineering Technology, Hanoi University of Industry, Vietnam (HaUI). She has been the head of the department since 2015. She has more than 18 years of academic experience in electronics, IoT, Smart Garden, and telecommunication. She has authored or coauthored more than 21 research articles in journals, books, and conference proceedings. She earned her Ph.D. in Information and Communication Engineering from Harbin Institute of Technology, Harbin, China, in 2013; her Master of Electronic Engineering from Le Quy Don Technical University, Hanoi, Vietnam, in 2006 and her Bachelor of Electronic Engineering from HCMC University of Technology and Education, Vietnam in 2004. She edited *Artificial*

Intelligence Trends for Data Analytics Using Machine Learning and Deep Learning Approaches and *Distributed Artificial Intelligence: A Modern Approach.* She is also a board member of the *International Journal of Hyperconnectivity and the Internet of Things* (IJHIoT), *Information Technology Journal*, *Mobile Networks and Application Journal*, and other journals and international conferences.

Contributors

Rashi Agarwal
Galgotias College of Engineering and Technology
Uttar Pradesh, India

Neha Bhati
Department of Computer Science and Engineering
ABES Institute of Technology
Ghaziabad, India

Arpit Kumar Chauhan
Department of Computer Science and Engineering
ABES Institute of Technology
Ghaziabad, India

G. Uma Devi
Department of Computer Science and Engineering
University of Engineering and Management
Jaipur, India

Niharika Dhingra
Department of Computer Science and Engineering
ABES Institute of Technology
Ghaziabad, India

Raj Kumar Goel
Department of Computer Science & Engineering
Noida Institute of Engineering & Technology
Greater Noida, Uttar Pradesh, India

Aishwarya Gupta
ABES Institute of Technology
Ghaziabad, India

Mahima Gupta
Department of Computer Science and Engineering
ABES Institute of Technology
Ghaziabad, India

Nidhi Gupta
Department of Computer Science and Engineering
RKGIT
Ghaziabad, India

Sonia Gupta
Department of Mechanical Engineering
GNIT
Greater Noida, India

Naziya Hussain
IPS Academy
Indore, India

Akarshita Jain
ABES Institute of Technology
Ghaziabad, India

Supriya Khaitan
Galgotias University
Uttar Pradesh, India

Rijwan Khan
Department of Computer Science and Engineering
ABES Institute of Technology
Ghaziabad, India

G. Kiruthiga
Department of Computer Science and Engineering
IES College of Engineering
Kerala, India

N. V. Kousik
School of Computer Science and Engineering
Galgotias University
Greater Noida, India

Amit Kumar
HMR Institute of Technology
New Delhi, India

Anuj Kumar
Department of Information Technology
ABES Institute of Technology
Ghaziabad, India

Suresh Kumar
AIACTR
Delhi, India

Pallavi Kumari
Department of Computer Science and Engineering
ABES Institute of Technology
Ghaziabad, India

Annu Mishra
Department of Computer Science
Birla Institute of Technology Mesra Extension (BIT Mesra Ext)
Noida, India

Sugandha Mittal
Department of Applied Science and Humanities
ABES Institute of Technology
Ghaziabad, India

Mohit Pandey
School of Computer Science and Engineering
Shri Mata Vaishnoishno Devi University
Katra, J&K, India

V. Radhika
Srinivasa Institute of Engineering and Technology
Cheyyeru, Amalapuram
Andhra Pradesh, India

Bipin Kumar Rai
ABES Institute of Technology
Ghaziabad, India

R. Arshath Raja
Research and Development
ICT Academy
Chennai, India

Preeti Rani
IPS Academy
Indore, India

Dhruv Rawat
HMR Institute of Technology
New Delhi, India

Shashank Sahoo
Ajay Kumar Garg Engineering College (AKTU)
Uttar Pradesh, India

Shubhangi Sankhyadhar
Department of Computer Application
ABES Engineering College
Ghaziabad, India

Ankur Seem
Department of Computer Science and Engineering
ABES Institute of Technology
Ghaziabad, India

Anisha Singh
ABES Institute of Technology
Ghaziabad, India

Murari Kumar Singh
Department of Computer Science and Engineering
IEC College of Engineering
Greater Noida, India

Narendra Singh
Department of Management Studies
G. L. Bajaj Institute of Management & Research
Greater Noida, India

Pushpa Singh
Department of Computer Science and Engineering
Delhi Technical Campus
Greater Noida, India

Raghuraj Singh
Computer Science and Engineering Department
Harcourt Butler Technical University
Kanpur, India

Rajnesh Singh
Department of Computer Science and Engineering
IEC College of Engineering
Greater Noida, India

Shailesh Singh
Department of Mechanical Engineering
GNIT
Greater Noida, India

Pankaj Tyagi
Department of Biotechnology
Noida Institute of Engineering & Technology
Greater Noida, India

Shweta Vishnoi
Department of Physics
Noida Institute of Engineering & Technology
Greater Noida, Uttar Pradesh, India

Satya Prakash Yadav
Department of Information Technology
ABES Institute of Technology (ABESIT)
Ghaziabad, India

Chandra Shekhar Yadav
Department of Biotechnology
Department of Computer Science & Engineering
Noida Institute of Engineering & Technology
Greater Noida, Uttar Pradesh, India

Vibhash Yadav
Rajkiya Engineering College
Banda, India

Vijay Yadav
Department of Computer Science and Engineering
Dr. A. P. J. Abdul Kalam Technical University
Lucknow, India

N. Yuvaraj
Research and Development
ICT Academy
Chennai, India

1 Distributed Artificial Intelligence

Annu Mishra

CONTENTS

1.1 INTRODUCTION

Evolutionary computing has been extensively investigated under varying environmental conditions, particularly with respect to interaction and controlling multiple agents. These agents can be autonomous bodies, such as software programs, applications, or robots. Multi-agent architecture may be used to investigate phenomena or to remedy issues that are difficult for human beings to examine and clear up. They may be utilized in various areas, ranging from computer games and informatics to economics and social sciences (Varro et al. 2005; Sun and Naveh 2004; Gutnik and Kaminka 2004; Kubera et al. 2010). Evolutionary computing has evolved as a paradigm for complex and composite problems. The basic idea is involvement of multi-agent to solve a problem instead of a single agent problem-solving technique. These agents are capable of discovering the solution by making their own decisions in the presence of multiple agents (Shoham and Leyton-Brown 2010). Traditional Artificial Intelligence (AI) systems have concentrated on gathering information, knowledge representation, and execution via a single intelligent agent, whereas in distributed artificial intelligence (DAI), these processes are distributed among the agents that are proficient for making independent as well as coordinated decisions.

According to Ponomarev and Voronkov (2017), a DAI can be characterized by three principle qualities (Sichman et al. 1994):

1. It is a technique for the dispersion of jobs between operators;
2. It is a technique for dispersion of forces;
3. It is a technique for communicating among the participating agents.

To satisfy the first quality, two aspects must be considered: first, the dissemination of knowledge and, second, the allocation of tasks among the agents.

1.2 WHY DISTRIBUTED ARTIFICIAL INTELLIGENCE?

Modeling and computational responsibilities have gained extra complexity as dimensions continue to grow. As a result, it is hard to address the use of centralized methods. Although motivations to apply multi-agent systems (MASs) for researchers from various disciplines are special, as discussed by Yu and Liu (2016), the principal benefits of using multi-agent technologies include (1) individuals or agents keep track of software-specific nature and environment; (2) interactions among agents can be modeled and supervised; and (3) any difficulty is managed by sublayers and/or subcomponents of the system. Therefore, MASs can be considered to be an optimal solution for a computational paradigm in a distributed environment as well as for vast data complexities. In addition, AI techniques can be utilized.

Pattison et al. (1987) state that agents take sensory input from the environment produces the output action as shown in Figure 1.1.

Before we study multi-agents further, let us first understand the term "Agent." An agent has the following characteristics:

- Self-sufficiency: agents work without the immediate mediation of people or others, and have an authority over their activities and inside state.
- Social capacity: agents collaborate with different agents (and conceivably people) through a specialist correspondence language.
- Reactivity: agents see their condition (which might be the physical world, a client by means of a graphical UI, an assortment of different specialists,

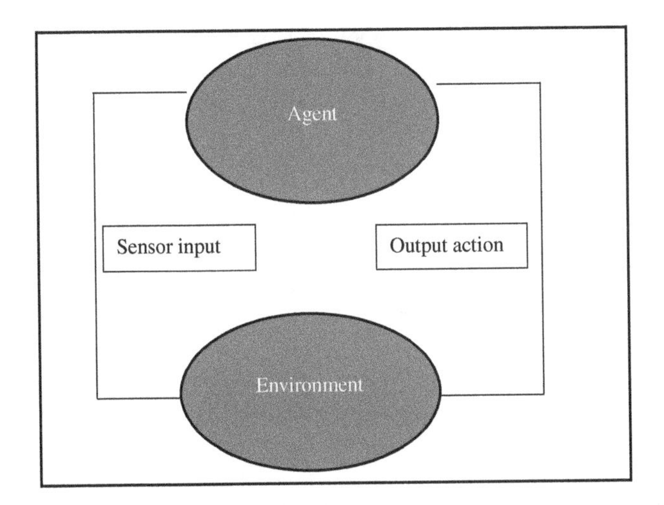

FIGURE 1.1 Distributed artificial intelligence structure. Source: Pattison et al. (1987).

the Internet, or maybe these joined), and react in an auspicious design to changes that happen in it.

- Pro-animation: agents do not just act in light of their condition; they can show objective coordinated conduct by stepping up to the plate.

The independent operator approach replaces a concentrated database and control PC with a system of operators, each supplied with a neighborhood perspective on its condition and the capacity and power to react locally to that condition (Andrews 1991). The general framework execution is not all inclusive arranged; however, it develops progressively through the dynamic cooperation of specialists. Hence, the framework does not shift back and forth between patterns of booking and execution. Thus, the computation rises up out of the simultaneous free choices of the neighborhood operators or the multi-agents.

A distributed approach additionally offers benefits when managing vulnerability, for example in the area of information. In open frameworks (Hewitt and de Jong 1983), it is frequently beneficial to have numerous central focuses of critical thinking action so as to have the option to manage more than one unexpected occasion at once. A distributed framework will not experience the ill effects of single point failure. This gives it a preferred position in circumstances where unwavering quality is of specific significance. In a distributed framework, the failure of a critical thinking hub or a single point will bring about an effortless debasement of execution instead of a total framework disappointment. Likewise, from a framework structure viewpoint, an appropriated arrangement might be the most practical. It might be that the all-out expense of the quantity of minimal effort equipment gadgets required to actualize a disseminated framework is lower than the expense of a huge gadget that is groundbreaking enough to play out a similar assignment. It might likewise be the situation that the measured quality upheld by a disseminated approach lessens programming costs. DAI leads to distributed problem-solving. This approach tackles a

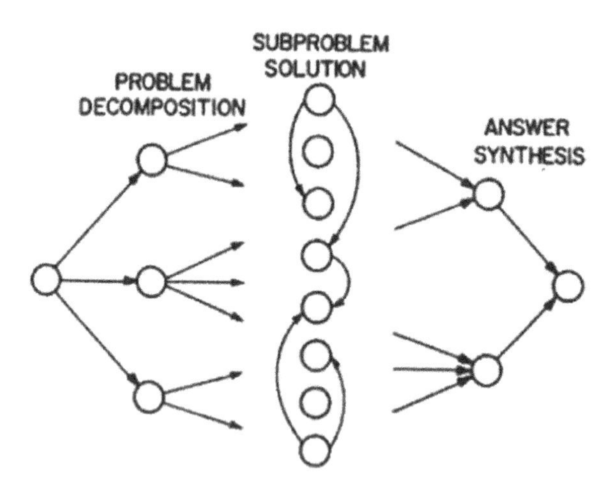

FIGURE 1.2 Phases of solving a problem in distributed environment.

specific issue that can be separated among various modules that coordinate in isolating and sharing information about the issue and its solution(s). Distributed problem-solving is a type of critical thinking that manages the issues by disseminating it among several problem solvers which communicate with each other in a cooperative manner and share the burden of transferring the partial solution to each other. The communication between partial problem solvers is expressly characterized in appropriated critical thinking situations. Multi-agent systems, by contrast, works together with loosely coupled problem solvers to tackle an issue. These autonomous problem solvers, or computational operators, are self-sufficient elements that have the capacity to brilliantly work under different natural conditions given their tangible and strong abilities (Lesser 1991). Figure 1.2 depicts the phases or steps of a solver. It can be observed that, first, the problem that arrives is fragmented into several sections. Fragmentation continues until the unit issue is obtained. In the next phase, solvers try to find a solution for each unit issue by interacting and sharing information and amenities. In the last and final phase, the subsolutions obtained are integrated to form a final solution to a problem.

1.3 CHARACTERISTICS OF DISTRIBUTED ARTIFICIAL INTELLIGENCE

1. Distributed artificial intelligence (DAI), also called decentralized artificial intelligence, is a subfield of artificial intelligence (AI) that is committed to the improvement of distributed reasoning for issues.
2. DAI is firmly identified by multi-agent systems and distributed problem solving (DPS).
3. The goals of DAI are to illuminate the thinking, arranging, learning, and recognition issues of AI, particularly when they require enormous amounts of information, by passing off the issue to independent agents.

4. DAI takes into consideration the interconnection and interoperation of different existing hereditaments system. By building an agent sheath around such legacy systems, they can be incorporated into the agents list.
5. DAI upgrades framework system execution, explicitly using the components of computational effectiveness, dependability, extensibility, viability, responsiveness, adaptability, and reuse.
6. DAI gives arrangements in circumstances where the ability of the subsystems are spatially and transiently disseminated.

1.4 PLANNING OF DAI MULTI-AGENTS

The agents participating in DAI communicate in several ways regarding their neighboring agents in order to understand their behaviour.

Specialists generally structure an arrangement that indicates all of their future activities and communications in order to accomplish a specific goal. Before execution begins, the zones of the inquiry space that will be crossed and the course that every specialist should take at every choice point in the activity are identified. Multi-operator plans are normally worked to maintain a strategic distance from conflicting or clashing activities, especially concerning the utilization of rare assets. Multi-agents arrange contrasts from hierarchical organizing and meta-level data trade regarding the degree of detail to which it each operator's exercises are determined (Peterson et al. 1990). With this methodology, specialists know ahead of time precisely what moves they will make, what moves their associates will make, and what connections will happen. Therefore, such arrangement of the agents can store some amount of association memory and prove dynamicity at times.

There are two types of multi-agent planning:

- centralized multi-agent planning
- distributed multi-agent planning

In centralized multi-agent planning, there is normally a coordinator who, on receipt of all neighborhood plans from every agent, investigates them so as to distinguish potential irregularities and clashing connections (for example, the struggle between agents over constrained assets). The planning operator or the coordinator at that point endeavors to alter these fractional or the partial plans and joins them into a special plan where clashing interactions are removed.

In Figure 1.3, Georgeff (1983) epitomizes such a methodology where clashing connections are distinguished and assembled into basic areas. In the final special plan, correspondence directions are embedded to synchronize all the participating agents so that they can collaborate effectively. Cammarata, McArthur, and Steeb (1983) utilized concentrated multi-agent arranging in a mimicked air traffic control space. In their model, the agents (in this case, planes) appear to be in a potential clash situation, for example two planes are heading toward a crash; choose a random plane (plane 1 or plane 2) among them which would behave as coordinator and make a decision to fix the problem of crashing. The scheme of the supplemental agents is

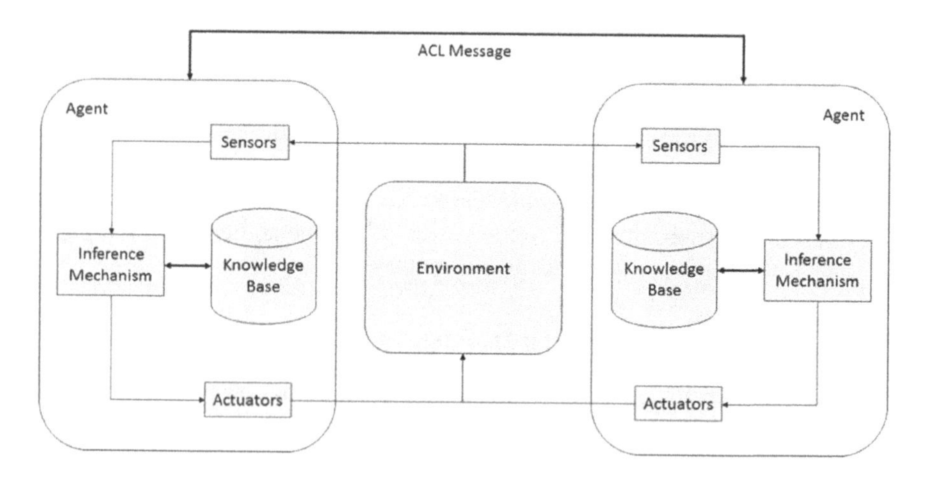

FIGURE 1.3 Centralized multi-agent planning. Source: Georgeff (1983).

forwarded to this coordinator. The coordinator at that point will endeavor to revise his own plan of flight in such a way that fix the strife.

In distributed multi-agent arrangement, the motivation is to share a structure/model with every agent. The multi-agents communicate among each other so as to fabricate the path to communicate, revise their individual plans and their structures of others unless all contentions are evacuated. An epitome of this methodology is Lesser et al. (1991), exact and agreeable (FA/C) approach. In this methodology, loosely coupled participating agents form a high level (most likely deficient) plans, results, and speculations. These speculations are shared. Next, they refine these shared results until they all unite on some final complete plan. Nearby irregularities are distinguished and only those parts of an agent's speculation that are very close to the non-global particulars are coordinated in the repository or databases. This methodology was utilized for Distributed Vehicle Monitoring Testbed(DVMT), which was a framework for testing coordination procedures. Another representation of appropriated multi-agent planning is mentioned in Dufree et al. (1989) halfway worldwide arranging (PGP) approach. In this approach, nearby agents in collaboration with one another plan their moves, which as a result, are adjusted constantly on the basis of incomplete final or non-local plans (worked by interchanging non-global nearby plans). In this manner, the agents are continually paying special attention to potential upgrades to amass coordination.

1.5 COORDINATION AMONG MULTI-AGENTS

Most of the architecture is based on the fact of whether the agents must communicate cooperatively or competitively. There are numerous reasons explaining why coordination among agents is required. The following sections explain this in more detail.

1.5.1 Forestalling Mobocracy or Confusion

Because of the decentralization in MAS-based frameworks, mobocracy can set in effectively. Never again does any agent have to have a global perspective on the whole organization. For example, the head of an organization cannot keep track of all the representatives of the organization (Durfee et al. 1989).They recruit other managers at different levels to supervise the employees. Similarly, in the multi-agent-based system, coordination is executed among the agents so that one single agent does not suffer the overhead of planning, execution, and knowledge transfer.

1.5.2 Meeting Overall Requirements

There typically are uniform imperatives that a group of agents must fulfill in order to be considered effective. For instance, the group of agents building a structure of a software may need to work within prerequisites and a predetermined budget. Also, some agents involved in network management may need to react to certain problems almost immediately, whereas some may take hours to make a particular decision. Thus, these agents need to coordinate among themselves to meet the constraints and protocols imposed on them.

1.5.3 Distributed Skill, Resources, and Data

When we consider a typical scenario of an operation room in a hospital, we can imagine the involvement of everybody, for example cardiologist, anaesthetist, nurse, etc. Everybody has their own expertise and area in which to work and total involvement of these characters may lead to a successful operation. Similarly, agents may have different limits, plans, knowledge, and explicit data. Of course, these agents may rely on distinct sources of information, amenities (for instance taking care of intensity, memory), steady quality levels, commitments, limitations, charges for organizations, etc. In such circumstances, these agents must work together on a global plan.

This distributed approach can be well understood using BOINC (Berkeley Open Infrastructure for Network Computing), an open software platform for distributed computing.

The server of BOINC has a set of multi-agents. Each of these agents has their specific task to perform, for example testing, processing, cleaning up files, etc., as depicted in Figure 1.4. Every agent checks the status of the other agents present in the server and delivers a few activities and changes the condition of result generated, working in an unbounded circle (Hayes-Roth 1985).

As a rule, the framework comprises of BOINC server, a majority of clients, playing out the undertakings and responsibilities of the server and, perhaps, extra components as GRID-affiliated systems (Ponomarev and Voronkov 2017). Interestingly, the distributed insight in the BOINC model can tackle a wide scope of errands. At present, it is effectively utilized in various territories and, yet, it is carefully incorporated and restricted by configuration design.

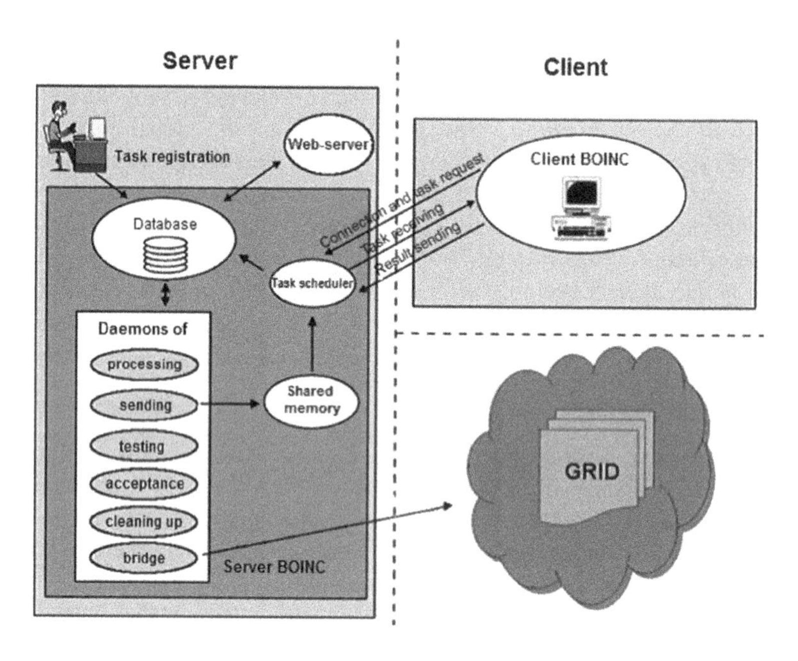

FIGURE 1.4 Multi agent systems and decentralized artificial superintelligence. Source: Ponomarev, Voronkov (2017).

1.5.4 Dependency among the Agents

Most agent's goals are interdependent on other agents. This can be seen in the trivial block world problem as depicted in Figure 1.5.

The most straightforward approach to understanding the problem for agents is to assume the subobjective of plunging B on C while the subsequent subgoal would be A over B and C to accomplish A-B-C. Obviously, the intermediate objectives are related: the subsequent operator needs to pause for the principal operator to finish its subobjective prior performing its errand. In systems involving this type of interdependency constantly performing in MAS, exercises need to be coordinated.

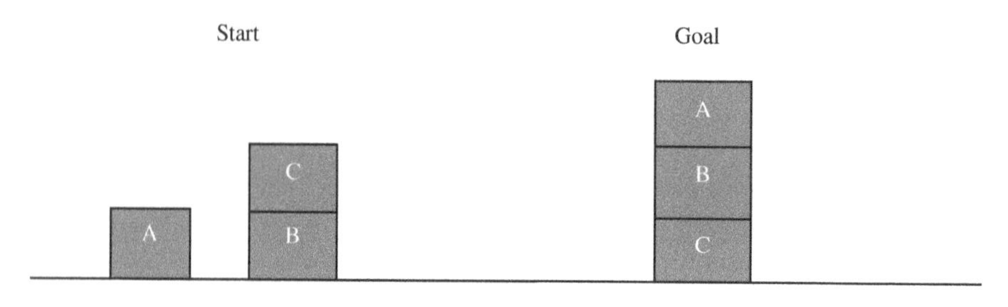

FIGURE 1.5 The trivial block world problem.

1.5.5 Efficiency

In some circumstances, a single agent may solve the problem but coordinating among other agents and sharing the knowledge may lead to a quicker completion of the task. Thus, coordination is required to increase the efficiency of every single agent.

1.6 COMMUNICATION MODES AMONG THE AGENTS

To accomplish coordination, operators may need to speak with each other. Singh and Huhns (1994) call attention to the fact that specialists may accomplish coordination without correspondence if they have models of each others' practices. Coordination can be accomplished chiefly through association. In any case, to encourage coordination, where operators need to coordinate through correspondence, it is fundamental that they publish their objectives, expectations, consequences, and predicaments to different operators (Gilbert and Conte 1995).

Communication methods among the agents are discussed in the following paragraphs.

Blackboard architecture: The fundamental idea behind blackboard architecture, also known as tuple space communication, is the traditional classroom blackboard, which acts as a knowledge pool that can be updated by several groups or individuals. Thus, it gives a platform to contribute collectively from problem definition to solution.

Blackboard architectural design as shown in Figure 1.6, is basically used for those problems for which there is no definite solution available. In blackboard architecture subsystems, knowledge is gained to generate either partial or complete solutions.

Knowledge sources are similar to an expert writing on a classroom blackboard. Similarly, DAI also requires information from different sources to load its knowledge base to deliver partial or exact solutions. The information needed to solve a problem is kept in different knowledge source and stored independently. These knowledge sources can be used during problem solving or solution construction. We can use these knowledge sources differently for several problems, which can be considered as a set of rules and procedures and logics.

The repository is the blackboard itself, which absorbs all the information written on it by the expert. Knowledge sources produce changes to the blackboard that lead

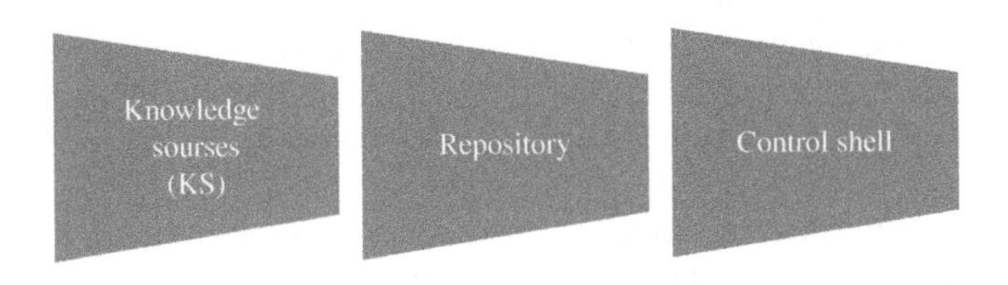

FIGURE 1.6 Blackboard architecture.

gradually to an answer, or a lot of worthy arrangements, to the issue. Collaboration among these knowledge sources happen exclusively through changes on the slate.

The control shell monitors all the activities to solve the given problem or reach to the final state. It contains the list of registered knowledge sources (KSs) and their state. The control shell is required because every agent has a shared repository to write and this may lead to certain muddled scenario.

The fundamental way to deal with critical thinking in the blackboard system is to partition the issue into loosely coupled tasks. These subtasks generally relate to regions of specialization inside the task. For a specific application, the solution designer characterizes a solution space (set of solutions) and information expected. This solution space is isolated into many investigation levels of incomplete or complete solutions, and the information is isolated into particular information sources that help in the completion of subtasks. Data on the investigation levels is globally open on the blackboard, making it a medium of communication between different knowledge sources. For the most part, an information source utilizes data on one degree of investigation as to its information and yields data on another level. The choice to utilize a specific information source is made powerfully by utilizing the most recent data contained in the board information structure (the present arrangement state). This specific way to deal with issue decay and information application is truly adaptable and functions admirably in different application areas. How the problem is broken into subproblems has a large effect to the lucidity of the methodology, the speed with which arrangements are discovered, the assets required, and even the capacity to imagine the issue.

This architecture is applicable for weather forecasting, pattern recognition, and much more. Heresay II and Douglas Hofstadter's Copycat are some examples of blackboard architecture.

Remote procedure call (RPC): This mode of communication as you may be aware, is often used in a server-based architecture. The nodes at different spaces are facilitated to execute the instruction at remote space without having their physical presence. General working of RPC is shown in Figure 1.7. RPC facilitates client with pellucid reach to a server. Amenities or assistance that the server offers to the client in distributed environment should encompass computational utility, file system access, efficient operation in a distributed environment, and all other features supported by the traditional RPC. The essential stipulation of RPC is that it offer the client with dependable, pellucid ingress to a server. A remote procedure call to a server is precisely the identical to the patron software as a confined procedure call. This is a totally effective idea. The software programmer no longer needs to be concerned about accessing a distant laptop when executing an RPC function.

Figure 1.7 shows the drift of a RPC call. Step 1 explicates the procedure call made by the client application. It is indistinguishable from any confined or non-global procedure. This procedure impetrated by the client is known as a stub. The stub compresses

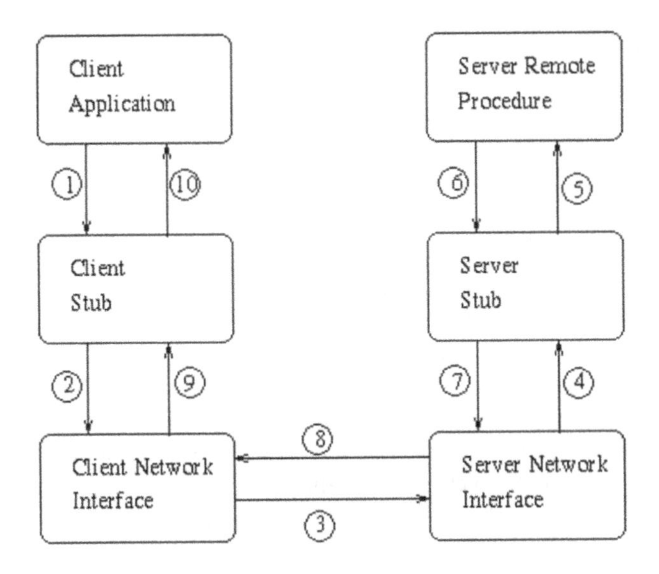

FIGURE 1.7 Elucidation of the drift in a RPC call.

the parameters given to it, crams them with extra statistics and information that may be required by the server, and sends the complete packet towards the customer network interface. Here it can be noted that the client application may or may not know about the existence of the stub as it is generated by the RPC library and it thus gives an semblance of non-global procedure call to the client application. The network interface then performs a series of steps on a protocol to infalliably relay the packet spawned by the stub to the precise server. In step 4, the network interface belonging to the server calls the server stub. The server stub unloads the received request and processes the operation on the server. After this, the data is redirected to the server stub, which bundles it with the additional information. This bundle is forwarded to client network interface again via the server network interface. The consumer network interface thereafter courses the packet to the client stub, unlades the accepted packet, and returns the preferred information to the customer application.

1.7 CATEGORIES OF RPC

RPC can be divided into three categories:

1. Batch mode RPC
2. Broadcast RPC
3. Callback RPC

Batch mode RPC: To reduce the overhead attached with sending each request independently and wait for the response, batch mode RPC queues the request and then sends it in a batch or group to the server. This is a provision for sending numerous client non-blocking calls in one batch.

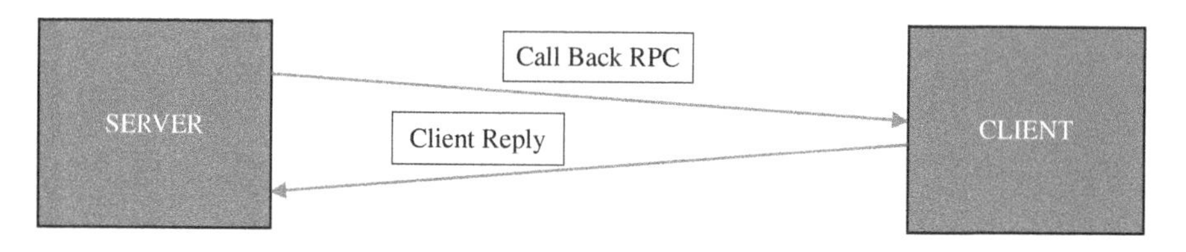

FIGURE 1.8 Callback RPC.

Broadcast RPC: In broadcast RPC, a message is despatched to all rpcbind daemons in a particular network. The preregistered request service then forwards the request to the server. The main variations among broadcast RPC and traditional RPC calls include:

1. Traditional RPC expects only one solution, whereas broadcast RPC expects multiple answers from every responding system
2. Broadcast RPC works best on connectionless protocols that assist broadcasting, which includes user datagram protocol
3. Broadcast RPC filters out all failed responses
4. The size of broadcast requests is limited via the maximum transfer unit (MTU) of the nearby community, which is 1500 bytes

Callback RPC: In callback RPC shown in Figure 1.8, the agent that acts as a server sends a call to the client agent for further execution of instruction or to gain some input from the client. The client then sends a reply to the server so the server can resume the functioning of knowledge transfer and other information passing.

Message passing: When we pass messages from client to server and vice versa the data must be in such format that the client and server both could understand. Well-known message designs for RPC are JSON and XML. Such correspondence is called JSON-RPC and XML-RPC for RPC that utilizes JSON and XML individually. Some keys of JSON objects sent by the client include:

- method—name of the service
- params—array arguments to be send
- id—it is an integer used to identify the request that could be received in synchronous or asynchronous ways

Some keys of JSON objects sent by the server include:

- result—contains the result or value of the service called; the result is null for any kind of error
- error—if there is an error, this key will provide the error code or error message
- id—it identifies the request it is responding to and is an integer value as well

FIPA standards: The Foundation for Intelligent Physical Agents (FIPA) was set up to create and set PC programming benchmarks for heterogeneous and associating agent- and operator-based frameworks.

FIPA was established as a Swiss non-profit organization in 1996 with the goal-oriented objective of characterizing a full arrangement of guidelines for both actualizing frameworks inside which operators could execute (specialist stages) and indicating how agents themselves ought to convey and interoperate in a standard way. Inside its lifetime the association's participation incorporated a few scholarly establishments and an enormous number of organizations including Hewlett Packard, IBM, BT (formerly British Telecom), Sun Microsystems, Fujitsu, and some more. Various norms were proposed, notwithstanding, regardless of a few specialist stages receiving the FIPA standard for operator correspondence it never prevailing with regards to picking up the business bolster which was initially imagined. The Swiss association was broken down in 2005 and an IEEE gauges council was set up in its place. The most generally embraced of the FIPA measures are the Agent Management and Agent Communication Language (FIPA-ACL) determinations.

1.8 PARTICIPATION OF MULTI-AGENTS

Multi-agents participating under DAI may exhibit full cooperation or partial cooperation in knowledge sharing. The fully cooperating agents solve typical and dependent problems. They even can switch their result or final state in order to cooperate the other agents. They can be regarded as the well-wishers who help the other agents to acquire their goal and thus maintain the sustainability and amity between themselves. The partial cooperative agents, on the other hand, may or may not reflect this idea of coherence. They are antagonistic in nature and may hamper other agents for their own benefit.

1.8.1 Fully Cooperative Architecture

The fully cooperative architecture is an arrangement of agents like in the mesh topology of the network architecture, shown in Figure 1.9. In this architecture, each agent participating has a connection with every other agent in the system. Unfortunately, this arrangement may face problems dealing with huge data and scaling to complex domains that would involve large numbers of agents. If we have an environment with S states, this would result in S^N states. This fulmination in the number of states can be an overhead for the service. The maximum delay in this type of architecture from receive of request to its response generation is 50 msec to 100 msec. However, this type of architecture involves a high cost for maintenance and communication links. It should be noted that most of the MAS (multi-agent systems) possess partial cooperative architecture. Fully cooperative frameworks support to peer-to-peer processing.

Team learning may be divided into two broad categories: homogeneous and purely heterogeneous team learning. Homogeneous learners develop a single agent behavior that is used by every agent on the team. Purely heterogeneous team learners develop a unique behavior for each agent; such approaches hold the promise of

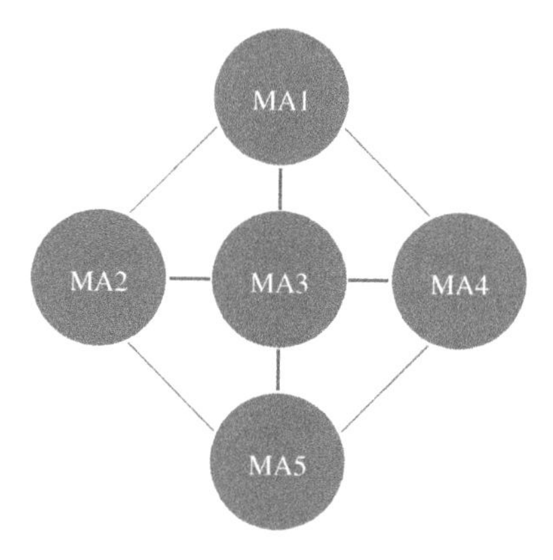

FIGURE 1.9 Fully cooperative architecture.

better solutions through agent specialization, but they must cope with larger search spaces. There are approaches in the middle ground between these two categories: for example, divide the team into groups in which group mates share the same behavior. We refer to these as hybrid team learning methods.

1.8.2 Partial Cooperative Architecture

The partial cooperative architecture is a two-level organization of the agents. In this arrangement some agents also called as the "councillors" are connected in a way to form a fully connected graph. These fully connected graphs are then connected to other fully connected graphs, which are formed in a similar manner. Each set of graphs has a councillor or controller that makes decisions about communication. Figure 1.10 illustrates the partial architecture of the agents.

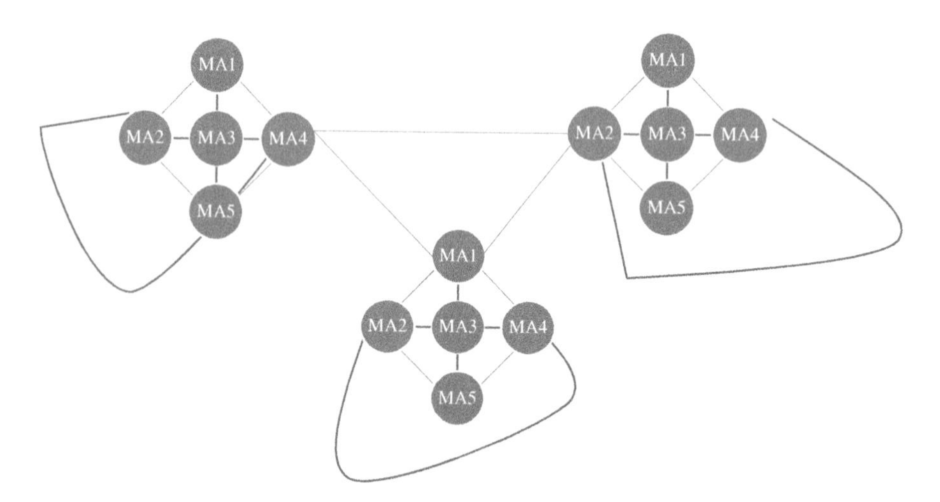

FIGURE 1.10 Partially cooperative architecture.

1.9 APPLICATIONS OF DAI

DAI has many applications because of its distributive nature. In recent years, many applications have been built based on this idea, which not only has brought a revolution but also has ushered in new technology and ideas. Some of the quotidian applications are listed below.

1.9.1 Electricity Distribution

As discussed by Chaib-draa (1995), a modern application from the ARCHON project, called Cooperating Intelligent Systems for Circulation System Management (CIDIM), is a guide for control engineers (CEs) who ensure the power supply to the users. CIDIM helps CEs in areas such as deficiency conclusion, client-driven reclamation arranging, and security investigation, just as consequently gathering a significant part of the data CEs collate physically by reference to independent systems. CIDIM is comprised of ten intelligent agents, some containing regular projects and some containing master frameworks. Seeking a multi-agent approach, CIDIM permits each unmistakable capacity to be executed utilizing the most proper model, regardless of whether it is a master system, database, or regular programming.

1.9.2 Telecommunications Systems

Utilizing DAI strategies in broadcast communications frameworks appears to be inescapable when we think about two patterns in the structure of such frameworks: dispersion of usefulness and joining of knowledge programming that executes complex administrations and dynamic (Chaib-draa 1995). DAI writing incorporates various methodologies that address the media communications field. A striking model is a framework called LODES (Large-Internetwork Observation and Diagnostic Expert System), which has tried to tackle operational systems. LODES identify and analyze issues in a portion of a LAN. Different LODES framework duplicates—each going about as a specialist—can screen and oversee diverse system segments. LODES incorporate segments that let every operator help out different specialists. In spite of the fact that LODES was grown basically as an examination testbed, it has been used in operational systems with some success. LODES' designers picked a circulated approach to deal with the physical and practical appropriation of systems. A dispersed methodology likewise empowers critical thinking, encouraging the correspondence of the consequences of the examination instead of requiring all of the data for the procedure.

1.9.3 Database Technologies for Service Order Processing

Singh and Huhns (1994) also characterized a distributed agent engineering for workflow management that functions over Carnot's environment. Their framework is comprised of four specialists that interact to produce the ideal conduct, and a database that includes the significant information and the application programs. The four operators are:

- the graphical interaction agent
- the transaction scheduling agent
- the schedule processing agent
- the schedule repairing agent

Applications are executed by the schedule processing agent. During processing, if the agent experiences a startling condition, for example, a task failure, it tells the transaction scheduling agent, which then approaches the schedule repairing agent for guidance on the most proficient method to fix the issue. This guidance can include how to restart a schedule process, how to prematurely end the transaction, and other tasks. In the end, the graphical interaction agent asks the frameworks to support clients or the users. Singh and Huhns (1994) prototyped a model that executes over a featured condition to enable a telecommunications organization to offer an assistance that requires coordination among numerous activities, networks, and system components condition monitoring.

1.9.3.1 Concurrent Engineering

The Palo Alto Collaborative Testbed (PACT) is a concurrent engineering infrastructure including various local sites, subsystems, and controls. Through PACT, agents—from Stanford University, Lockheed Palo Alto ResearchLabs, and Enterprise Integration Technologies—look at the mechanical and sociological issues of building enormous scope-distributed concurrent engineering systems. PACT tests have investigated fabricating a larger structure along three measurements:

- cooperative advancement of interfaces, conventions, and design
- sharing of information among frameworks that keep up their own particular information bases and thinking components
- computer-helped support for arranging and settling on choices in simultaneous building ventures

The PACT design depends on interfacing projects, or specialists, that epitomize building apparatuses. The specialist association depends on three things:

- shared ideas and wording for conveying information across disciplines
- a typical language for moving information among agents
- a correspondence and control language that empowers operators to demand data and administrations

1.9.3.2 Weather Monitoring

Sun-powered illumination and temperature figures are utilized in an assortment of various control frameworks. One of these frameworks is atmosphere control, where sun irradiance influences artificial lighting.

Weather forecast results for a neighborhood often are purchased from business climate figure offices and agencies. Shockingly, conjectures are never as exact as one might hope. The Danish Meteorological Institute (DMI) has claimed that they have a 97% exactness rate. The precision is excellent, yet the interval is nevertheless too wide. Business climate gauges are commonly produced in a 10 × 10 km grid. It is sometimes possible to predict upcoming climate events based on previous weather events. Certain models have been developed using DAI up to 24 hours before (Wollsen and Jorgensen 2015). These models make use of artificial neural networks.

1.9.3.3 Intelligent Traffic Control

Artificial Intelligence (AI) and Internet-of-Things (IoT), called AI-fuelled Internet-of-Things (AIoT), is able to prepare enormous amounts of information produced from a large number of gadgets and can deal with complex issues in social frameworks. For example, observation cameras can provide continuous data for traffic light control and report conditions of both automated and manual traffic. Liu, Liu, and Chen (2017) proposed a traffic light control arrangement that utilized disseminated multi-agent Q learning to improve the general execution of the entire control framework. By utilizing a multi-agent Q learning calculation, they advanced both automated and manual traffic.

1.10 CONCLUSION

Distributed artificial intelligence (DAI) is an extent of artificial intelligence that manages communication among wise and smart agents. It endeavors to build smart agents that can take decisions and thus help themselves in achieving the objectives in a world inhabited by various intelligent agents. DAI focuses on acquiring and analyzing information to plan strategies required for finding solutions to a given problem. The use of DAI in various fields has proved to be a revolutionary upgrade of the system. It has helped to solve complex problems that require a number of resources, power, and information through a distributed approach. However, there are several difficulties associated with DAI as it manages multiple agents. Coordination is one of the prominent issues. However, DAI is one of the most powerful methods that can be used to solve complex and large tasks.

REFERENCES

Andrews, G.R., "Paradigms for process interaction in distributed programs", *ACM Computing Surveys*, 23: 49–90, 1991.

Cammarata, S., McArthur, D., Steeb, R., "Strategies of cooperation in distributed problem solving", in A.H. Bond, L. Gasser (ed.), *Readings in Distributed Artificial Intelligence*, 102–105, 1988, Elsevier.

Chaib-draa, B., "Industrial applications of distributed AI", *Communications of the ACM*, 38(11): 49–53, 1995.

Durfee, E.H., Lesser, V.R., Corkill, D.D., "Trends in cooperative distributed problem solving", *IEEE Knowledge and Data Engineering*, 1(1): 63–83, 1989.

Georgeff, M.P., "Strategies in heuristic search", *Artificial Intelligence*, 20(4), 393–425, 1983.

Gilbert, N., Conte, R., *Artificial Societies: The Computer Simulation of Social Life*, UCL Press, London, 1995.

Gutnik, G., Kaminka, G., "Towards a formal approach to overhearing: Algorithms for conversation identification", in *Proceedings of the Third International Joint Conference on Autonomous Agents and Multiagent Systems, 2004. AAMAS 2004*, 78–85, IEEE, July 2004.

Hayes-Roth, B., "A blackboard architecture for control", *Artificial Intelligence*, 25: 251–321, 1985.

Hewitt, C., Jong, Peter de., *Analyzing the Roles of Descriptions and Actions in Open Systems*, Massachusetts Institute of Technology, Cambridge Artificial Intelligence Lab, 1983.

Kubera, Y., Mathieu, P., Picault, S., "Everything can be Agent!", in *Proceedings of the Ninth International Joint Conference on Autonomous Agents and Multi-Agent Systems (AAMAS' 2010)*, 2010, Toronto, Ontario, Canada, 1547–1548, hal-00584364.

Lesser, V.R., "A retrospective view of FA/C distributed problem solving", *IEEE Transactions on Man, Machine, and Cybernetics*, 21: 1347–1363, 1991.

Liu, Y., Liu, L., Chen, W.P., "Intelligent traffic light control using distributed multi-agent Q learning", in 2017 *IEEE 20th International Conference on Intelligent Transportation Systems* (ITSC), 1–8, IEEE, October 2017.

Pattison, H.E., Corkill, D.D., Lesser, V.R., Huhns, M., "Distributed artificial intelligence, instantiating descriptions of organizational structures", *Distributed artificial intelligence*, volume1, 59–96, 1987.

Peterson, L., Hutchinson, N., O'Malley, S., Rao, H., "The x-kernel: A platform for accessing internet resources", *Computer*, 23(5): 23–33, 1990.

Ponomarev, S., Voronkov, A.E., "Multi-agent systems and decentralized artificial superintelligence", 2017. arXiv preprint arXiv:1702.08529.

Shoham, Y., Leyton-Brown, K., *Multiagent Systems: Algorithmic, Game-Theoretic, and Logical Foundations*, 1st edn., 2010, Cambridge University Press. www.masfoundations.org/mas pdf.

Sichman, J.S., Demazeau, Y., Conte, R., Castlefranchi, C., "A social reasoning mechanism based on dependence networks", in *ECAI 94 11th European* Conference *on Artificial Intellegence*, 189–192, John Wiley and Sons, California, 1994.

Singh, M.P., Huhns, M.N., "Automating workflows for service order processing: Integrating AI and database technologies", *IEEE Expert*, 9(5): 19–23, 1994.

Sun, R., Naveh, I., "Simulating organizational decision-making using a cognitively realistic agent model", *Journal of Artificial Societies and Social Simulation*, 7(3), 2004.

Varro, G., Schurr, A., Varró, D., "Benchmarking for graph transformation", in 2005 *IEEE Symposium on Visual Languages and Human-Centric Computing* (*VL/HCC*'05), 79–88, IEEE, September 2005.

Yu, N.P., Liu, C.C., "Multiagent systems. Advanced solutions in power systems: HVDC, FACTS, and artificial intelligence", 903–930, 2016, John Wiley & Sons.

Wollsen, M.G., Jorgensen, B.N., "Improved local weather forecasts using artificial neural networks", In *Distributed Computing and Artificial Intelligence, 12th International Conference*, 75–86, Springer, Cham, 2015.

2 Intelligent Agents

Rashi Agarwal, Supriya Khaitan, and Shashank Sahu

CONTENTS

2.1 INTRODUCTION

The word software agent was first introduced by John McCarthy and Oliver G. Selfridge in the mid-1950s. Today, software agent is very diverse and hot topic of research in artificial intelligence, robotics, distributed computing, human-computer interaction, intelligent and adaptive interfaces, and information retrieval, etc. We all are familiar with the word an agent, e.g., estate agent. The commonplace meaning of an agent is "an element or an entity having the power to act on behalf of another." These agents are acting on behalf of other entities and fix appointments without reference to the owners. These agents are very proactive and reactive with respect to their behavior; e.g., these agents proactively advertise the "for sale" property in the local press or the same agent can display a high amount of both proactivity and reactivity at different times.

Here, we are talking about two essential properties of programming specialists, also known as software agents, that are autonomous/self-governing and situated in an environment. The foremost property, acting naturally overseeing, infers that pros are self-sufficient and choose their own decisions. This is one of the properties that separate between programming operators from objects. At the point, when we believe a structure to be decentralized containing different autonomous specialists. The ensuing property does not validate the idea of an authority, particularly since all product items can be seen as masterminded in a circumstance. Regardless, programming operators will, when all is said and done, be used where condition is testing. Even more unequivocally, a specialist's surroundings are dynamic, whimsical, and tricky. The earth is dynamic in that they change rapidly. By "rapidly," we suggest that the specialist can't acknowledge that the earth will remain static while it is endeavoring to achieve a goal.

These conditions are unordinary in that it cannot predict the future states of condition. Consistently, this is in light of the fact that a person cannot have dynamic information about nature. Agents are regularly arranged in unique situations that change quickly. Specifically, this implies an agent must react to huge changes in its condition. For the model, an agent controlling a robot playing soccer can make arrangements based on the current situation of the ball and of different players, yet it must be set up to adjust or relinquish its arrangements as the environment change in a critical manner. As it were, agents should be reactive, reacting in an opportune way to changes in their environment.

Another key property of an agent is that they seek after objectives, they are proactive. One property of objectives is that they are persevering; this is valuable in that it makes agents increasingly hearty. An agent will keep on endeavoring to accomplish an objective regardless of failed endeavors. Objects are reactive and having an implicit goal. However, agents are not proactive in having multiple goals. This property differentiates agents with objects. Also, these agents have very interesting characteristics like learning, cooperation, and mobility. Therefore, the word agent can be summed up as seen in Figure 2.1: "An agent is a computational entity which can act on the behalf of other entities in an autonomous way, performs some actions with some degree of proactively or reactively and exhibit some key attribute of learning, cooperation and mobility."

A study of agents is related with anything that can take decisions or an action with the best of its past and current perceptions. An agent can be a person, firm, machine, or software. An artificial intelligence system is made of an agent and its environment as shown in Figure 2.2. The agent takes decisions within the boundary of its

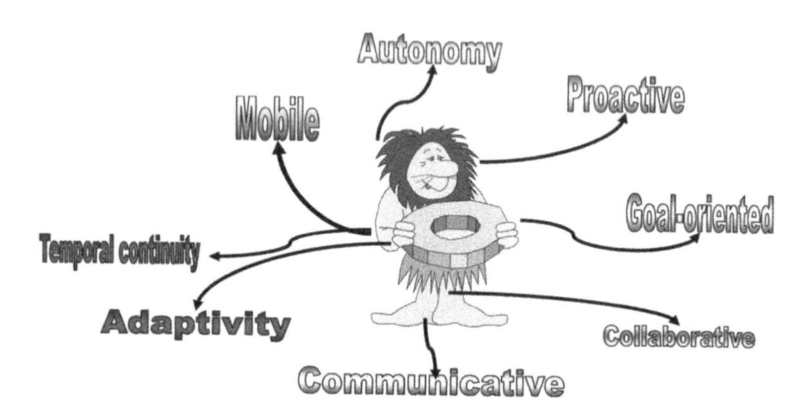

FIGURE 2.1 An agent's characteristics.

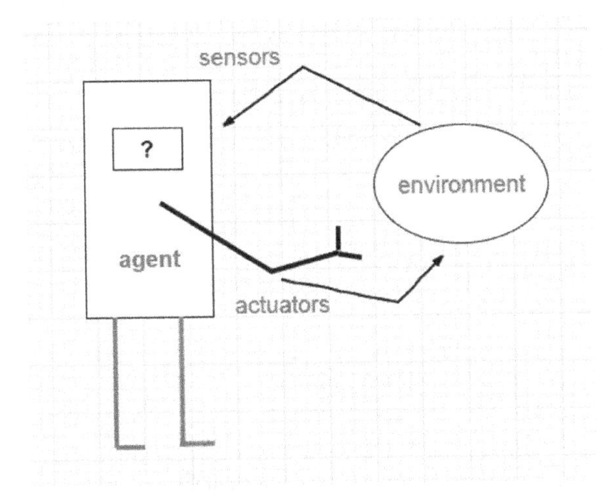

FIGURE 2.2 An agent.

environment. Software developed with the help of agents has two parts: agent and environment. Two major parts of an agent are given below:

- Sensors: Perceiving the environment
- Actuators: Acting upon the environment

Nikola Kasbov defined intelligent agents as a system (see Figure 2.3) that should have the following characteristics:

- Be able to accommodate new algorithms to solve a problem
- Be able to adopt changes in real-time environment

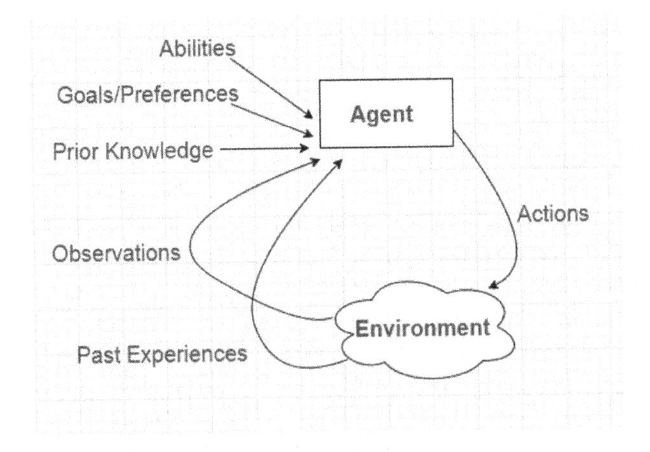

FIGURE 2.3 Intelligent agent.

TABLE 2.1
Different Agents with Sensors and Actuators

Agent	Sensors	Actuators
Human	Eyes, ears	Hand, legs, vocal tract
Robot	Camera, infrared range finder, NLP	Various motors
Software	Keystroke, file contents	Display output devices

- Be able to analyze its behavior in terms of success rate and error rate
- Be able to learn from previous experience

$$\text{Agent} = \text{Architecture} + \text{Agent Program}$$

where
Architecture = Machine where agent is executed
Agent Program = Functional implementation of an agent.

The study of the rational agent and its environment creates an AI system. Sensors are used by agents to sense the environment and actuators act on their environment. An AI agent has properties such as intention, knowledge, and belief. An agent can be human, software or robot. It perceives the environment through sensors and can act upon that environment with actuators (see Table 2.1). An agent completes their tasks in a cycle of perceiving, thinking, and acting.

2.2 NEED FOR EVOLVING AGENTS IN EVOLUTIONARY SOFTWARE SYSTEMS

Systems are required to be adaptive in nature in order to understand the user's needs and the changes that are necessary. A program's functionality has limited flexibility to handle unforeseen events. They are able to act upon conditions for that they were designed. They are able to act according to programming of the software. This rigidity of a program works well in the applications where functionality requirements are not changing with respect to time. This rigidity in the computer programs may negatively affect functionality of the system if its requirements are changing with respect to time. A generic process model is used at design time. It is difficult to tune generic model according to organization need that changes with respect to time (Van Der Aalst et al., 2009). Software engineering delivers static approach of functionality. It is a challenge to develop the software which evolves over a period of time (Eisenbach et al., 2002). Globalization of business also demands frequent changes in software development (Alaranta et al., 2004). Zimmermann (2015) emphasized need for architectural decisions to accommodate rapid changes occurring in the environment. In case of growing need for computer system today, there is a requirement for a system to be able to adjust with a dynamic and unpredictable environment. Therefore, a new approach is needed in software development.

2.2.1 Change of Requirements

Software deployed at a client site works fine for some duration, but as time passes the requirements in the terms of client/customer changes and the working functionalities are not able to satisfy the customer. It is natural that requirements are keeping changing, we cannot stop it. To accommodate these changes in the software, human intervention is required (A. G. Sutcliffe & Maiden, 1993). It is a difficult task to manage human intervention whenever the changes occurred in the functionality. An important phase of software development is requirement gathering. It collects user requirements and communicates these requirements to developers. User requirements cannot be fixed because it is changing with respect to time and software programs are required to include changes. For conventional software systems, new changes are included in programs with the help of a developer that increases maintenance costs and resources. This process makes it difficult to maintain a budget for implementing changes that incur an additional cost. Changes are needed to optimize software so that software can perform better for tomorrow. Software developed using a traditional software development approach is not able to modify functionality of the software itself and that reflects the additional cost for software as well as developer becomes busy most of the time in incorporating new changes occurred as requirements with time.

Changes in requirements of software also increase risk in success of the software. On the other hand, it gives the opportunity for improving usability in the software (McGee & Greer, 2012). Changes may be occurred because of various sources like market and organizational changes. Changes in the requirements are not only affecting in software specification but it is also affecting other components of software development. Therefore, it is necessary to provide flexible and effective changes to the management system (Mu K. D. et al., 2011). Changes in the environment impose constraints on functionality of the software. Software is not able to achieve the goal for that it was developed. Software functionality should be able to adapt these changes. Generally, adaptation of changes reflects the reorganization of software architecture (Hoogendoorn et al., 2011). The client is not able to understand that incorporating changes in the software is not easy. It may also reduce the efficiency of working software. A trust buildup is needed between client and developer for incorporating changes in the software (Chudge & Fulton, 1996).

There are constantly evolving changes in the requirements to meet business goals. Frequent changes in the environment also reflect that it is difficult to find out reasons for the change (Kavakli & Loucopoulos, 2006). Changes in the requirements can occur during software development as well as after software development. Changes may be of different types varies from error correction to change in the behavior of functionality (Lin & Poore, 2008). Changes affect product quality. It also affects the testing of the software because we need to re-test the code that has been changed (Heindl & Biffl, 2008). Requirements are a fundamental activity of software development. It acts as reducing the gap between business and its services (Zhao et al., 2010). There are unknown requirements which are arising in the domain (A. Sutcliffe et al., 2018). Requirements deal with involvement of product engineering, service engineering, and software engineering. Requirements needed for integration are missing and unclear (Berkovich et al., 2014). Large organizations are struggling for management of requirements that may come from informal communications also. There should be suitable methods to adapt and balance these changes in organizations (Heikkilä et al., 2017).

2.2.2 Need for an Evolving System

It is a difficult and tedious job to incorporate new changes through a manual process. Extra time is needed to reflect on the functionality of the software. This problem can be solved by incorporating changes in the software without a programmer/developer. Today, the dynamic world needs it. One approach to handle this problem is that software evolves over time (Van Lamsweerde, 2000). Accommodating the requirements in functionalities in the software is an important phase (Sajid et al., 2010). And these requirements are evolutionary in nature because customer objectives are evolutionary. Perini (2007) emphasized that agents can be developed using agent-oriented software engineering (AOSE) approach. Several methodologies have been proposed for AOSE, each one is addressing software development using agents. An effort toward standardization of AOSE methodologies is in progress, in parallel research is also addressing challenging issues, such as designing agents with autonomic properties i.e., evolving agent. Garcia et al. (2004) discussed use of multiagent systems in software engineering and also addressed integration of agents with an object-oriented approach. Evolving software systems are the best suitable systems to handle a dynamic and unpredictable environment.

There is a requirement of a system that becomes adaptive by understanding changes that are occurring in the requirements of the user. Software developed with the help of agents has two parts: Agent and Environment. Environment represents the outside world with which agents are interacting. Markets of the business impose evolution of currently working software (Stone, 2014). Market affects internal and external resources of the business.

Advancements in technology and standardization create a need for an existing system to update. New technology is also generating information that affects the functionality of the software (Frömel & Kopetz, 2016). Distributed real-time systems have a set of system work that is performed in collaborative and distributed fashion. A major problem in such collaboration is a need for dynamic evolution because it changes the properties of the system (Gezgin et al., 2013). Education is changing over time and it also varies with culture. Education has an impact on the work and technology of the organization. Software supporting learning needs to evolve accordingly to support enhanced learning (Sancho, 2006). A recommender system detects customer's preferences in a proactive manner. Due to the change in the customer's preferences, rules need to be evolved (Zhou & Hirasawa, 2017).

Software evolution is needed in various aspects of applications such as socio-technical, search-based web page, mining of unstructured repositories, and gathering of software requirements. Attention should be paid to inter-dependent software components. Due to the rapid growth of the internet and internet resources, it is comfortable for the user to find related information. It reflects software should be in pace with new information. New requirements emerge when a user is using the software. There are various new requirements which are basically changes in business requirements, errors corrections, adding of new equipment, and improvement in performance and reliability of the system. The complexity of large-scale software yields volatility in requirements. It makes evolution in the software necessary. There is also a risk in the evolution of software-related schedule and budget for new capabilities. Evolution should be carefully incorporated, and evolution should be around new information which has been captured by observation. Evolution should be achieving the ultimate objectivities of the user. Daily maintenance tasks are labor-intensive. It motivates that software should evolve itself.

There are various reasons for change such as missing requirement identified, a bug in the coding, misinterpreted requirement, change in marketplace, and change in legislation. If requirements are finalized at an early stage of software development, then it is more likely that requirements change later and software should be able to cope with. Prioritization of requirements may be implemented during evolution of the software. Software may be evolved by adding and enhancing functionality through a developer. Software may be evolved itself also. But software evolved itself is more effective than software evolved through changing the code. Software that is maintained for a long time needs attention to revisit the architecture of the software (Kelly, 2006). It is necessary to design each element of a set of design for long-time evolution of the software.

Generally, interaction is considered as one of the factors that contribute to the complexity of the software. Software architecture consists of many interacting components that create their own thread of control. By understanding the needs of interacting components, software industries have developed many tools and techniques to model the interactions of the complex software. From the last two decades, the use of agents and multiagent systems have been grown up in the research community. There are many reasons for agent development, of which one is that the agent is autonomous. Today many systems are composed of passive objects. These passive objects have the states on which, an operation can be performed. These objects are just reflecting the static structure of the design of the system. However, some objects are interacting at run time. We can call that a semi-autonomous object, but these objects are not able to make decisions on their own.

There are several reasons to consider agents as a new direction for software engineering. It is natural to have interactive components in the system. Some components are working as passive components and other components are working as active components. Systems consist of data that is distributed over various components. Therefore, control of processing should be distributed using various components. Components also demand cooperation with each other. Components of the system should be able to be flexible for various types of outside components and able to interact with them. All of these qualities are reflected in an agent.

Programmers are generally familiar with object-oriented objects. It is difficult for them to consider a new idea of an agent. Objects have state and perform actions using various methods. Objects are similar to agents. But there are differences between object and agent. Basic differences between object and agent lie in terms of autonomy. Encapsulation is principal characteristic object-oriented programming. It indicates that an object has control over its state. But if an object is declared as public, then the object does not have control over its behavior. Objects have methods that are available for other objects to use. In the case of object-oriented programming, objects share a common goal. In the case of agent development, no such common goal exists. Nor it happens that an agent executes an action because of another agent wants it. One of the biggest differences between an agent and an object is that an agent reflects autonomous behavior. On the other hand, an object in object-oriented programming does not.

An agent is always active. An agent is continuously observing the environment in a loop fashion. An agent is updating its internal state based on the observation. Agents create their own thread for control. In the case object-oriented, objects are active when other objects require their services. However, there is a concept of an active object. We can say an active object is an agent, but the active object does exhibit flexible autonomous behavior. There are three distinctions (Wooldridgey & Ciancarini, 2001) between agent and object:

(a) Agents have a strong notion of autonomy. They are able to make decisions themselves for performing an action on the request.
(b) Agents exhibit flexible behavior such as reactive, proactive, and social. Objects do not show such type of behavior.
(c) Agent systems are inherently multi-threaded. Each agent has at least one thread of control.

There are basically two approaches to the development of agent-based systems:

(i) Extends the existing object-oriented methodologies for the development of agents.
(ii) Adapts other technology or knowledge engineering.

2.2.3 Software System

Software is developed on the basis of requirements which are gathered from the customer. Analysis of requirements is done by an analyst and then it is forwarded to the design and coding phase for implementation. After the implementation, testing is carried out and then it is deployed at a customer site. The customer uses the software and gets outcome by functionalities developed in the software (Gefen & Schneberger, 1996; Bosch, 2000).

Requirement engineering is an important phase in software development. It explores real goals that need to be implemented as functions in the software system.

Requirement engineering also explores constraints needed to be imposed on the functionality of the software system under development. Requirement engineering finds out the relationship among factors that are contributing to the precise specifications of software to be developed. A correct and consistent specification provides consistent behavior of software which is expected by customer/client. Other phases of the software development cycle start when requirements are collected and analyzed. Software developed with Software Development Life Cycle (SDLC) is deployed at a client site and a client starts working on the software. A continuous stream of requirement changes creates a major problem during software development (Navarro et al., 2010). Requirement changes affect all software development phases. It increases the budget and development time of the software project. It is difficult to get consistency in software requirement specifications. Domain knowledge has also impact on effectiveness of requirement engineering. However, teams of both ignorant and aware of the domain are more effective in requirement idea generation (Niknafs & Berry, 2017). Requirements are also impacting return on investment. Businesses are advocating flexibility in requirements collection (Jarke & Lyytinen, 2010). Customer requirements are fluctuating in the modern era. Requirements engineering should be able to accommodate these fluctuations in software requirements specifications. Demand for customized functionality in the software is increasing day by day (Universit & Group, 2012). It shows that current requirement gathering techniques need be to look into and research on new and elaborative techniques are going on.

Software development has the pressure of excessive budget and schedule (Nan & Harter, 2009). However, the new way of software development like pair programming is going on to reduce software development costs (Moise Kattan & Goldman, 2017).

2.2.4 Evolving Software System

Evolving software systems are able to observe the environment and able to take decisions accordingly. After the observations, the system understands new changes in the observations and updates itself according to new changes. Now the evolving system interacts with the environment after adaptation of new changes occurred in the environment and it is able to make better decisions on behalf of the application user. An evolving software system evolves itself without human intervention. This way, an evolving system is able to cope with a dynamic and unpredictable environment.

Evolving software systems are distributed, heterogeneous, decentralized, and independent. They are able to operate in a dynamic and unpredictable environment. Evolving software systems (ESS) prepares themselves for good decision in the future. For a mission-critical system, normal software systems are not able to maintain their availability and reliability for a long time (Mens, Magee, & Rumpe, 2010). However, on the other hand, ESS systems adjust themselves for future compatibility. Local properties and software structure affect the evolution of software systems. ESS systems are working as human. They reflect human intelligence and human reasons. ESS system needs various technologies for its implementation. It ranges from natural science to social science. ESS utilizes industrial experience for further evolution.

The purpose of ESS is to reduce the gap between a real-time scenario and functionality behavior designed in the software. It leads to smart management. An evolving system uses software agents as components of the system. Software agents have different responsibilities and roles in an ESS system. All agents cooperate to each other to fulfill the goal of the evolving system. The evolving system is also used in the area of big data analysis. Architecture of evolving systems uses advanced level design for big data applications (Gorton & Klein, 2015). Evolving systems are designed with architectures that evolve over a period of time (Woods, 2016). Architecture has evolutionary intelligent components.

The evolving software system may have various architecture alternatives for different functionalities of the software (Koziolek et al., 2013). The importance of requirements traceability in the evolving system has been advocated by Mäder & Egyed (2015). Elements of an evolving system should be identified uniquely and there should be linkage among all elements of the evolving system (Wenzel, 2014). Software architecture plays an important role in the sustainability of an evolving system. Classical architectures cannot cope with frequent changes occurring due to maintenance and other aspects. Architecture that supports evolving capability becomes a core element of the system that evolves over time (Naim et al., 2017).

2.3 AGENTS

Evolving software systems are designed and developed using software agents. A software agent is a component of software that has the capability of performing the tasks for another entity which can be software, hardware, or a human entity or user. Agents reflect autonomy and intelligence in their behavior. However, an agent in its basic form may perform pre-defined tasks such as data collection and transmission leaving little room for autonomy and intelligence. Another dimension of agenthood is sociality. Sociality refers to interaction and collaboration with other agents and non-agent entities. Aoftware evolving agent is a software program that has the capability to learn changes that occur in the requirements itself to fulfill user needs. Evolving agent (also called an intelligent agent) is gaining widespread applications in trading markets (Sueyoshi & Tadiparthi, 2007). An agent is a software component that perceives the outside world/environment and acts accordingly. Software developed with the help of agents is adaptive and able to adjust themselves according to changes in the environment. Software developed using agents provides a new way for the development of programs. This way of programming is for open environments such as distributed domains; electronic commerce and web-based systems. Agents have the ability to learn the requirements of users autonomously and able to cooperate with other agents to fulfill the requirements of the user.

An agent has the following properties:

- Autonomy: Agent is able to take decision according to changes occur in the environment.

- Reactivity: Agent is able to respond to changes happening in activities of the external environment.
- Pro-activeness: Agent is able to plan in advance for better decisions in the future and towards achieving the goal.
- Social ability: Agent is able to communicate with other agents towards achieving the goal.

Agents can react according to the environment without human intervention (Banerjee & Tweedale, 2007). It gives a new emerging field: Agent-Oriented Software Engineering (AOSE) (Zambonelli & Omicini, 2004). AOSE consists of agents that are working towards achieving the desired goal.

Agent-based computing has a novel of abstractions and required adaptive methodologies. It advocates the development of new tools to support an agent-based system. There are two views points of engineering about the agent: one with a strong artificial intelligence view in which proactivity of agent and intelligence of agent are addressed; and in the second view, the software engineering view, an agent is a reactive or proactive thread for its environment. A complex system could not be developed with one agent only. It has many agents that have different roles in performing tasks of the complex system. When multiagents are working in a system, an interaction protocol is needed for interaction between agents. Agents also need local context in which agents will live.

Agent-based systems reflect characteristics of agents like autonomy and reactivity. Agent-Oriented Software Engineering (AOSE) provides enriched methodologies for abstraction of agents. An agent-based system is suitable for a wide class of scenarios and applications. The agent consists of artificial intelligence which is an important characteristic of an agent. AOSE is suitable for a high-level and complex system. Some examples of components that can be modeled with AOSE are autonomous network processes, computing based-sensors, personal digital assistants, and robots. On one hand, software systems like internet applications, peer-to-peer communications, sensor networks, and pervasive computing may be designed using agent methodologies. There some issues in implementing agents and multiagent systems. These issues include:

(i) How to move from agent-based design to concrete agent code
(ii) Agent architecture abstraction should be independent of implementation
(iii) What the communication architectures for interaction among agents are
(iv) What are available tools for implementing agents are

There are two major categories of methodologies: (1) object-oriented methodologies, and (2) belief, desire, intention (BDI)–based methodologies

It is still to decide which methodology to follow based on the performance and maintenance of the system. AOSE multiagent infrastructure demands the following various components:

(i) Ruling interactions
(ii) Middleware layer support communications and coordination activities

(iii) Active layer of communications instead of passive layer
(iv) Support of execution of interaction protocol
(v) Providing help to other agents
(vi) Proactively controlling interactions
(vii) Routing messages facilities
(viii) Synchronization of interactions

Agents in multiagent systems (MAS) interact with other agents. An agent first finds another agent and keeps track of the service agents that are around it. The agent understands the characteristics and services of other agents. Then it routes the message accordingly. There may be a facilitator agent in the communication infrastructure. A facilitator agent receives the messages to be delivered to other agents and accordingly, it re-routes the messages. Communication infrastructure supported by AOSE should be able to identify internal agent behaviors and inter-agent interaction protocol. Communication infrastructure should be able to handle global interactions. Problems with the communication infrastructure are that it needs to retune sometime for application domain and global behavior of the agent. There should be embedded law in communication infrastructure for communication. It is needed to prevent malicious behavior. Enforcement of law also helps in re-configuration of protocol according to the environment. Separation of communication infrastructure from agent development reduces complexity in agent analysis and design. Inter-agent and intra-agent issues are handled separately. Design of agent should be adaptive and able to support distributed computing. While implementing agent, its infrastructure also needs to be implemented. Now agents are no longer homogenous. They also need standardization for designing the agent. Open issues in agent-oriented software engineering include:

(i) Designing of MAS for mobility and ubiquity
(ii) Agent design for dynamic and complex systems
(iii) Relationship of the agent with social networks
(iv) Self-organization
(v) Performance model for measuring activities of the agent

An adaptive system using agent technology has been effective for leaning market mechanisms (Seppecher et al., 2019) and applications of self-adaptiveness for data management found promising (Rafique et al., 2019). It is more effective to apply software agents in new situations (Jaeger et al., 2019).

2.3.1 Evolving Agents

To design an agent, it is necessary to have architecture (Jennings, 2000). Use of evolving agents has been advocated by several authors (Rosenblatt, 1958; Samuel, 1959; Fogel, 2006). Evolving agents are able to take suitable decisions by observing the environment (Hanna & Cagan, 2009). Evolving agents need evolving architecture. Evolving architecture is able to adapt to new changes occurring in the environment (Nunes et al., 2009). The authors also addressed separate handling of agent's designing and development. Xiao Xue et al. (2006) emphasized agent-oriented software development for solving modern complex problems. The authors present an agent structure which is called C4I (command, control, communication, computer and information). It is used as information processing in a naval warship. The authors concluded the requirement of a detail agent structure that plays a key role in agent design and development. Various frameworks of agent-oriented requirements engineering have been discussed by Singh et al. (2008) and the authors proposed parameters for evaluating the frameworks. They emphasized that customer requirements are dynamic in nature and it can be addressed using agents. Further research is needed in this area. A spiral model framework based on agents for requirement engineering is proposed by Gaur et al. (2010a). It uses story cards to specify the requirements, which are useful in an agent's development. They emphasized further study is needed to satisfy the customer. Software development using agents is a recent paradigm for designing and development of software (Gaur Vibha et al., 2010b; Huiying and Zhi 2009). Agents may have goals and tasks that software agents would try to accomplish to fulfill the goals of the system (Gaur et al., 2010b). Huiying and Zhi (2009) presented graphical symbols that describe the requirements of agents. A combined framework using i* model and UML activity is presented by Bhuiyan et al. (2007). A graphical architecture for web services is presented by (Zhu & Shan, 2005). The architecture is written using agent modeling language CAMLE and abstract using specification language SLABS. They illustrated how agents can be used in web services applications. An emphasis on adding goals is given by Khallouf & Winikoff (2005) in interaction diagrams and protocols. They presented the importance of goals with the help of case study and concluded that adding a goal in the design of the agent is both usable and effective. Vilkomir et al. (2004) explored a combination of i* framework and Z formal methods for requirements collection for agents. They used i* framework for an early phase of requirements and further requirements are specified using Z notations. Agents may also play roles to perform their designated tasks (Chan & Sterling, 2003). A component-based agent architecture has been proposed by Qu Youtian et al. (2009). A semi-automated change propagation framework based on agents is proposed by Dam et al. (2006) for evolving software systems. Agent-based development for evolving software systems is promising and has been advocated by Pour (2002). Evolving agents are a core part of an evolving software system. Evolving agents observe new changes and incorporates these changes in the software without any human intervention.

2.3.2 Agent Architecture

Evolution of software systems has been emphasized by Paderewski-Rodríguez et al. (2004). Two techniques: adaptive and inheritance have been presented in the evolutionary framework of the paper. In an adaptive technique, the agent modified itself according to the environment. In inheritance technique, an agent is developed. A similar evolving model that consists of two types of agents: student agent and teacher agent has been proposed by Wang X et al. (2007) for the system that answers students' queries. The answering system consists of an agent which has a 5-dimensional template <A, I, S, T, K> where A defines the name which is unique, I defines the interface, S shows the agent's status, T shows the agent's behavior, and K defines the source of knowledge. Vandewoude and Berbers (2002) demonstrates dynamic communications with the help of asynchronous messages using ports which are connected to components. A general architecture of multiagent for the website has been proposed by Jonker C et al. (2001). These agents are intelligent. Authors recognize classes for multiagents, including personal assistant agents (software agents), customers (human agents), employees (human agents), and website agents (software agents). Stefano et al. (2004) discussed the adaptability in the development of software and proposed that classes be replaced at run time. A component which is called checker in developed software observes replacement time and replaces the class, compile the class and run the class. A structure or directory is required to store classes to be connected with the software. The architecture for intelligent virtual environments that can be used for training has been presented in Mendez and Antonio (2008). It is in peer-to-peer style. This offers service-oriented behavior using publish-subscribe style. It is necessary to adapt changes over a period of time in life (Q. Wang et al., 2003). Authors categorized types of changes which are Runtime Platform changes, Interactive Context changes, and customer requirements changes and emphasized requirements of evolutionary software. An agent model based on a BDI model is suggested in Meng (2009) that communicates with other agents using ports. A new methodology, "Gaia," for analysis and design of agent proposed by Spanoudakis and Moraitis (2010). The authors defined the architecture which has three levels. It consists of roles that are defined for software agents. Similarly, an agent methodology called "Tropos" with the help of a case study eCulture System proposed by Bresciani et al. (2004). BDI concept is implemented by Tropos methodology. The methodology presented early requirements module in the life cycle of software development. Two types of organizations of agents: "Mechanical" and "Organic" described in Isern et al. (2011). A multiagent system (MAS) based on simulation has five basic steps: exact verification, pointwise simulation, modeling, tuning, and approximate verification (May & Zimmer, 1996). Architecture that provides provision for machine-to-machine communication (M2M) with the help of 2G-RFID and Internet of Things (IoT) technology, discussed by Oluyomi, Karunasekera, and Sterling (2008). The structure of BDI agent could be: <Agent> {<Beliefs> Constraints; Data Structures; <Desires> Values; Condition; Functions; <Intentions> Methods; Procedures;} shows an implementation approach (Meng, 2009). An approach of model-driven architecture is implemented in agent-oriented development (Jawahar & Nirmala, 2015). The structure of the meeting agent uses the multiagent concept efficiently (Huamonte et al., 2005). Meeting Agent worked for both synchronous and asynchronous algorithms. The authors found that an asynchronous is less efficient than a synchronous one. There are big challenges in designing of agent behavior. A sequential technique that moves from core building blocks to enhancing blocks is suitable for agent behavior. A persistence memory may be used for designing evolving behavior (De Jong, 2008; Krzywicki et al., 2014). The architecture of agents defines modules that are contributing to the completion of a task. It also defines relationships among modules for designing and developing an agent. Selecting the right architecture gives a proper way to solve the complex problem of designing and developing an agent. It reduces the time of development and enhances the efficiency of the solution. Authors emphasized the development of evolving architecture in software development (Rosenblatt, 1958; Samuel, 1959; Fogel, 1995). Agents are making suitable decisions while evolving for a solution (Hanna & Cagan, 2009). There is a need for software development paradigms that motivates adaptable architecture for the environment that changes with respect to time (Nunes et al., 2009). The importance of system architecture in dynamic environments gives commercial importance. Some approaches related to the architecture of agent systems have been discussed by Bratman et al. (1988), Doyle (1992), Shoham (1993), and Rao & Georgeff (1991).

There is a lack of a well-defined structure that helps in designing agent systems (Jennings et al., 1998). A road map on agent research and its development is given by Jennings et al. (1998). Various architectures and theories in this regard are specified by Wooldridge & Jennings (1995) and Wooldridge et al. (2000). Intelligent agents are also useful in trading markets (Samuel, 1959). An agent has various properties (Sahu et al., 2016), including autonomy, reactivity, proactiveness, and social ability. A popular architecture for designing an agent is belief, desire, and intention (BDI) (Rao & Georgeff, 1995; Amalanathan et al., 2015). However, the BDI model is lacking in learning capability. BDI is not able to adapt the changes based on past experience. It is an important requirement for an agent to be evolved (Phung et al., 2005). The lack of learning in BDI architecture is also mentioned by Guerra-Hernández et al. (2005). A good research from theory to implemented applications in the area of intelligent agent has been emphasized (Hindriks et al., 1999). BDI framework demonstrated a specification of designing agents that are intelligent. However, less work has been done for refining such specifications for implementing BDI using known programming languages (Singh & Asher, 1991; Hoek et al., 1994; Rao, 1996; Linder et al., 1996; Hindriks et al., 1999). Adaptive behavior can be created using reusable agent architecture (Decker & Sycara, 1997).

Intelligent is also applied in big data (Bhargava et al., 2016). There is also interprocess synchronization using agents (Bhargava & Sinha, 2012). Agents are used in mobiles for resource management (Ogwu et al., 2006) and in decision support systems with multi-criteria (Wen, 2009). A suitable architecture is needed for implementation of agents. Some architectures are proposed by Bratman et al. (1988), Doyle (1992), Shoham (1993), Hoek et al. (1994), and Rao & Georgeff (1991). A roadmap of an agent's architecture is presented by David & Fogel (2006). There is a lack of agent architecture for developing an evolving agent (David & Fogel, 2006).

Following is a description of some methodologies (Wooldridgey & Ciancarini, 2001) used in agent system development.

(i) AAII methodology: Australian AI Institute (AAII) methodology has two models: internal and external models. The external model represents agents and relationships among agents. The internal model is concerned with the internal working of the agents which is related to the BDI paradigm. Inheritance relationships are defined among the agents. The external model itself consists of two further models: agent model and interaction model.

1. A brief description of this methodology is given below:
   a) Identify the roles to develop agent class hierarchy
   b) Identify the responsibility and goal of each agent
   c) Determine a plan for each goal to achieve it
   d) Determine the information needed for each goal and plan. It means determine the belief structure of the system
2. The process is iterated to develop a functional agent-oriented system.

(ii) Gaia Methodology: This methodology suggests systematic movement from requirements to design. Design is in the detail and it can be implemented directly. Analyst moves from abstract to concrete concepts. Gaia methodology is based on the process of organizational design. It consists of two types of concepts: abstract and concrete concepts. Abstract concept is used by analysts to captures the requirements of the system. It is not necessary for abstract concepts to be realized later. Concrete concepts are used in designing part of the system. They are directly implemented in the system. A role has four attributes: responsibilities, permissions, activities, and protocols. Responsibility is the functionality of the role. Each role has permissions i.e., rights. Permissions identify resources associated with each role. Activities are actions that the agent performs for itself. Protocols of the agents define interaction mechanisms. It is needed for interacting with other agents.

(iii) Agent UML: Researches extended the UML notations for agent-oriented development. Proposed UML notations are expressing concurrent threads for interactions and a notation of "role" for specifying the role of the agent.

(iv) DESIRE: It is a framework for designing and formal specifications for agent-based systems. DESIRE supports compositional systems. It has a graphical editor and tools for the development of agent systems.

(v) Cassiopeia: This methodology supports the behavior view of agents. This methodology has three steps.
   (a) Identify basic behaviors of the system
   (b) Identify relationships between basic behaviors
   (c) Identify overall organization behavior

(vi) Agents in Z: In this approach, an agent is developed using Z language. The framework defines the hierarchical structure of the agents. The framework first defines entities that have attributes. Secondly it defines objects that have capabilities. Thirdly it defines agents that have goals and are active. Lastly it defines autonomous agents that have motivations.

A good methodology should be able to differentiate between agents and objects clearly. In the object-oriented methodology, many aspects of agents cannot be captured such as pro-activeness, dynamic reaction and cooperation, and negotiation with other self-interested agents. The most active agent-oriented software engineering is a formal method. Formal methods are working in three areas: specification of the system, directly programming the system, and verification of the system. Here a number of agent methodologies have been discussed, but there is still a need for research in the direct implementation of agents.

2.3.3 Application Domain

The grocery industry has manual and repetitive work. It requires automation to reduce labor. There is a requirement for an evolving agent to solve this problem that can evolve itself to handle the changes occurring in the repetitive work (Hanna & Cagan, 2009). Software agents are also used in the marketing of supermarkets (Rykowski, 2005). A personalized recommender system was developed that suggests new products to customers (Lawrence et al.,2001). Architecture Agent 7 (AGNT7) of evolving agent has been experimented on an application for purchasing items (Sahu et al., 2017). An agent that provides a list of prices of items for purchasing has been discussed by Benedicenti et al. (2004). A self-evolving agent for simulation is suggested by Kang et al. (2017). Software agents are also used as middleware for client-side performance (Ivanović et al., 2017). Agile software can use software agents for identification, assessment, and monitoring of risks (Odzaly et al., 2017). An adaptive multiagent tool has been developed that evolves ontologies from the text (Sellami et al., 2013). A framework using agents simulates health states to predict epidemics automatically (Miksch et al., 2014). Agent systems can be helpful in self-adaptive systems (Weyns & Georgeff, 2010). An agent framework for analysis of cyber events for various networks is proposed by Kendrick et al. (2018). Agents are utilized for individual tracking in a constrained environment (Zaghetto et al., 2017).

2.3.3.1 Types of Agents

There are several reasons to consider agents as a new direction in software engineering. It is natural to have interactive components in the system. Some components are working as passive components and other components are working as active components. Systems consist of data that are distributed over various components. Therefore, control of processing should be distributed using various components. Components also demand cooperation with each other. Components of the system should be flexible for various types of outside components and be able to interact with them. All of these qualities are reflected in an agent.

Agents can be grouped into different classes according to capability and degree of intelligence:

- Simple Reflex Agents: Simple reflex agent action is defined only on the current perception, ignoring perception history. It is based on the condition-action rule. If a condition is true, then take the action, else not, as shown in Figure 2.4.
- Model-Based Reflex Agents: Model-based reflex agents keep track of internal state which is updated during current perception based on percept history, as shown in Figure 2.5.
- Goal-Based Agents: Goal-based agent action is defined in terms of distance from the goal see Figure 2.6. The agent has to select one option among multiple possibilities to reduce the distance. These agents are more flexible and require planning based on search. The behavior of these agents can be changed easily.

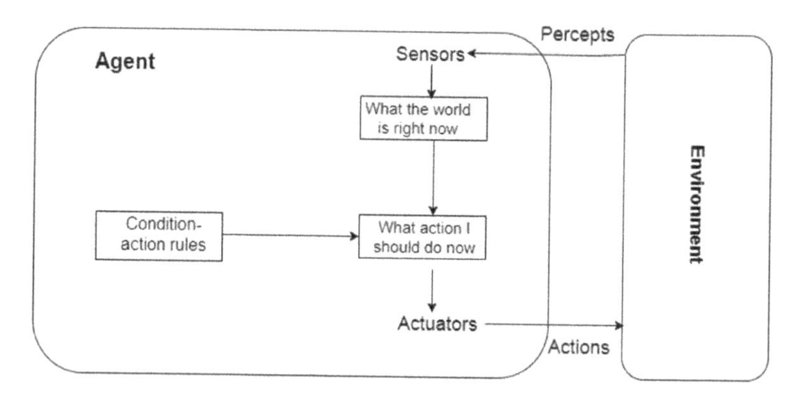

FIGURE 2.4 Simple reflex agent.

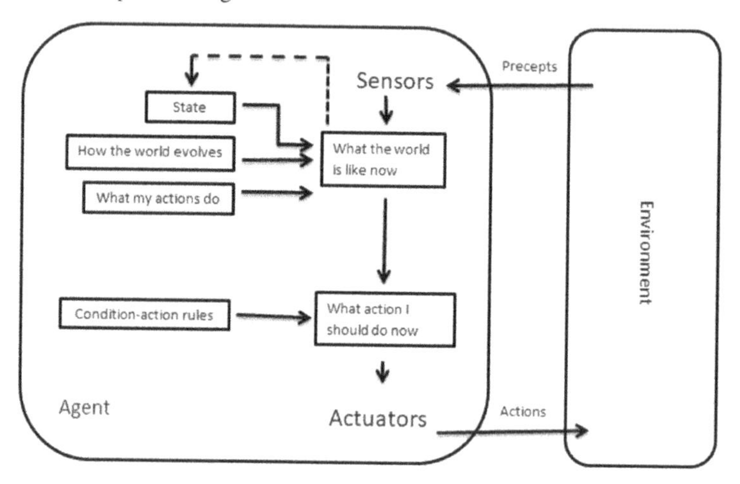

FIGURE 2.5 Model-based reflex agents.

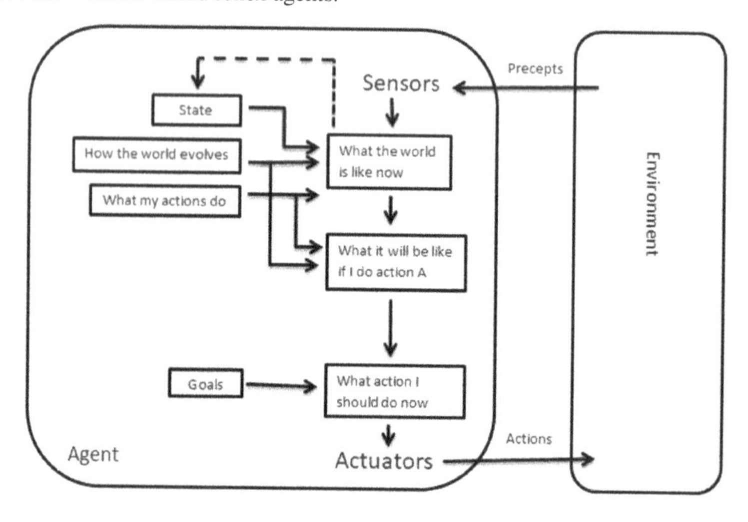

FIGURE 2.6 Goal-based agent.

- Utility-Based Agents: The term "utility" is used to describe how "happy" the agent is. The action of a utility agent is defined in terms to maximize the utility as shown in Figure 2.7.
- Learning Agents: The agent that has capabilities of learning from his history is defined as a learning agent (see Figure 2.8). A learning agent's initial actions are based on some knowledge and then adapt through learning.

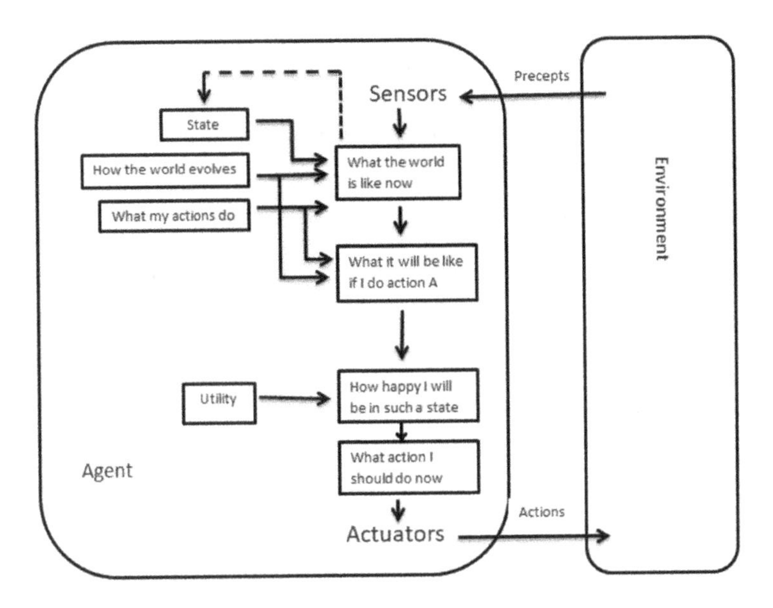

FIGURE 2.7 Utility-based agent.

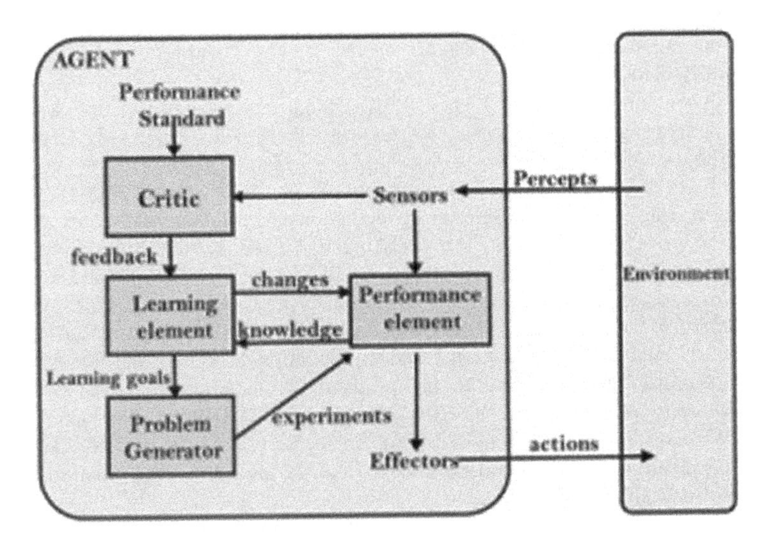

FIGURE 2.8 Learning agent.

REFERENCES

Alaranta, M., Valtonen, T., & Isoaho, J. (2004). Software for the changing E-Business. In *Digital Communities in a Networked Society* (Vol. 139, pp. 103–115). https://doi.org/10.1007/1-4020-7907-9_9

Amalanathan, A., Acouncia, S. M., Vairamuthu, S., & Vasudevan, M. (2015). A framework for E-Governance system using linked data and belief-desire-intention agent. *Indian Journal of Science and Technology*, *8*(15). https://doi.org/10.17485/ijst/2015/v8i15/54038

Banerjee, D., & Tweedale, J. (2007). Reactive (Re) planning agents in a dynamic environment. In *Intelligent Informational Processing III* (Vol. 228, pp. 33–42). Boston, MA: Springer US. https://doi.org/10.1007/978-0-387-44641-7_4

Benedicenti, L., Chen, X., Cao, X., & Paranjape, R. (2004). An agent-based shopping system. *Canadian Conference on Electrical and Computer Engineering*, *2*, 703–705. https://doi.org/10.1109/CCECE.2004.1345210

Berkovich, M., Leimeister, J. M., Hoffmann, A., & Krcmar, H. (2014). A requirements data model for product service systems. *Requirements Engineering*, *19*(2), 161–186. https://doi.org/10.1007/s00766-012-0164-1

Bhargava, D., & Sinha, M. (2012). Design and implementation of agent based inter process synchronization manager. International Journal of Computer Applications, *46*(21), 17–22. https://doi.org/10.5120/7065-9670

Bhargava, D., Poonia, R., & Arora, U. (2016). Design and development of an intelligent agent based framework for predictive analytics. In *Proceedings of the 10th INDIACom; 2016 3rd International Conference on Computing for Sustainable Global Development, INDIA Com 2016*, Bharati Vidyapeeth's College of Engineering, New Delhi, India (pp. 3715–3718).

Bhuiyan, M., Zahidul Islam, M. M., Krishna, A., & Ghose, A. (2007). Integration of agent-oriented conceptual models and UML activity diagrams using effect annotations. In *Proceedings - International Computer Software and Applications Conference* (Vol. 1, pp. 171–178). IEEE. https://doi.org/10.1109/COMPSAC.2007.130

Bosch, J. (2000). *Design and Use of Software Architectures: Adopting and Evolving a Product-Line Approach*. Addison-Wesley Professional. Pearson Education.

Bratman, M. E., Israel, D. J., & Pollack, M. E. (1988). Plans and resource-bounded practical reasoning. *Computational Intelligence*, *4*(3), 349–355. https://doi.org/10.1111/j.1467-8640.1988.tb00284.x

Broy, M., Fuhrmann, S., Huemer, J. *et al.* (2012). Requirement-based function design in the automotive engineering process. *ATZelektronik Worldwide*, *7*(6), 54–59. https://doi.org/10.1365/s38314-012-0137-7

Bresciani, P., Perini, A., Giorgini, P., Giunchiglia, F., & Mylopoulos, J. (2004). Tropos: an agent-oriented software development methodology. *Autonomous Agents and Multi-Agent Systems*, *8*(3), 203–236. https://doi.org/10.1023/B:AGNT.0000018806.20944.ef

Chan, K., & Sterling, L. (2003). Specifying roles within agent-oriented software engineering. In *Proceedings - Asia-Pacific Software Engineering Conference, APSEC* (Vol. 2003-January, pp. 390–395). IEEE. https://doi.org/10.1109/APSEC.2003.01254393

Chudge, J., & Fulton, D. (1996). Trust and co-operation in system development: applying responsibility modelling to the problem of changing requirements. *Software Engineering Journal*, *11*(3), 193–204. https://doi.org/10.1049/sej.1996.0025

Dam, K., Winikoff, M., & Padgham, L. (2006). An agent-oriented approach to change propagation in software evolution. In Software Engineering Conference Australia, Sydney, NSW, Australia (p. 10). IEEE.

De Jong, K. A. (2008). Evolving intelligent agents: a 50 year quest. *IEEE Computational Intelligence Magazine*, *3*(1), 12–17. https://doi.org/10.1109/MCI.2007.913370

Decker, K. S., & Sycara, K. (1997). Intelligent adaptive information agents. *Journal of Intelligent Information Systems*, *9*(3), 239–260. https://doi.org/10.1023/A:1008654019654
Doyle, J. (1992). Rationality and its roles in reasoning. *Computational Intelligence*, *8*(2), 376–409. https://doi.org/10.1111/j.1467-8640.1992.tb00371.x
Eisenbach, S., Sadler, C., & Shaikh, S. (2002). Evolution of distributed java programs.In *International Working Conference on Component Deployment* (pp. 51–66). Berlin: Springer.
Fogel, D. B. (1995). Evolutionary computation: toward a new philosophy of machine intelligence. Ed. J. B. Anderson, Piscataway, NJ: IEEE Press, 274.
Fogel, D. B. (2006). *Evolutionary Computation : Toward a New Philosophy of Machine Intelligence*. John Wiley & Sons.
Frömel, B., & Kopetz, H. (2016). Interfaces in evolving cyber-physical systems-of-systems. In *Lecture Notes in Computer Science* (Including Subseries Lecture Notes in Artificial Intelligence and Lecture Notes in Bioinformatics) (Vol. 10099 LNCS, pp. 40–72).Cham: Springer. https://doi.org/10.1007/978-3-319-47590-5_2
Garcia, A. F., de Lucena, C. J. P., & Cowan, D. D. (2004). Agents in object-oriented software engineering. Software: *Practice and Experience*, *34*(5), 489–521. https://doi.org/10.1002/spe.578
Gaur, V., Soni, A., & Bedi, P. (2010a). An agent-oriented approach to requirements engineering. In 2010 *IEEE* 2nd International Advance Computing Conference (IACC) (pp. 449–454). https://doi.org/10.1109/IADCC.2010.5422878
Gaur, V., Soni, A., & Bedi, P. (2010b). An application of multi-person decision-making model for negotiating and prioritizing requirements in agent-oriented paradigm. In *DSDE 2010—International Conference on Data Storage and Data Engineering* (pp. 164–168). https://doi.org/10.1109/DSDE.2010.59
Gefen, D., & Schneberger, S. L. (1996). The non-homogeneous maintenance periods: a case study of software modifications. In International Conference on Software Maintenance, Proceedings (pp. 134–141). IEEE. https://doi.org/10.1109/ICSM.1996.564998
Gezgin, T., Henkler, S., Rettberg, A., & Stierand, I. (2013). Contract-based compositional scheduling analysis for evolving systems. In *IFIP Advances in Information and Communication Technology* (Vol. 403, pp. 272–282). Berlin: Springer. https://doi.org/10.1007/978-3-642-38853-8_25
Gorton, I., & Klein, J. (2015). Distribution, data, deployment: software architecture convergence in big data systems. *IEEE Software*, *32*(3), 78–85. https://doi.org/10.1109/MS.2014.51
Guerra-Hernández, A., El Fallah-Seghrouchni, A., & Soldanof, H. (2005). Learning in BDI multi-agent systems. *Computational Logic in Multi-Agent Systems*, *3259*(Ml), 218–233. https://doi.org/10.1007/978-3-540-30200-1_12
Hanna, L., & Cagan, J. (2009). Evolutionary multi-agent systems: an adaptive and dynamic approach to optimization. *Journal of Mechanical Design*, 131(1), 011010. https://doi.org/10.1115/1.3013847
Heikkilä, V. T., Paasivaara, M., Lasssenius, C., Damian, D., & Engblom, C. (2017). Managing the requirements flow from strategy to release in large-scale agile development: A case study at Ericsson. *Empirical Software Engineering*, *22*(6), 2892–2936. https://doi.org/10.1007/s10664-016-9491-z
Heindl, M., & Biffl, S. (2008). Modeling of requirements tracing. In *Lecture Notes in Computer Science* (Including Subseries Lecture Notes in Artificial Intelligence and Lecture Notes in Bioinformatics) (Vol. 5082 LNCS, pp. 267–278). Berlin: Springer. https://doi.org/10.1007/978-3-540-85279-7_21
Hindriks, K. V., De Boer, F. S., Van der Hoek, W., Meyer, J. J. C., & Ch. Meyer, J.-J. (1999). Agent programming in 3APL. *Autonomous Agents and Multi-Agent Systems*, *2*(4), 357–401. https://doi.org/10.1023/A:1010084620690

Hoek, W., Linder, B., & Meyer, J.-J. C. (1994). A logic of capabilities. In *International Symposium on Logical Foundations of Computer Science* (pp. 366–378). Berlin: Springer. https://doi.org/10.1007/3-540-58140-5_34

Hoogendoorn, M., Jonker, C. M., & Treur, J. (2011). A generic architecture for redesign of organizations triggered by changing environmental circumstances. *Computational and Mathematical Organization Theory*, *17*(2), 119–151. https://doi.org/10.1007/s10588-011-9084-8

Huamonte, J., Smith, K., Thumma, D., & Bowles, J. (2005). The use of roles to model agent behaviors for model driven architecture. Proceedings *of IEEE SoutheastCon, 2005*, 594–598. https://doi.org/10.1109/SECON.2005.1423311

Huiying, X., & Zhi, J. (2009). An agent-oriented requirement graphic symbol representation and formalization modeling method. In 2009 WRI World Congress on Computer Science and Information Engineering.

Isern, D., Sánchez, D., & Moreno, A. (2011). Organizational structures supported by agent-oriented methodologies. *Journal of Systems and Software*, *84*(2), 169–184. https://doi.org/10.1016/j.jss.2010.09.005

Ivanović, M., Vidaković, M., Budimac, Z., & Mitrović, D. (2017). A scalable distributed architecture for client and server-side software agents. *Vietnam Journal of Computer Science*, *4*(2), 127–137. https://doi.org/10.1007/s40595-016-0083-z

Jaeger, C. B., Hymel, A. M., Levin, D. T., Biswas, G., Paul, N., & Kinnebrew, J. (2019). The interrelationship between concepts about agency and students' use of teachable-agent learning technology. *Cognitive Research: Principles and Implications*, *4*(1), 14. https://doi.org/10.1186/s41235-019-0163-6

Jarke, M., & Lyytinen, K. (2010). High impact requirements engineering. *Wirtschaftsinformatik*, *52*(3), 115–116. https://doi.org/10.1007/s11576-010-0218-2

Jawahar, S., & Nirmala, K. (2015). Quantification of learner characteristics for collaborative agent based e-learning automation. *Indian Journal of Science and Technology*, *8*(14). https://doi.org/10.17485/ijst/2015/v8i14/73119

Jennings, N. R. (2000). On agent-based software engineering. *Artificial Intelligence*, *117*(2), 277–296. https://doi.org/10.1016/S0004-3702(99)00107-1

Jennings, N. R., Sycara, K., & Wooldridge, M. (1998). A roadmap of agent research and development. *Autonomous Agents and Multi-Agent Systems*, *1*(1), 7–38. https://doi.org/10.1023/A:1010090405266

Jonker, C. M., Lam, R. A., & Treur, J. (2001). A reusable multi-agent architecture for active intelligent websites. *Applied Intelligence*, *15*(1), 7–24. https://doi.org/10.1023/A:1011248605761

Kang, D. O., Bae, J. W., & Paik, E. (2017). Incremental self-evolving framework for agent-based simulation. In *Proceedings*—2016 International Conference on Computational Science and Computational Intelligence, CSCI 2016 (pp. 1428–1429). IEEE. https://doi.org/10.1109/CSCI.2016.0282

Kavakli, E., & Loucopoulos, P. (2006). Experiences with goal-oriented modeling of organizational change. *IEEE Transactions on Systems, Man and Cybernetics Part C: Applications and Reviews*, *36*(2), 221–235. https://doi.org/10.1109/TSMCC.2004.840066

Kelly, D. (2006). A study of design characteristics in evolving software using stability as a criterion. *IEEE Transactions on Software Engineering*, *32*(5), 315–329. https://doi.org/10.1109/TSE.2006.42

Kendrick, P., Criado, N., Hussain, A., & Randles, M. (2018). A self-organising multi-agent system for decentralised forensic investigations. *Expert Systems With Applications*, *102*, 12–26. https://doi.org/10.1016/j.eswa.2018.02.023

Khallouf, J., & Winikoff, M. (2005). Towards goal-oriented design of agent systems. In *Proceedings - International Conference on Quality Software* (Vol. 2005, pp. 389–394). IEEE. https://doi.org/10.1109/QSIC.2005.68

Koziolek, H., Schlich, B., Becker, S., & Hauck, M. (2013). Performance and reliability prediction for evolving service-oriented software systems: industrial experience report. *Empirical Software Engineering, 18*(4), 746–790. https://doi.org/10.1007/s10664-012-9213-0

Krzywicki, D., Faber, Byrski, A., & Kisiel-Dorohinicki, M. (2014). Computing agents for decision support systems. *Future Generation Computer Systems, 37*, 390–400. https://doi.org/10.1016/j.future.2014.02.002

Lawrence, R. D., Almasi, G. S., Kotlyar, V., Viveros, M. S., & Duri, S. S. (2001). Personalization of supermarket product recommendations. *Data Mining and Knowledge Discovery, 5*(1–2), 11–32. https://doi.org/10.1023/A:1009835726774

Leung, R. W. K., Lau, H. C. W., & Kwong, C. K. (2003). On a responsive replenishment system: a fuzzy logic approach. *Expert Systems, 20*(1), 20–32. https://doi.org/10.1111/1468-0394.00221

Lin, L., & Poore, J. H. (2008). Pushing requirements changes through to changes in specifications. *Frontiers of Computer Science in China, 2*(4), 331–343. https://doi.org/10.1007/s11704-008-0034-7

Linder, B., Hoek, W., & Meyer, J.-J. C. (1996). Formalising Motivational Attitudes of Agents (pp. 17–32). Berlin: Springer. https://doi.org/10.1007/3540608052_56

Mäder, P., & Egyed, A. (2015). Do developers benefit from requirements traceability when evolving and maintaining a software system? *Empirical Software Engineering, 20*(2), 413–441. https://doi.org/10.1007/s10664-014-9314-z

May, E. L., & Zimmer, B. A. (1996). The evolutionary development model for software. *Hewlett-Packard Journal, 47*(4), 39–45.

McGee, S., & Greer, D. (2012). Towards an understanding of the causes and effects of software requirements change: two case studies. *Requirements Engineering, 17*(2), 133–155. https://doi.org/10.1007/s00766-012-0149-0

Mendez, G., & de Antonio, A.. (2008). A modifiable agent-based software architecture for intelligent virtual environments for training. In Seventh Working IEEE/IFIP Conference on Software Architecture (WICSA 2008) (pp. 319–322). IEEE. https://doi.org/10.1109/WICSA.2008.35

Meng, Y. (2009). Agent-based reconfigurable architecture for real-time object tracking. *Journal of Real-Time Image Processing, 4*(4), 339–351. https://doi.org/10.1007/s11554-009-0116-2

Mens, T., Magee, J., & Rumpe, B. (2010). Evolving software architecture descriptions of critical systems. *Computer, 43*(5), 42–48. https://doi.org/10.1109/MC.2010.136

Miksch, F., Urach, C., Einzinger, P., & Zauner, G. (2014). A flexible agent-based framework for infectious disease modeling. In *Lecture Notes in Computer Science* (Including Subseries Lecture Notes in Artificial Intelligence and Lecture Notes in Bioinformatics) (Vol. 8407 LNCS, pp. 36–45). Berlin: Springer. https://doi.org/10.1007/978-3-642-55032-4_4

Moise Kattan, H., & Goldman, A. (2017). Software development practices patterns. In *Proceedings of the International Conference on Agile Software Development: Agile Processes in Software Engineering and Extreme Programming* (XP 2017), Cologne, Germany, 22-26 May, 2017 (pp. 298–303). Cham: Springer. https://doi.org/10.1007/978-3-319-57633-6_23

Mu, K. D., Liu, W., Jin, Z., Hong, J., & Bell, D. (2011). Managing software requirements changes based on negotiation-style revision. *Journal of Computer Science and Technology, 26*(5), 890–907. https://doi.org/10.1007/s11390-011-0187-y

Naim, S. M., Damevski, K., & Hossain, M. S. (2017). Reconstructing and evolving software architectures using a coordinated clustering framework. *Automated Software Engineering, 24*(3), 543–572. https://doi.org/10.1007/s10515-017-0211-8

Nan, N., & Harter, D. E. (2009). Impact of budget and schedule pressure on software development cycle time and effort. *IEEE Transactions on Software Engineering, 35*(5), 624–637. https://doi.org/10.1109/TSE.2009.18

Navarro, I., Leveson, N., & Lunqvist, K. (2010). Semantic decoupling: reducing the impact of requirement changes. *Requirements Engineering, 15*(4), 419–437. https://doi.org/10.1007/s00766-010-0109-5

Niknafs, A., & Berry, D. (2017). The impact of domain knowledge on the effectiveness of requirements engineering activities. *Empirical Software Engineering, 22*(1), 80–133. https://doi.org/10.1007/s10664-015-9416-2

Nunes, I., Nunes, C., Kulesza, U., & Lucena, C. (2009). Developing and Evolving a Multiagent System Product Line: An Exploratory Study (pp. 228–242). Berlin: Springer. https://doi.org/10.1007/978-3-642-01338-6_17

Odzaly, E. E., Greer, D., & Stewart, D. (2017). Agile risk management using software agents. *Journal of Ambient Intelligence and Humanized Computing, 9*(3), 823–841. https://doi.org/10.1007/s12652-017-0488-2

Ogwu, F. J., Talib, M., & Aderounmu, G. A. (2006). An analytical survey of mobile agent for resource management in a network. *Journal of Statistics and Management Systems, 9*(2), 427–439. https://doi.org/10.1080/09720510.2006.10701215

Oluyomi, A., Karunasekera, S., & Sterling, L. (2008). Description templates for agent-oriented patterns. *Journal of Systems and Software, 81*(1), 20–36. https://doi.org/10.1016/j.jss.2007.06.020

Paderewski-Rodríguez, P., Torres-Carbonell, J. J., Rodríguez-Fortiz, M. J., Medina-Medina, N., & Molina-Ortiz, F. (2004). A software system evolutionary and adaptive framework: application to agent-based systems. *Journal of Systems Architecture, 50*(7), 407–416. https://doi.org/10.1016/j.sysarc.2003.08.012

Perini, A. (2007). Agent-oriented software engineering. In *Wiley Encyclopedia of Computer Science and Engineering* (pp. 1–11). Hoboken, NJ: John Wiley & Sons, Inc. https://doi.org/10.1002/9780470050118.ecse006

Phung, T., Winikoff, M., & Padgham, L. (2005). Learning within the BDI framework: an empirical analysis. In *Proceedings of Knowledge-Based and Intelligent Information and Engineering Systems* (pp. 282–288). https://doi.org/10.1007/11553939_41

Pour, G. (2002). Integrating agent-oriented enterprise software engineering into software engineering curriculum. 32nd Annual Frontiers in Education, 3, 8–12. https://doi.org/10.1109/FIE.2002.1158672

Qu, Y., Wang, C., Lili, Z, & Huilai, L. (2009). Research for an intelligent component-oriented software development approaches. *Journal of Software, 4*(10), 1136–1144. https://doi.org/10.4304/jsw.4.10.1136-1144

Rafique, A., Van Landuyt, D., Truyen, E., Reniers, V., & Joosen, W. (2019). SCOPE: self-adaptive and policy-based data management middleware for federated clouds. *Journal of Internet Services and Applications, 10*(1), 2. https://doi.org/10.1186/s13174-018-0101-8

Rao, A., & Georgeff, M. P. (1991). Modeling rational agents within a BDI-architecture. *Readings in Agents*, 473–484. https://doi.org/10.1.1.51.5675

Rao, A. S. (1996). Decision procedures for prepositional linear-time belief-desire-intention logics. In *Lecture Notes in Computer Science* (Including Subseries Lecture Notes in Artificial Intelligence and Lecture Notes in Bioinformatics) (Vol. 1037, pp. 1–48). Berlin: Springer. https://doi.org/10.1007/3540608052_57

Rao, A. S., & Georgeff, M. P. (1995). BDI agents: from theory to practice. In *Proceedings of the 1st International Conference on Multi-Agent Systems (ICMAS-95)* (pp. 312–319). https://doi.org/10.1.1.51.9247

Rosenblatt, F. (1958). The perceptron: a probabilistic model for information storage and organization in the brain. *Psychological Review, 65*(6), 386–408. https://doi.org/10.1037/h0042519

Rykowski, J. (2005). Active advertisement in supermarkets using personal agents. In *Challenges of Expanding Internet: E-Commerce, E-Business, and E-Government* (pp. 405–419). Boston, MA: Kluwer Academic Publishers. https://doi.org/10.1007/0-387-29773-1_27

Sahu, S., Agarwal, R., & Tyagi, R. K. (2016). AGNT7 for an intelligent software agent. *Indian Journal of Science and Technology, 9*(40). https://doi.org/10.17485/ijst/2016/v9i40/89126

Sahu, S., Agarwal, R., & Tyagi, R. K. (2017). A novel OLDA evolving agent architecture. *Journal of Statistics and Management Systems, 20*(4), 553–564. https://doi.org/10.1080/09720510.2017.1395175

Sajid, A., Nayyar, A., & Mohsin, A. (2010). Modern trends towards requirement elicitation. In *Proceedings of the 2010 National Software Engineering Conference on - NSEC '10* (pp. 1–10). New York, NY: ACM Press. https://doi.org/10.1145/1890810.1890819

Samuel, A. L. (1959). Some studies in machine learning using the game of checkers. *IBM Journal of Research and Development, 3*(3), 210–229. https://doi.org/10.1147/rd.33.0210

Sancho, J. M. (2006). Evolving landscapes for education. IFIP International Federation for Information Processing, *195*, 81–100. https://doi.org/10.1007/0-387-31168-8_6

Sellami, Z., Camps, V., & Aussenac-Gilles, N. (2013). DYNAMO-MAS: a multi-agent system for ontology evolution from text. *Journal on Data Semantics, 2*(2–3), 145–161. https://doi.org/10.1007/s13740-013-0025-1

Seppecher, P., Salle, I., & Lang, D. (2019). Is the market really a good teacher? *Journal of Evolutionary Economics*, 299–335. https://doi.org/10.1007/s00191-018-0571-7

Shoham, Y. (1993). Agent-oriented programming. *Artificial Intelligence, 60*(1), 51–92. https://doi.org/10.1016/0004-3702(93)90034-9

Singh, M. P., & Asher, N. M. (1991). Towards a Formal Theory of Intentions (pp. 472–486). Berlin: Springer. https://doi.org/10.1007/BFb0018460

Singh, Y., Gosain, A., & Kumar, M. (2008). Evaluation of agent oriented requirements engineering frameworks. In *Proceedings - International Conference on Computer Science and Software Engineering*, CSSE 2008 (Vol. 2, pp. 33–38). IEEE. https://doi.org/10.1109/CSSE.2008.1555

Spanoudakis, N., & Moraitis, P. (2010). Modular JADE agents design and implementation using ASEME. In *Proceedings* - 2010 IEEE/WIC/ACM International Conference on Intelligent Agent Technology, IAT 2010 (Vol. 2, pp. 221–228). IEEE. https://doi.org/10.1109/WI-IAT.2010.136

Di Stefano, A., Pappalardo, G., & Tramontana, E. (2004). An infrastructure for runtime evolution of software systems. In Proceedings. *ISCC 2004*. Ninth International Symposium on Computers and Communications (IEEE *Cat. No.04TH8769*) (Vol. 2, pp. 1129–1135). https://doi.org/10.1109/ISCC.2004.1358691

Stone, M. (2014). The new (and ever-evolving) direct and digital marketing ecosystem. *Journal of Direct, Data and Digital Marketing Practice*. https://doi.org/10.1057/dddmp.2014.58

Sueyoshi, T., & Tadiparthi, G. R. (2007). Intelligent agent technology: an application to US wholesale power trading. In 2007 IEEE/WIC/ACM International Conference on Intelligent Agent Technology (IAT'07) (pp. 27–30). IEEE. https://doi.org/10.1109/IAT.2007.44

Sutcliffe, A. G., & Maiden, N. A. M. (1993). Bridging the requirements gap: policies, goals and domains. In *Proceedings of 1993 IEEE 7th International Workshop on Software Specification and Design* (pp. 62–65). https://doi.org/10.1109/IWSSD.1993.315514

Sutcliffe, A., Sawyer, P., Stringer, G., Couth, S., Brown, L. J. E., Gledson, A., & Leroi, I. (2018). Known and unknown requirements in healthcare. *Requirements Engineering*, pp. 1–20. https://doi.org/10.1007/s00766-018-0301-6

Van Der Aalst, W. M. P., Pesic, M., & Schonenberg, H. (2009). Declarative workflows: balancing between flexibility and support. *Computer Science - Research and Development*, *23*(2), 99–113. https://doi.org/10.1007/s00450-009-0057-9

Van Lamsweerde, A. (2000). Handling obstacles in goal-oriented requirements engineering. *IEEE Transactions on Software Engineering*, *26*(10), 978–1005. https://doi.org/10.1109/32.879820

Vandewoude, Y., & Berbers, Y. (2002). Run-time evolution for embedded component-oriented systems. In International Conference on Software Maintenance, 2002. *Proceedings* (pp. 242–245). https://doi.org/10.1109/ICSM.2002.1167773

Vilkomir, S. A., Ghose, A. K., & Krishna, A. (2004). Combining agent-oriented conceptual modelling with formal methods. In 2004 Australian Software Engineering Conference. Proceedings (pp. 147–155). IEEE. https://doi.org/10.1109/ASWEC.2004.1290467

Wang, Q. W. Q., Huang, G. H. G., Shen, J. S. J., Mei, H. M. H., & Yang, F. Y. F. (2003). Runtime software architecture based software online evolution. In *Proceedings 27th Annual International Computer Software and Applications Conference*. COMPAC 2003 (pp. 230–235). https://doi.org/10.1109/CMPSAC.2003.1245346

Wang, X., He, S., & Shi, Y. (2007). Design of intelligent answering system based on agent technology. In 2007 First IEEE International Symposium on Information Technologies and Applications in Education (pp. 333–336). https://doi.org/10.1109/ISITAE.2007.4409299

Wen, Y.-F. (2009). An agent-based decision support system for solving the dynamic location and distribution problem using multi-criteria decision models. *Journal of Statistics and Management Systems*, *12*(1), 1–12. https://doi.org/10.1080/09720510.2009.10701369

Wenzel, S. (2014). Unique identification of elements in evolving software models. *Software and Systems Modeling*, *13*(2), 679–711. https://doi.org/10.1007/s10270-012-0311-7

Weyns, D., & Georgeff, M. (2010). Self-adaptation using multiagent systems. *IEEE Software*, *27*(1), 86–91. https://doi.org/10.1109/MS.2010.18

Woods, E. (2016). Software architecture in a changing world. *IEEE Software*, *33*(6), 94–97. https://doi.org/10.1109/MS.2016.149

Wooldridge, M., & Jennings, N. R. (1995). Intelligent agents: theory and practice. *The Knowledge Engineering Review*, *10*(2), 115–152. https://doi.org/10.1017/S0269888900008122

Wooldridge, M., Jennings, N. R., & Kinny, D. (2000). The Gaia methodology for agent-oriented analysis and design. *Autonomous Agents and Multi-Agent Systems*, *3*(3), 285–312. https://doi.org/10.1023/A:1010071910869

Wooldridgey, M., & Ciancarini, P. (2001). Agent-oriented software engineering: the state of the art. In *Lecture Notes in Computer Science* (Including Subseries Lecture Notes in Artificial Intelligence and Lecture Notes in Bioinformatics) (Vol. 1957 LNCS, pp. 1–28). Berlin: Springer. https://doi.org/10.1007/3-540-44564-1_1

Xue, X., Zeng, J., & Liding, L. (2006). Towards an engineering change in agent oriented software engineering. In First International Conference on Innovative Computing, Information and Control - Volume I (ICICIC'06) (Vol. 3, pp. 225–228). IEEE. https://doi.org/10.1109/ICICIC.2006.542

Zaghetto, C., Aguiar, L. H. M., Zaghetto, A., Ralha, C. G., & Vidal, F. de B. (2017). Agent-based framework to individual tracking in unconstrained environments. *Expert Systems With Applications*, *87*, 118–128. https://doi.org/10.1016/j.eswa.2017.05.065

Zambonelli, F., & Omicini, A. (2004). Challenges and research directions in agent-oriented software engineering. *Autonomous Agents and Multi-Agent Systems*, *9*(3), 253–283. https://doi.org/10.1023/B:AGNT.0000038028.66672.1e

Zhao, B., Cai, G.-J., & Jin, Z. (2010). Semantic approach for service oriented requirements modeling. In *Intelligent Information Processing V SE - 8* (Vol. 340, pp. 35–44). Berlin: Springer. https://doi.org/10.1007/978-3-642-16327-2_8

Zhou, H., & Hirasawa, K. (2017). Evolving temporal association rules in recommender system. *Neural Computing and Applications*, 1–15. https://doi.org/10.1007/s00521-017-3217-z

Zhu, H., & Shan, L. (2005). Agent-oriented modelling and specification of web services. In *Proceedings - International Workshop on Object-Oriented Real-Time Dependable Systems, WORDS* (pp. 152–159). https://doi.org/10.1109/WORDS.2005.14

Zimmermann, O. (2015). Architectural refactoring: a task-centric view on software evolution. *IEEE Software*, *32*(2), 26–29. https://doi.org/10.1109/MS.2015.37

3 Knowledge-Based Problem-Solving

How AI and Big Data Are Transforming Health Care

Raj Kumar Goel, Shweta Vishnoi, Chandra Shekhar Yadav, and Pankaj Tyagi

CONTENTS

3.1 INTRODUCTION

Big data begins with distributed and decentralized control over large and multifaceted autonomous sources and attempts to explore the complex and growing relationships between data. Big data is growing expeditiously in all areas, including science, technology, sports, physics, biology, and medicine/biomedicine, due to the intense development of network, data storage, and data acquisition functions [1]. Big data is used to identify consumer behavior patterns and develop more advanced, progressive, and innovative business solutions. In the health care industry, big data refers to predictive analytics and machine learning platforms that offer sustainable solutions such as treatment plans and personalized medical service implementation [2].

In parallel with the increasing availability of big data, progress has been made in AI techniques that allow machines and computers to detect and act alone or to increase human activities. AI has great expertise in health care, as in other sectors, in uncovering new sources of development, changing the way people work, and empowering people to drive growth in the workplace [3]. AI and machine learning can liberate potential health professionals from routine work and protect lives through effective early identification [4].

The health care industry has experienced practical change at various points from the stakeholders involved using big data analytics and associated technologies. Historically, the health care sector has produced a wealth of data, drawn from record keeping, insurance companies, regulatory and compliance requirements, medical imaging, and patient health care [5]. Although a large amount of data is stored on paper, the recent trend is to quickly digitize this plenty of data. Big data relates to electronic patient records that are so extensive, harmonized, and complex that they cannot be handled using traditional software and/or hardware in the health care industry. It is very difficult to maintain this data using usual data management operations and methods. Big data in the health care industry involves gathering plenty of information from various health care agencies and then saving, managing, analyzing, viewing, and providing effective decision-making techniquest [6].

The basic challenge for health care big data applications is to research a lot of data and gain relevant information and insight for future action. In many cases, the process of knowledge extraction should be very efficient and almost real-time, as it is almost impossible to store all the observed data. It has been observed that very valuable health data is usually available in unstructured or semi-structured form [7]. It is not easy to collect valuable information using traditional data techniques and tools in complex, dynamic, and specific features.

While synthesizing and analyzing big data and relationship patterns and trends, health care contributors and different health care stakeholders can grow more comprehensive and meaningful diagnoses and treatments, leading to superior quality of care that is economical, people-centered. This leads to better overall outcomes [8].

The whole world seems to be connected to the arrival of the IoT. Everything is connected to something. Of course, health care cannot be unaffected, and smart health system comes into play. Wearable, portable, or implanted sensors form a body area network that transfers data at tremendous speeds. This means that large volumes of data, often referred to as big data, is saved and analyzed. In the age of AI, researchers should focus on machine learning tools to process this large volume of Medicare/health care data [9,10].

As a result, scientific and technological advances in information handling, processing, and networking for decision support and forecasting systems must go hand in hand with parallel changes globally that involve many stakeholders, including citizens and society. Only global acceptance can transform big data and the potential of AI into effective advances in health and medicine. This requires science, and scientists, society, and citizens must progress mutually. Data-driven science in machine learning and the vast area of AI have the potential to make significant changes in the health care industry. However, medicine is not like any other science: it is deeply and intimately connected to a vast and vast network of legal, ethical, regulatory, economic, and social functions [11–13].

In this chapter, we will look in depth at biomedical and health care applications using AI, the Internet of Things (IOT), and big data. The challenges of big data, scope, and opportunities in health care informatics also will be discussed. The rest of this chapter is organized thus.

Section 3.2 deals with the role of AI, big data, and IoT in health care. The concept of image-based diagnosis will be discussed in Section 3.3. Section 3.4 describes big data analytics processes using machine learning. The analysis and discussion of the study is in Section 3.5 and subsequently concludes the work.

3.2 THE ROLE OF AI, BIG DATA, AND IOT IN HEALTH CARE

Big data analytics has a tremendous ability to change medical models and business into intelligent and efficient care. This allows anonymous collection of health information to allow secondary use of data. Authentic decision making can also be facilitated by recognizing patterns and decoding associations. In clinical practice, big data analysis can contribute to the primary detection of diseases, the precise prognosis of disease courses and deviations from healthy conditions, the detection of symptoms, and the uncovering of fraud [14]. Management and analysis of large health data is critical to ensure patient-centered care. As the type and size of data sources have grown over the past two decades, obsolete data management tools have become inadequate.

Modern and innovative big data tools and technologies are needed that can complement the ability to manage large health data. Forecast for a research study on global big data expenditure in health care to reach an average compound interest rate (CAGR) of 42% in 2014-2019 [15–17].

Figure 3.1 shows the Google trends from 2004 to 2019 for the analysis of big data, AI, and IoT in health care. Google Trends is a distinctive, freely available online portal from Google Inc. that enables consumers to collaborate with Internet hunting

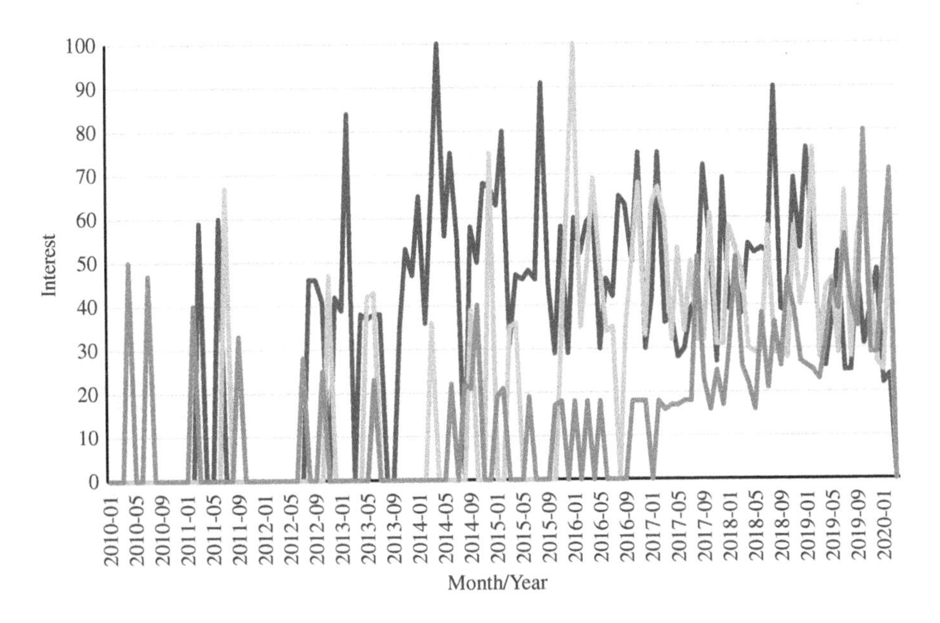

FIGURE 3.1 Google trends for AI, big data, and IoT in health care from 2004 to 2019.

data, providing an in-depth view of human activities and health-related events. Google flu and dengue trends are now being used to broadcast the spread of flu and dengue-like diseases. Google Trend has been used in a number of publications [18, 19]. Figure 3.1 shows that the term "big data in health care" first was used in early 2013. The growing attention in this topic can be linked to a widespread McKinsey & Company report that was published in early 2013 [20].

AI and big data in medicine are still in the early development stage but are growing rapidly. Although AI algorithms manage big data effectively and access data quickly to solve healthcare issues, their development is an open question to answer in the future. However, AI algorithms will not grow well unless IoT devices generating large amounts of structured data actively help them to do so. AI has become more powerful due to big data generated by IoT and historical medical images. IoT-generated big data has different characteristics than common big data because of the variety of data it collects. Figure 3.2 depicts the relationships among AI, IoT, and big data.

The interconnection of health care systems with IoT and the emergence of big data require AI to automate the clinical processes and provide rapid data analysis to identify current, new, or future issues [21,22]. System AI should be activated so that the machine can automatically forecast the outcomes from its conjugation of experience and determine them. In order to activate systems AI, natural language processing (NLP), knowledge representation, automated thinking, and machine learning must exist in the system. NLP helps to make computers/machines intelligent in the same way as humans are in understanding language [23].

3.3 IMAGE-BASED DIAGNOSIS

With megapixels of megapixel data to the results of X-rays, CAT scans, MRIs, and other test methods, combining high-resolution images can be a challenge even for the experienced clinician. Artificial intelligence has already shown that it can be a valuable partner for radiologists and pathologists to increase their productivity and improve their accuracy [24]. Advanced AI algorithms, especially deep learning,

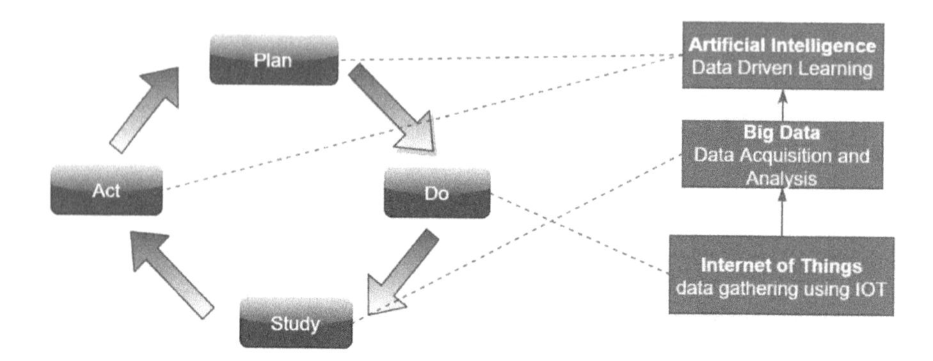

FIGURE 3.2 Relationships among AI, IoT, and big data.

have shown remarkable advances in image recognition tasks. Numerous medical experts, including those in ophthalmology, dermatology, radiology, pathology, and neurology, rely on image-based diagnostics [25]. Historical picture databases are mostly stored by radiology departments in an image collection and communication system, which usually contains thousands of examples of training networks [26].

The convolutional neural network trained on more than one lakh clinical images has achieved dermatologist-level precision in diagnosing skin malignancies. In a comparison of AI algorithm prediction against 21 dermatologists evaluating a set of photographs and dermoscopy images, the performance of deep learning algorithms was better than that of ordinary dermatologists. With the emergence of deep convolutional neural networks, AI may be useful in detecting prostate cancer from biopsies, identifying breast cancer metastases in lymph nodes, and mitosis in breast cancer [27].

3.4 BIG DATA ANALYTICS PROCESS USING MACHINE LEARNING

Modern machines with large computational capabilities can analyze the volumetric data center and use approximate correlation models. Big-data technique has the ability to accurately affect machine learning skills and enable instantaneous decision making to improve overall operational performance and lessen redundant costs [28]. Figure 3.3 shows the process of bigdata analytics in the Medicare sector. Machine learning has its own unique importance in sensing various parameters and correlating them with diseases. It is an effective method in which computers use machine learning algorithms to analyze huge amounts of data presented in a nonlinear manner, categorize patterns, and make verifiable and validated predictions.

Existing data and information from past experiences are predicted during supervised learning. In addition, patterns of unknown effects can be forecasted from the data during unsupervised learning. [29]. Machine learning, deep learning, and cognitive computing are all different steps toward higher levels of AI, but they are not same. Deep learning is a subset of machine learning. It uses artificial neural networks, which simulate human brain connections that are "trained" with the time to answer questions with almost 100% precision [30,31].

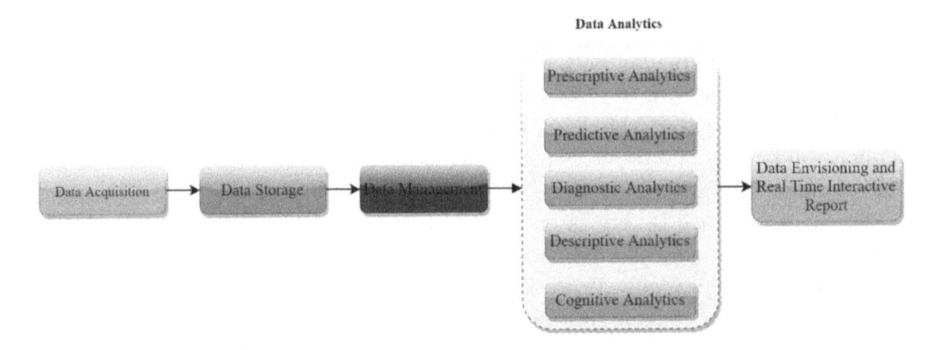

FIGURE 3.3 Process of big data analytics in the Medicare sector.

Big data in health care can be in an unstructured, semistructured, or structured format [32] and can be obtained from primary sources and secondary sources. Clinical decision support systems, computerized physician order entry (CPOE), electronic patient records (EPRs), etc. are just a few examples of primary sources and insurance companies, laboratories, government sources, health maintenance organizations (HMOs), pharmacies, etc. [33] are just a few examples of secondary sources.

EPRs, image processing, social media, smartphones, and web databases are the core sources of big data in the Medicare sector. The digitization of medical history in the past has provided hospitals with a basis for medical records of patients. EPRs data is received from doctor records, ECGs, lab reports, scans, health sensor devices, X-rays, medical prescriptions, etc. [34]. This dataset is the foundation of personalized medicine and extensive cohort studies for hospitals. In image processing, medical images are also one of the main reasons for data analysis. Photo acoustic imaging, computed tomography (CT), mammography fluoroscopy, molecular imaging magnetic resonance imaging (MRI), ultrasound, positron emission tomography-computed tomography (PETCT), and X-rays are just some examples of well-recognized imaging procedures within medical systems [35].

Twitter, YouTube, Facebook, Instagram, and LinkedIn are the most popular social media platforms for collecting health care data. Social media offers health care professionals with advanced tools to share information, discuss policies and practice health issues, interact with the public, promote healthy behaviors, and communicate with patients, colleagues, students, and nursing staff [36–38].

Data from medical social journals is generally used to analyze the spread/transmission of diseases. For example, gathering information on Twitter about a person affected by the flu is faster than the traditional method [39]. Applications are the most vital data sources in the field of Medicare self-management. Currently, smartphones have health related applications like the adjustment bit, pedometers tjat produce a lot of step data, the number of steps climbed, and the number of calories burned [40]. Another mood panda application is used to measure individual moods as well as everything related to emotional, mental, and physical aspects as well as social and environmental outlook of everyday life. We also have mobile applications such as the glucose buddy, diabetes connect mysugr, and sugar sense for diabetes management iBGstar (sugar and glucose meter app) to monitor blood sugar. These applications generate large amounts of data 24/7, which contributes to health research. Popular websites that document health data include 23 and Me and uBiome. 23 and Me is a DNA study service that provides evidence and tools for individuals to get to know and research their DNA. uBiome is a microbiome sequencing facility that provides people with evidences and tools to discover their microbiome [41].

The role of storage is very essential and effective in big data. In Medicare, data is growing rapidly, so a streamlined and heavy storage is necessary, such as cloud computing technology [42]. It provides elasticity and competence to access data, create awareness, accelerate the ability for scalable analysis solutions and increase value. This technology is a robust technology for storing large amounts of data and for carrying out comprehensive and complex data processing. This eliminates the need to get expensive computer hardware, software, and dedicated storage space [43]. The need for cloud for big data research in health care is understandable for the following reasons: it may be important to invest in big data research and increase the ability to have an accomplished and affordable infrastructure. Big data in health organizations is a mixture of internal and external bases, as they often store the patient's most sensitive data internally. Gigantic amounts of big data made by public providers and third parties may be outside [44]. Data management in health care comprises cleaning, organizing, retrieval, data governance, and data mining. It also includes methods for checking and verifying scrap data or lost values and such data must be eliminated [45].

Data analytics is a splendid way to transform raw data into information. In the health care industry, big data analytics can be done by five different analytics named as descriptive, predictive, diagnostic, prescriptive, and cognitive. Current research in industry and academia suggests that by integrating big data into analytics, retailers can achieve a return on investment of almost 15% to 20%. Descriptive analysis examines former performance based on historic data. Descriptive analysis is also called unsupervised learning. It briefly describes what is happening in health management and clearly explain the effects of parameters on the system [46,47].

Diagnosis analytics uses historic data to estimate and identify the core reason of the problem. Predictive analytics analyzes both historic and real-time data, also referred as supervised learning. This may merely predict what will happen later, because all forecasting is probabilistic in nature. It cannot predict the future. What does it expect? What are the futuristic trends? Prescriptive analytics automatically integrates big data and suggests before deciding for multiple feasible solutions. This precious information taken and processed by strategic decision maker. This analytic recommends that what we should do. What is the optimum result and how it can be achieved easily [48–50]?

Cognitive analytics is an innate development of data processing and visual analysis. Cognitive analysis takes humans out of the loop and works fully automatically. Cognitive analytics builds on advances in many areas and integrates computer and cognitive science approaches [51]. In cognitive analytics, extensive data come from heterogenous sources and having structured, semistructured, and unstructured data. Moreover, it uses knowledge structures including taxonomy and ontology to facilitate reasoning and inference. For cognitive analytics, it is very important to extract low-level as well as high-level information. Unlike all other analytics, cognitive analytics makes several responses to a query and allots a certain level of trust for every suggestion [52].

Data visualization provides the results of analyzing health data from numerical to graphical or pictorial form so that complex data is easy to understand and decision making can be improved. It can be used to understand the relationship between the model and the data. IBM Watson analysis, Graphviz R and Cytoscape are visualization tools used in health care. These tools can be used for predictive analytics, advanced analytics, reporting, instantaneous analysis, dashboards, and large data visualization [53,54].

3.5 DISCUSSION

Only transparency is not the sole trait of AI. The social function taken over by the AI algorithm should be predictable for those who govern it. The barrier to AI in the health care sector is the issue of privacy and information security. Medical history data can be used to teach AI, and appropriate security measures should be taken to prevent sharing this data with third parties. There should be reliable protective measures to defend against cyberattacks [55–57]. This is an important issue in every sector, but it is very important in the health care industry because this sector is directly related to human life, and cyberattacks can actually cause deaths. Wearable devices must be protected against external attacks. But then there are a number of questions: what security is considered credible, who evaluates the credibility, and who is responsible for an incident [58].

Regardless of how effective it is for the advancement of medical science; big data is critical to the success of all health organizations and can only be used to solve privacy and security issues. To ensure a safe and reliable big data environment, the limits of existing solutions must be determined and guidelines for future research must be developed. The biggest challenges for big data in health care are data security, data loss, data privacy, efficient handling of vast amounts of medical imaging data, privacy, confidentiality, integrity and security of information, misuse of health data or failure to protect health information, and unstructured data. Accurately understanding medical tips that extract useful information is of vital importance [59,60].

The Open Web Application Security Program (OWASP) has proposed preventative measures to protect various IoT features. Security practices include only accepting strong passwords, checking user interfaces for any known vulnerabilities in software tools, and using https with a firewall to handle high-level insecure user interfaces. In addition, the software or firmware installed on the device must be constantly updated through an encrypted transfer mechanism. software update files must be securely downloaded from a genuine and firewall protected server and signed and verified before installation [61,62].

3.6 CONCLUSION

AI is basically how we make machines smarter, although machine learning is the procedure of implementing computational systems that support it. AI mechanisms can forecast postoperative outcomes, reduce relapses, and improve patient flow. In other words, doctors can use these solutions to quickly and accurately manage medical diagnoses. In order to make the correct decision for the right patient by doctors at the right time, doctors/clinicians need knowledgeable data, real-time analysis, and predictive modeling to improve health care. Real-time health analysis helps streamline and automate the process of gathering and measuring abundant health data by meeting the needs of care and cost.

In the future, the main task will be to solve the problem of using "big data" in every particular phase by determininig which hardware functions advanced methods are required, as well as specific implementation techniques and methods for realizing the best combination of "big data" with techniques such as IoT, AI, cloud capability, and distributed processing.

REFERENCES

1. Kambatla, K., Kollias, G., Kumar, V. and Grama, A., 2014. Trends in big data analytics. *Journal of Parallel and Distributed Computing*, 74(7), pp. 2561–2573.
2. Wang, Y., Kung, L. and Byrd, T.A., 2018. Big data analytics: Understanding its capabilities and potential benefits for healthcare organizations. *Technological Forecasting and Social Change*, 126, pp. 3–13.
3. Qiu, J., Wu, Q., Ding, G., Xu, Y. and Feng, S., 2016. A survey of machine learning for big data processing. *EURASIP Journal on Advances in Signal Processing*, 2016(1), p. 67.
4. Makridakis, S., 2017. The forthcoming artificial intelligence (AI) revolution: Its impact on society and firms. *Futures*, 90, pp. 46–60.
5. Manogaran, G., Thota, C., Lopez, D., Vijayakumar, V., Abbas, K.M. and Sundarsekar, R., 2017. Big data knowledge system in healthcare. In Chintan Bhatt, Nilanjan Dey, Amira S. Ashour (ed.), *Internet of Things and Big Data Technologies for Next Generation Healthcare* (pp. 133–157). Springer, Cham.
6. Fang, R., Pouyanfar, S., Yang, Y., Chen, S.C. and Iyengar, S.S., 2016. Computational health informatics in the big data age: a survey. *ACM Computing Surveys (CSUR)*, 49(1), pp. 1–36.
7. Krawczyk, B., 2016. Learning from imbalanced data: open challenges and future directions. *Progress in Artificial Intelligence*, 5(4), pp. 221–232.
8. Ebner, K., Bühnen, T. and Urbach, N., 2014, January. Think big with big data: Identifying suitable big data strategies in corporate environments. In *2014 47th Hawaii International Conference on System Sciences* (pp. 3748–3757). IEEE, Waikoloa, HI.
9. Dimitrov, D.V., 2016. Medical internet of things and big data in healthcare. *Healthcare Informatics Research*, 22(3), pp. 156–163.
10. Jha, R., Bhattacharjee, V. and Mustafi, A., 2020. IoT in healthcare: A big data perspective. In Prasant Kumar, Pattnaik Suneeta Mohanty, Satarupa Mohanty (ed.), *Smart Healthcare Analytics in IoT Enabled Environment* (pp. 201–211). Springer, Cham.
11. Lovis, C., 2019. Unlocking the power of artificial intelligence and big data in medicine. *Journal of Medical Internet Research*, 21(11), p. e16607.
12. He, J., Baxter, S.L., Xu, J., Xu, J., Zhou, X. and Zhang, K., 2019. The practical implementation of artificial intelligence technologies in medicine. *Nature Medicine*, 25(1), pp. 30–36.
13. Rus, D., 2020. Artificial intelligence: a vector for positive change in medicine. In Bernard Nordlinger, Cedric Villani, Daniela Rus (ed.), *Healthcare and Artificial Intelligence* (pp. 17–20). Springer, Cham.
14. Pramanik, M.I., Lau, R.Y., Demirkan, H. and Azad, M.A.K., 2017. Smart health: Big data enabled health paradigm within smart cities. *Expert Systems with Applications*, 87, pp. 370–383.
15. Palanisamy, V. and Thirunavukarasu, R., 2019. Implications of big data analytics in developing healthcare frameworks—A review. *Journal of King Saud University—Computer and Information Sciences*, 31(4), pp. 415–425.
16. Feldman, K., Davis, D. and Chawla, N.V., 2015. Scaling and contextualizing personalized healthcare: A case study of disease prediction algorithm integration. *Journal of Biomedical Informatics*, 57, pp. 1–9.
17. Waller, M.A. and Fawcett, S.E., 2013. Data science, predictive analytics, and big data: A revolution that will transform supply chain design and management. *Journal of Business Logistics*, 34(2), pp. 77–84.

18. Zhang, Y., Milinovich, G., Xu, Z., Bambrick, H., Mengersen, K., Tong, S. and Hu, W., 2017. Monitoring pertussis infections using internet search queries. *Scientific Reports*, 7(1), pp. 1–7.
19. Nuti, S.V., Wayda, B., Ranasinghe, I., Wang, S., Dreyer, R.P., Chen, S.I., et al., 2014. The use of Google trends in health care research: A systematic review. *PLoS One*, 9, pp. 1–49.
20. Wamba, S.F., Akter, S., Edwards, A., Chopin, G. and Gnanzou, D., 2015. How "big data" can make big impact: Findings from a systematic review and a longitudinal case study. *International Journal of Production Economics*, 165, pp. 234–246.
21. Chen, M., Mao, S. and Liu, Y., 2014. Big data: A survey. *Mobile Networks and Applications*, 19(2), pp. 171–209.
22. Plageras, A.P., Psannis, K.E., Stergiou, C., Wang, H. and Gupta, B.B., 2018. Efficient IoT-based sensor BIG data collection—processing and analysis in smart buildings. *Future Generation Computer Systems*, 82, pp. 349–357.
23. Kaur, J. and Mann, K.S., 2018. AI based healthcare platform for real time, predictive and prescriptive analytics using reactive programming. *Journal of Physics*, 933(1), p. 012010. IOP Publishing.
24. Wang, S., Yang, D.M., Rong, R., Zhan, X., Fujimoto, J., Liu, H., Minna, J., Wistuba, I.I., Xie, Y. and Xiao, G., 2019. Artificial intelligence in lung cancer pathology image analysis. *Cancers*, 11(11), p. 1673.
25. Yu, K.H., Beam, A.L. and Kohane, I.S., 2018. Artificial intelligence in healthcare. *Nature Biomedical Engineering*, 2(10), pp. 719–731.
26. Litjens, G., Kooi, T., Bejnordi, B.E., Setio, A.A.A., Ciompi, F., Ghafoorian, M., Van Der Laak, J.A., Van Ginneken, B. and Sánchez, C.I., 2017. A survey on deep learning in medical image analysis. *Medical Image Analysis*, 42, pp. 60–88.
27. Fujisawa, Y., Otomo, Y., Ogata, Y., Nakamura, Y., Fujita, R., Ishitsuka, Y., Watanabe, R., Okiyama, N., Ohara, K. and Fujimoto, M., 2019. Deep-learning-based, computer-aided classifier developed with a small dataset of clinical images surpasses board-certified dermatologists in skin tumour diagnosis. *British Journal of Dermatology*, 180(2), pp. 373–381.
28. Popovič, A., Hackney, R., Tassabehji, R. and Castelli, M., 2018. The impact of big data analytics on firms' high value business performance. *Information Systems Frontiers*, 20(2), pp. 209–222.
29. Amershi, S. and Conati, C., 2007, January. Unsupervised and supervised machine learning in user modeling for intelligent learning environments. In *Proceedings of the 12th International Conference on Intelligent User Interfaces* (pp. 72–81), Honolulu, HI.
30. L'heureux, A., Grolinger, K., Elyamany, H.F. and Capretz, M.A., 2017. Machine learning with big data: Challenges and approaches. *IEEE Access*, 5, pp. 7776–7797.
31. Bini, S.A., 2018. Artificial intelligence, machine learning, deep learning, and cognitive computing: What do these terms mean and how will they impact health care? *The Journal of Arthroplasty*, 33(8), pp. 2358–2361.
32. Nambiar, R., Bhardwaj, R., Sethi, A. and Vargheese, R., 2013. A look at challenges and opportunities of big data analytics in healthcare. In *Proceedings of the 2013 IEEE International Conference Big Data* (pp. 17–22), Silicon Valley, CA. doi:10.1109/BigData.2013.6691753.
33. Al-Jarrah, O.Y., Yoo, P.D., Muhaidat, S., Karagiannidis, G.K. and Taha, K., 2015. Efficient machine learning for big data: A review. *Big Data Research*, 2, pp. 87–93. doi:10.1016/j.bdr.2015.04.001

34. Yang, J.J., Li, J., Mulder, J., Wang, Y., Chen, S., Wu, H., Wang, Q. and Pan, H., 2015. Emerging information technologies for enhanced healthcare. *Computers in Industry*, 69, pp. 3–11.
35. Dunne, R.M., O'Neill, A.C. and Tempany, C.M., 2017. Imaging tools in clinical research: Focus on imaging technologies. In David Robertson, Gordon H. Williams (ed.), *Clinical and Translational Science* (pp. 157–179). Academic Press, London, UK.
36. Borgmann, H., Cooperberg, M., Murphy, D., Loeb, S., N'Dow, J., Ribal, M.J. and Wirth, M., 2018. Online professionalism—2018 update of European association of urology (@ Uroweb) recommendations on the appropriate use of social media. *European Urology*, 74(5), pp. 644–650.
37. Crilly, P., Hassanali, W., Khanna, G., Matharu, K., Patel, D., Patel, D. and Kayyali, R., 2019. Community pharmacist perceptions of their role and the use of social media and mobile health applications as tools in public health. *Research in Social and Administrative Pharmacy*, 15(1), pp. 23–30.
38. Islam, S.M.S., Tabassum, R., Liu, Y., Chen, S., Redfern, J., Kim, S.Y. and Chow, C.K., 2019. The role of social media in preventing and managing non-communicable diseases in low-and-middle income countries: Hope or hype? *Health Policy and Technology*, 8(1), pp. 96–101.
39. Shafqat, S., Kishwer, S., Rasool, R.U., Qadir, J., Amjad, T. and Ahmad, H.F., 2018. Big data analytics enhanced healthcare systems: A review. *The Journal of Supercomputing*, 76, pp. 1–46.
40. Zhang, S., Wu, Q., van Velthoven, M.H., Chen, L., Car, J., Rudan, I., Zhang, Y., Li, Y. and Scherpbier, R.W., 2012. Smartphone versus pen-and-paper data collection of infant feeding practices in rural China. *Journal of Medical Internet Research*, 14, p. e119.
41. Almalki, M., Gray, K. and Sanchez, F.M., 2015. The use of selfquantification systems for personal health information: Big data management activities and prospects. *Health Information and Science Systems*, 3, pp. 1–11. doi:10.1186/2047-2501-3-S1-S1.
42. Hashem, I.A.T., Yaqoob, I., Anuar, N.B., Mokhtar, S., Gani, A. and Khan, S.U., 2015. The rise of "big data" on cloud computing: Review and open research issues. *Information Systems*, 47, pp. 98–115.
43. Hurwitz, J.S., Bloor, R., Kaufman, M. and Halper, F., 2010. *Cloud computing for dummies*. John Wiley & Sons, Hoboken, NJ.
44. Auffray, C., Balling, R., Barroso, I., Bencze, L., Benson, M., Bergeron, J., Bernal-Delgado, E., Blomberg, N., Bock, C., Conesa, A. and Del Signore, S., 2016. Making sense of big data in health research: Towards an EU action plan. *Genome Medicine*, 8(1), p. 71.
45. Archenaa, J. and Anita, E.A.M., 2015. A survey of big data analytics in healthcare and government. *Procedia Computer Science*, 50, pp. 408–413. doi:10.1016/j.procs.2015.04.021.
46. Beheshti, S.M.R., Tabebordbar, A., Benatallah, B. and Nouri, R., 2017, April. On automating basic data curation tasks. In *Proceedings of the 26th International Conference on World Wide Web Companion* (pp. 165–169), Perth, Australia.
47. Storey, V.C. and Song, I.Y., 2017. Big data technologies and management: What conceptual modeling can do. *Data & Knowledge Engineering*, 108, pp. 50–67.
48. Xu, Y., Sun, Y., Wan, J., Liu, X. and Song, Z., 2017. Industrial big data for fault diagnosis: Taxonomy, review, and applications. *IEEE Access*, 5, pp. 17368–17380.
49. Vernon, D., Metta, G. and Sandini, G., 2007. A survey of artificial cognitive systems: Implications for the autonomous development of mental capabilities in computational agents. *IEEE Transactions on Evolutionary Computation*, 11(2), pp. 151–180.

50. Lepenioti, K., Bousdekis, A., Apostolou, D. and Mentzas, G., 2020. Prescriptive analytics: Literature review and research challenges. *International Journal of Information Management*, 50, pp. 57–70.
51. Gudivada, V.N., 2017. Data analytics: Fundamentals. In Mashur Chowdhury, Amy Apon, Kakan Dey (ed.), *Data Analytics for Intelligent Transportation Systems* (pp. 31–67). Elsevier, Amsterdam, Netherlands.
52. Eastman, C. and Computing, D., 2001. New directions in design cognition: Studies of representation and recall. In C. Eastman, W. Newstetter, M. McCracken (ed.), *Design Knowing and Learning: Cognition in Design Education* (pp. 147–198). Elsevier Science, Kidlington, UK.
53. Senthilkumar, S.A., Rai, B.K., Meshram, A.A., Gunasekaran, A. and Chandrakumarmangalam, S., 2018. Big data in healthcare management: A review of literature. *American Journal of Theoretical and Applied Business*, 4(2), pp. 57–69.
54. Zhu, Z., Hoon, H.B. and Teow, K.L., 2017. Interactive data visualization techniques applied to healthcare decision making. In Mehdi Khosrow-Pour, Steve Clarke, Murray E. Jennex, Annie Becker, Ari-Veikko Anttiroiko (ed.), *Decision Management: Concepts, Methodologies, Tools, and Applications* (p. 1157).
55. Truong, T.C., Zelinka, I., Plucar, J., Čandík, M. and Šulc, V., 2020. Artificial intelligence and cybersecurity: Past, presence, and future. In Subhransu Sekhar Dash, C. Lakshmi, Swagatam Das, Bijaya Ketan Panigrahi (ed.), *Artificial Intelligence and Evolutionary Computations in Engineering Systems* (pp. 351–363). Springer, Singapore.
56. Grimsman, D., Chetty, V., Woodbury, N., Vaziripour, E., Roy, S., Zappala, D. and Warnick, S., 2016, July. A case study of a systematic attack design method for critical infrastructure cyber-physical systems. In *2016 American Control Conference (ACC)* (pp. 296–301). IEEE, Boston, MA.
57. Indre, I. and Lemnaru, C., 2016, September. Detection and prevention system against cyber attacks and botnet malware for information systems and internet of things. In *2016 IEEE 12th International Conference on Intelligent Computer Communication and Processing (ICCP)* (pp. 175–182). IEEE, Cluj-Napoca, Romania.
58. Iliashenko, O., Bikkulova, Z. and Dubgorn, A., 2019. Opportunities and challenges of artificial intelligence in healthcare. In *E3S Web of Conferences* (Vol. 110). EDP Sciences, Russia.
59. Abouelmehdi, K., Beni-Hessane, A. and Khaloufi, H., 2018, Big healthcare data: Preserving security and privacy. *Journal of Big Data*, 5, p. 1. https://doi.org/10.1186/s40537-017-0110-7
60. Islam, S.R., Kwak, D., Kabir, M.H., Hossain, M. and Kwak, K.S., 2015. The internet of things for health care: A comprehensive survey. *IEEE Access*, 3, pp. 678–708.
61. Riadi, I., Umar, R. and Sukarno, W., 2018. Vulnerability of injection attacks against the application security of framework based bebsites open web access security project (OWASP). *Journal Informatika*, 12(2), pp. 53–57.
62. Kellezi, D., Boegelund, C. and Meng, W., 2019, December. Towards secure open banking architecture: An evaluation with OWASP. In *International Conference on Network and System Security* (pp. 185–198). Springer, Cham.

4 Distributed Artificial Intelligence for Document Retrieval

Ankur Seem, Arpit Kumar Chauhan, Rijwan Khan, and Satya Prakash Yadav

CONTENTS

4.1 INTRODUCTION

Document retrieval (also known as information retrieval) is a computerized user query process that searches for the most relevant data and arranges all of the relevant results in priority order. This is an evolving field that began in the 1970s. Many models for document retrieval exist. Document retrieval methods are regularly analyzed using standard tests [1]. For example, someone creates a new method for document retrieval, then applies it to real-world problems, then compares that with the existing method. The result will only be considered to be positive if it is an improvement over the existing method. The most recent and popular advancement in document retrieval applies artificial intelligence (AI) to create intelligent machines which can work and respond like humans. More recently, distributed AI (DAI), which is a combination of AI and blockchain technology, has been used in document retrieval. Using DAI, records are stored in blocks and then these blocks are linked with each other using cryptography [2,3]. Each block has a cryptographic hash for the block and the main advantage is nobody owns the data.

4.2 PROPOSED RESEARCH

Distributed artificial intelligence (DAI) techniques will be used to make a more precise system. Standard test sets and judgments from a text retrieval conference (TREC) will be used for analysis and learning. This method will increase efficiency, decrease processing time, and ensure the safety and privacy of data [4].

4.2.1 Improving Precision

We will explore three approaches in this chapter:

- A general-purpose ranking method
- The use of document structure to improve precision
- The combination of general-purpose ranking and document structure techniques

4.3 GENERAL-PURPOSE RANKING

Exactness enhancements increased using structure-weighted recovery will be compared with a standard. That standard is the exactness of the picked unweighted positioning capacity. Many functions have been published, so many taxonomies of ranking functions exist. These taxonomies account for upwards of 100,000 ranking functions [5].

It is not feasible to test 100,000 functions, but a coordinated inquiry is possible. Hereditary programming is acceptable to use for this type of search. With the assistance of a genetic algorithm, genetic programming (GP) uses a theoretical sentence structure tree to speak with people.

The confirmations to consolidate (the iotas in the ranking function) ought to be "available easily" in a transformed record (for proficiency reasons). This incorporates word frequencies, record length, assortment size, and the most elevated report term recurrence, among others.

Very large enhancements are normal with this method, although up to this point they have not been utilized. Past analyses to consolidate a constrained measure for a "tf.idf" like proof on the cystic fibrosis assortment neglected to display huge improvement. In the meantime, tests to learn progressively broad positioning capacities for Hypertext Markup Language (HTML) have neglected to better BM25 (where BM means "best matching") without utilizing archive structures [6]. Earlier outcomes are in opposition to the desired outcome. They propose "tf.idf" and BM25 are ideal positioning capacities for inspected assortments and cannot be improved.

Close assessment of the earlier outcomes shows why they are sudden. The picked-up positioning capacity ought to be in any event a blend of proof and administrators utilized in the gauge work. Except if this is the situation the standard couldn't be scholarly and it is sensible to accept it won't be bettered. At the end of the day, if a previously existing positioning capacity f() is a mix of proof (α) and administrators (β), it can't beat a GP learned capacity being the mix of proof (ε) with administrators (φ) if α is a subset of ε, and β is a subset of φ. This is because the GP could learn f(). Furthermore, the learning can be seeded with f() ensuring at any rate f(). The decision of proof is significant. Past trials had negative outcomes because the decision of proof was deficient.

4.4 STRUCTURE-WEIGHTED RANKING

TREC reports are dispersed in Extensible Markup Language (XML). The Wall Street Journal (WSJ) records, for instance, incorporate structured markup for the title, first passage, and content, as different info fields. Term events can be weight dependent on which of these structures they happen to have [7,8]. This strategy isn't new. Exactness enhancements have just appeared, and an attempt has been made for changed weight picking strategies. In any case, there is no precise technique for picking the loads, which might be the reason picking loads was a repetitive inquiry at INEX (Initiative for the Evaluation of XML Retrieval).

The loads can be encoded in a cluster, which for each reported structure weight one exhibit component is used. Ideal record structure weighting won't decrease mean normal accuracy when contrasted with recovery without structure weighting. On the off chance that all archive structures are weighted equivalent and 1, the likeness unweighted recovery happens. Execution of weighted recovery (with ideal loads) can in this way consistently in any event equivalent to that of unweighted recovery over a huge arrangement of questions [9]. Furthermore, genetic algorithms are a settled improvement procedure, so it should be conceivable to put an exhibition upper bound on this strategy.

4.5 THE STRUCTURE-WEIGHTED/LEARNED FUNCTION

First we should learn and enhance common purpose ranking function, after which we should learn and enhance structure weight for such a function. Better again, the purposed function and structure weighted function can be learned to be related, and along these lines any cooperation of strategies would be abused at the time of learning. In earlier tests, word frequencies in some of the HTML labels were utilized in the GP learning test. This brought improvement for the unweighted recovery with Okapi BM25 [10,11]. Development is likewise expected if this strategy is applied to non-web archives.

4.6 IMPROVING RECALL AND PRECISION

There are now a significant number of procedures to improve review, including pertinence input, stemming, and thesaurus.

4.6.1 Stemming

At the time of indexing, each word in the archive assortment is changed over according to the algorithm in a stem. For stemming, the words with similar stem are then recorded as though they were a similar term. The expectation is to combine the postings with all words with a typical form. For instance, postings for "treatment," "treating," and "treat" would be converted into the stem "treat." At the point when a client scans "treats," the word "treats" is stemmed in "treat" and postings for the stem are recovered, discovering archives not containing the asked for term by the user [12]. It should, along these lines, be conceivable to figure out how to stem utilizing GP.

Some of the investigations on stemming raise some question on the effectiveness of the stemming due to its error rate. We can hope for the improvement in the stemming algorithms to increase the suggested average.

4.6.2 Relevance Feedback

A client plays out a pursuit, the outcome is returned for judging, at that point, with information on the decisions, the search is reconsidered and the new outcomes recovered [13]. The strategies previously talked about could be utilized for significant input. After the underlying deciding round, the first inquiry and a lot of decisions are known – all that is expected to beome familiar with a positioning capacity and structural loads. Previous procedures for significance criticism, for example, question development and term weighting could be utilized related to this learning strategy. Genetic algorithms have been utilized to understand the term loads.

4.6.3 Thesaurus

Thesaurus can be spoken to as a lot of bit-strings, one for every thesaurus passage. In a string, each part of it has a piece for every remarkable term in the record assortment. On the off chance that, in the given string piece, two bits were set, which are equivalent words.

You can educate these bit-strings with a genetic algorithm. A community is seeded with people with irregular bits and a usual genetic algorithm is used for learning great blends. Particular weight is applied to build accuracy independent of the impact on review.

A similar strategy can be utilized to learn equivalent words and phrases (for example, data recovery and information retrieved (IR)). Singular substance uncovering expressions would initially be recognized utilizing previous methods, the report assortment ordered to incorporate these terms, and the bit-string thesaurus understanding applied.

4.7 PRELIMINARY RESULTS

Beginning examination concerning learning archive structure loads was directed with internal items, Okapi BM25 positioning, and probability. The internal item work was direct "tf.idf" usage [14,15]:

$$w_{dq} = \sum_{t \in q} \left(tf_{td} * vidf_t\right) \cdot \left(tf_{tq} * vidf_t\right)$$

where

$$vidf_t = \log_2 \frac{N - n_t + 1}{n_t}$$

In Harman's opinion, the probabilistic role was:

$$w_{dq} = \sum_{t \in q} (C + pidf_t) \times \left(K + (1 - K) * \frac{tf_{td}}{\max(tf_d)} \right)$$

where

$$pidf_t = \log_2 \frac{N - n_t + 1}{n_t}$$

where $C = 1$, $K = 0.3$. The BM25 execution was devoted to Robertson et al.

For each situation, "N" is the number of archives, n_{ot} the number of reports wherein the term happens and tf_{td} is the number of events of term "t" in the record "d" (similarly tf_{tq} in the inquiry). The preparation set was the TREC WSJ (1987–1992), with subjects 151–200. Subjects with fewer than five decisions were disposed of. The population size was 50 and a few examinations were run for 25 generations. The assessment was against themes 101–150; results are shown in Table 4.1.

A rise of 5 percent is seen for unweighted recovery in an inward item and likelihood model (huge at 1 percent). Okapi shows no improvement. No analyzed method essentially outflanked unweighted BM25.

To put an improvement in upper bound, the preparation and assessment inquiries were pooled and the progression of examinations rush to advance structure loads overall questions. The outcomes were like those appearing In Table 4.1, this strategy yields an improvement of approximately 5 percent in the internal item and probability, but less than 1 percent in BM25.

TABLE 4.1
MAP Differences for Weighted Retrieval (W-MAP) and Unweighted Retrieval for Evaluation

Function	MAP	W-MAP	Imp	Imp%	P-Value
BM25	0.2289	0.2281	–0.0008	–0.35%	0.92
Probability	0.1675	0.1787	0.0112	6.69%	0.00
Inner Prod	0.1657	0.1735	0.0078	4.71%	0.00

Training on each isolated query showed improvement for BM25 from 10 to 12 times to less than 1 percent. The improvement for the average interior product average was 30 percent and 15 percent for probability and BM25. The result suggests that such a methodology could be effective for input about important feedback; however, further research should be performed [16]. Table 4.2 shows that the mean average precision (MAP) is improving. In Figure 4.1, the enhancements are shown from the very much to the least ordered in each query:

$$vidf_t = \log_2 \frac{N+1}{n_t}$$

There was no hypothetical avocation for doing as such and the trials have not been directed utilizing the previous definition. It ought to be noted that the exhibition of the internal item utilizing this latter definition is worse than when previously utilized. Trials to get familiar with a broadly useful positioning capacity were directed on a similar preparing set (WSJ, topics 151–200).

A few runs of around 100 people for 100 generations were led. The learning process was listed and loaded with the others positioning capacities including BM25. The assessment was against WSJ (points 101–150), assortments from TREC circles 4 and 5 (themes 301–350), and the cystic fibrosis assortment. An assorted arrangement of assessment assortments is essential, since positioning execution is known to change significantly from assortment to assortment. The assessment results are shown in Table 4.2 where Run 5 is the best-learned to date capability. Played out the other an equivalent number of times, yet outperformed at the 5 percent level, Run 5 consistently beat BM25. Capacity Run 5 has been demonstrated to be compact from assortment to assortment.

4.8 SCOPE FOR DISTRIBUTED AI IN THIS PROCESS

Before using DAI for document retrieval, we should first know about centralized AI for document retrieval. In centralized systems such as Google whenever a person enters a query there are some robots or algorithms called spiders or crawlers that go through all the websites and search for the words in the query. Spiders search only for the keywords and ignore other words such as connecting words. That is the reason that if you search for anything by changing the connecting words the results are the same. Then after searching the sites or pages Google ranks them according to the algorithms, keywords, the number of times the page is referred by other websites, and the pages. Then it shows the result to you. After that it records the result and query and the next time it is searched Google will take this record to give the most appropriate result [17]. Now, the problem is it stores the data on its server and this is called the centralized system. The problem with a centralized system is that it requires a huge amount of storage and computation power for storing the data and using it. Another problem is companies can also use the data, in today's world data is everything and by querying, we are giving data away for free. Here we can relate to the saying "if you are not paying for the product then you are the product." With

TABLE 4.2
Comparison of Different Ranking Functions on Different Document Collections

Function	WSJ	CysFib	CR	FR94	FT	Trec4	FBIS	LATimes	Trec5	Trec4+5
Run 5	0.2849	0.2860	0.1727	0.1820	0.2223	0.1702	0.2566	0.1881	0.1834	0.1625
BM25	0.2553	0.2728	0.1949	0.2079	0.2022	0.1767	0.2082	0.1996	0.1824	0.1658
Probability	0.1890	0.2781	0.1481	0.1311	0.1240	0.0857	0.1423	0.1352	0.1155	0.0855
Inner Prod	0.1497	0.2873	0.0992	0.0863	0.1203	0.0346	0.0665	0.0797	0.0473	0.0286
MAP Imp	0.0297	0.0133	−0.0223	−0.0259	0.0200	−0.0065	0.0484	−0.0115	0.0010	−0.0033
Imp%	11.63%	4.86%	−11.42%	−12.48%	9.91%	−3.70%	23.24%	−5.78%	0.54%	−2.00%
P-Value	0.00	0.04	0.91	0.95	0.19	0.83	0.00	0.50	0.17	0.33

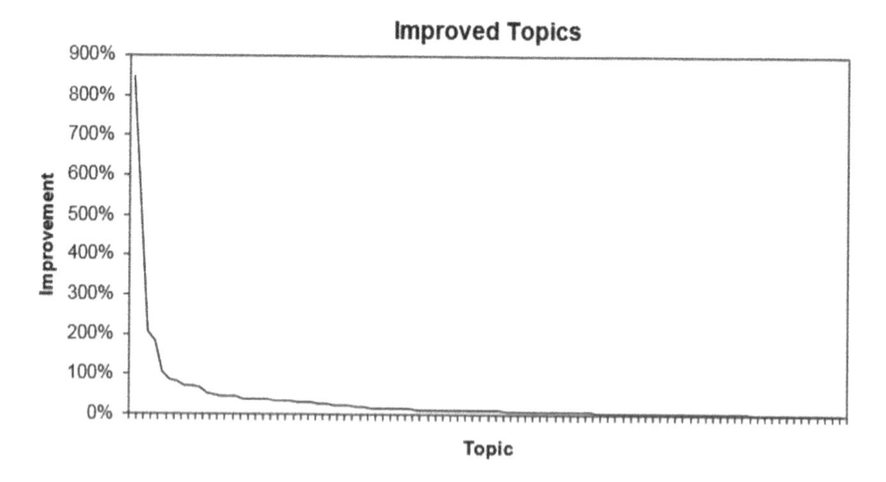

FIGURE 4.1 A few topics show large improvements while many topics show small improvements. Approximately 50 percent of topics show an improvement of over 10 percent.

this data, companies can influence our decisions, predict our decisions, and even manipulate our decision. To overcome these problems there is a concept of distributed or decentralized systems. In a distributed system, when any person searches for the query, the search engine tries to find it in the distributed ledger and then lists the result. The details about the search are stored in the distributed ledger after encrypting it. The distributed ledger is created or constructed with the help of all the nodes (each computer connected to the network) of the network. All the nodes work for the searching and updating of the distributed ledger.

4.9 BENEFITS OF DECENTRALIZED SEARCH ENGINES

Privacy: user searches remain anonymous; records of the user are not secretly recorded for financial gain.

Community-driven: centralized searched engines are owned by private companies or individuals, whereas decentralized search engines are owned by the community.

Shared resources: all people can have access to public ledgers; the only condition is that that people should have a copy of ledgers.

There is huge scope for distributed search engines but today there are only a few of them and the scope of improvement in these is very high.

4.10 DISCUSSION

This examination bases on one inquiry: How can conveyed manmade consciousness strategies be utilized to improve data recovery? As of now, demonstrated genetic algorithms and GP can be utilized to improve exactness and DAI will decrease power utilization, stockpiling required, and upgrade the security of protection. These calculations have the benefit over others (e.g., neural networks) in the emblematic outcomes. The positioning capacity can be analyzed. A resulting thesaurus may be printed as a thesaurus. More significantly, the outcomes can be moved to start with one report assortment then onto the next and can be relied upon to keep on performing admirably. Assessment of exactness and review is a self-assertive decision, and may not be the best decision. Without a doubt, classification can be improved with AI. How might it be utilized in an inquiry/answer framework? How could these strategies be utilized to improve the intelligent experience of a client? Could GP be utilized for file pressure? Maybe a clever storing instrument would improve throughput? DAI is significant for grouping, yet can hereditary procedures be utilized? Unanswered AI questions include: do other AI procedures (for example, molecule swarm enhancement) better fit this difficult area? What preferable encodings exist over those proposed in this? Have these methods been attempted previously? What effectiveness issues ought to be analyzed? The most significant unanswered inquiry is: In what future directions can (and should) this methodology be taken?

4.11 CONCLUSION

Knowledge management is an important aspect of human development. In the beginning, the amount of information was very limited so there was no need for an advanced information retrieval method. With the increasing amount of data, there is a need for improvement in information retrieval methods. We now have tools such as AI to manage this huge amount of data. With this technology, we can manage any amount of data, but to increase its efficiency and to decrease the amount of energy used, DAI for document retrieval is necessary. Only small work has been done in this field and there is huge scope yet to be covered in the hope we reach our destination as soon as needed.

REFERENCES

1. Hersh, William R. (2014). Information retrieval and digital libraries. In *Biomedical Informatics* (pp. 613–641). Springer, London.
2. Fan, W., Gordon, M. D., Pathak, P., Xi, W., & Fox, E. A. (2004). Ranking function optimization for effective web search by genetic programming: An empirical study. In *Proceedings of the 37th Annual Hawaii International Conference on System Sciences.* http://citeseerx.ist.psu.edu/viewdoc/summary?doi=10.1.1.11.2861
3. Fuhr, N., Gövert, N., Kazai, G., & Lalmas, M. (2002). INEX: Initiative for the evaluation of XML retrieval. In *Proceedings of the ACM SIGIR 2000 Workshop on XML and Information Retrieval.* http://www.dcs.gla.ac.uk/~mounia/Papers/xml-inex.pdf
4. Fuller, M., Mackie, E., Sacks-Davis, R., & Wilkinson, R. (1993). Structured answers for a large structured document collection. In *Proceedings of the 16th ACM SIGIR Conference on Information Retrieval* (pp. 204–213). https://dl.acm.org/doi/abs/10.1145/160688.160720
5. Harman, D. (1991). How effective is suffixing? *Journal of the American Society for Information Science*, 42(1), 7–15.

6. Harman, D. (1992). Ranking algorithms. In W. B. Frakes & R. Baeza-Yates (Eds.), *Information Retrieval: Data Structures and Algorithms* (pp. 363–392). Englewood Cliffs, NJ: Prentice-Hall.
7. Harman, D. (1993). Overview of the first TREC conference. In *Proceedings of the 16th ACM SIGIR Conference on Information Retrieval* (pp. 36–47). https://dl.acm.org/doi/abs/10.1145/160688.160692
8. Holland, J. H. (1975). *Adaptation in Natural and Artificial Systems.* Ann Arbor, MI: University of Michigan Press.
9. Kim, W., & Wilbur, W. J. (2001). Corpus-based statistical screening for content-bearing terms. *Journal of the American Society for Information Science and Technology*, 52(3), 247–259.
10. Kim, Y.-H., Kim, S., Eom, J.-H., & Zhang, B.-T. (2000). SCAI experiments on TREC-9. In *Proceedings of the 9th Text REtrieval Conference* (TREC-9) (pp. 392–399). http://citeseerx.ist.psu.edu/viewdoc/summary?doi=10.1.1.21.2888
11. Koza, J. R. (1992). *Genetic Programming: On the Programming of Computers Utilizing Natural Selection.* Cambridge, MA: MIT Press.
12. Losada, D. E., & Barreiro, A. (2001). A logical model for information retrieval based on propositional logic and belief revision. *Computer Journal*, 44(5), 410–424.
13. Løvbjerg, M., Rasmussen, T. K., & Krink, T. (2001). Hybrid particle swarm optimizer with breeding and subpopulations. In *Proceedings of the 3rd Genetic and Evolutionary Computation Conference.* www.semanticscholar.org/paper/Hybrid-Particle-Swarm-Optimiser-with-breeding-and-L%C3%B8vbjerg-Rasmussen/d16b37a8f383e41537d76b5085a9ba7d0c60920c?p2df
14. Oren, N. (2002). Reexamining tf.idf based information retrieval with genetic programming. In *Proceedings of the 2002 Annual Research Conference of the South African Institute of Computer Scientists and Information Technologists on Enablement through Technology (SAICSIT)* (pp. 224–234).https://citeseerx.ist.psu.edu/viewdoc/download?doi=10.1.1.102.8878&rep=rep1&type=pdf
15. Porter, M. (1980). An algorithm for suffix stripping. *Program*, 14(3), 130–137.
16. Pôssas, B., Ziviani, N., Meira, W., & Ribeiro-Neto, B. (2002). Set-based model: A new approach for information retrieval. In *Proceedings of the 25th ACM SIGIR Conference on Information Retrieval* (pp. 230–237). https://dl.acm.org/doi/abs/10.1145/564376.564417
17. Rapela, J. (2001). Automatically combining ranking heuristics for HTML documents. In *Proceedings of the 3rd International Workshop on Web Information and Data Management* (pp. 61–67). https://dl.acm.org/doi/abs/10.1145/502932.502945

5 Distributed Consensus

Satya Prakash Yadav

CONTENTS

5.1 INTRODUCTION

The distributed environment of a network gives the ability to each node to interact independently with the previous and node ahead of it. The blockchain technology utilizes this feature. The control of the network is not restricted to any center, thus using decentralization. Therefore, it is essential that the technology has to be safeguarded by a consensus mechanism [1]. A consensus mechanism regulates the working of the nodes, the number of nodes that can join the network, and how they are connected without giving way to any threat on the network.

5.2 NAKAMOTO CONSENSUS

Satoshi Nakamoto created the Nakamoto consensus for bitcoin in the blockchain technology. The consensus defines a group of rules and the concept of proof of work on the network; it guarantees a network of no trust. This consensus gave rise to the Byzantine Fault Tolerance (BFT) and peer-to-peer network, of which bitcoin became a part [2]. The consensus guarantees that any nodes can join the network or leave as and when they require. The participants of the network at not entitled to trust each other. The term fault tolerance stands for the ability of the system to withstand any kind of failure without affecting the overall working of the network as a whole. The BFT works in a distributed environment and has the power to remain unaffected by fault; it continues to follow a consensus that validates the transactions on the block. The failure of one or more nodes does not stop the overall working of the network. Before the introduction of bitcoin, the consensus model was working on a close or partial end loop. The popular model known as the partial Byzantine Fault Tolerance model (pBFT) used different nodes instead of what is followed in the Nakamoto model. Bitcoin network is large; it requires maintenance. This maintenance is done with the help of guidelines that follow cryptography and game theory rules—this helps to maintain an environment of no trust and the decentralization concept. The traditional pBFT worked in a small group of nodes that were closed [4]. The nodes could be 50 or less. The pBFT model uses a round robin mechanism to select a leader, which initiates communication. The limited number of nodes limits the leader selection process. The overhead calculated by communication was more and, thus, it was not used in bitcoin.

5.2.1 Nakamoto Consensus Working

Together these four concepts make the bitcoin network with its distributed platform a hit. The consensus is maintained to be without trust, and security is maintained by mining done by sincere miners. The Nakamoto consensus can be categorized into four parts [5].

5.2.1.1 Proof of Work

An important consensus mechanism is proof of work (PoW). It was firstly used by bitcoin, which made it popular enough to be adopted by other cryptocurrencies as a consensus mechanism.

5.2.1.1.1 Procedure

The proof of mining is referred to as mining, and the nodes that form the chain are called the miners. A complex mathematical problem in the form of a puzzle is presented to the miner. As soon as the miner solves that puzzle, it has the power to add a block to the chain as a reward. The puzzles with mathematical background have interesting features [6]:

- They are asymmetric, which implies that they take a lot of time to be solved but, once done, the answer can easily be verified.
- The puzzle has to be solved using a trial and error method. The approach is mainly guessing based. The whole process would require more computational power to be solved more quickly.
- The speed of block creation is dependent on the solution of the puzzle. If a puzzle is easily solved, then blocks are added faster to the coin otherwise vice versa.

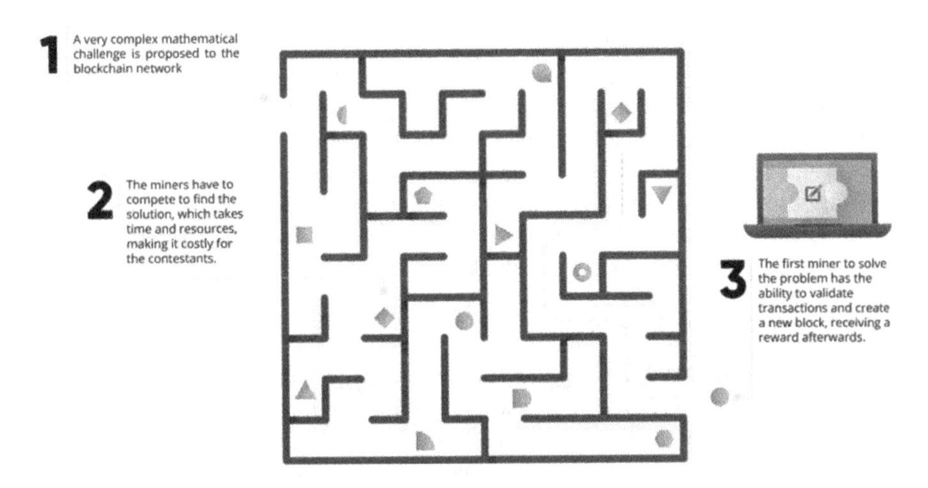

FIGURE 5.1 Proof of work [13].

Thus, resolving the puzzles requires faster computational cost for faster addition to the blockchain (Figure 5.1).

5.2.1.2 Block Selection

Block selection is an important step in the consensus. Here the miners compete against each other for the block. The block is selected based on a lottery system. The winner of the block can then mine the block, which means made transaction like read and write on that block. It is based on the PoW model [8]. The block selected, becomes a part of the chain. If it was pBFT then a leader would be selected through voting, the leader deiced which blocks will be added on the chain. The addition happens if they are approved by two-thirds of the total nodes available. Bitcoin does not make use of this voting process; it makes the miners solve a cryptographic puzzle. The puzzle is hash code and starts with 0 to reach the beginning of any node.

Once the puzzles are solved, the miners are required to add the block to along with chain which is already running. The process is random and because there is no leader, nothing can be predicted. The winner of the solution is the miner which has the most hashing power which is critical to the whole chain [9].

5.2.1.3 Scarcity

The original and old method of money transaction was the use of coins made of metals, mainly gold and silver. However, the system worked because there was a limit on the number of such coins. The effort that was put in to produce more coins each added to the cost and hence the scarcity factor was crucial. Bitcoin also works on the principle of scarcity. New bitcoins are added to the system based on the mining process (already discussed above). However, to avoid bulk production, the quantity is limited by halving the block reward every time it reaches 210,000 blocks that happens every four years—thus keeping a check on the production of blocks in the chain.

5.2.1.4 Incentive Structure

The consensus is drafted such that the increasing and decreasing blocks in the blockchain are balanced. The approach is based on game theory, which follows iteration. Incentives are given to the miners for securing the network. The award for agreeing to the rules is a bitcoin. As the value associated with a bitcoin falls, the network working also falls apart and it is the miner who suffers here. Thus following the above stated four points, the Nakamoto consensus has become successful. It is reliable even though trusts between the blocks are less.

5.2.2 Security of Bitcoin

The next question that arises is how the guarantee of blockchain security can be taken by following the simple rules discussed. With Bitcoin, a node can be made part of the network by anyone easily. Anonymity of the nodes makes it harder to detect the target and owner. The anonymity is not about one node but a group of nodes that are connected in a chain manner, the culprit of any malicious behavior cannot be detected easily. The uncertainty of the system has discouraged a lot of investors from putting their money in the technology, because if anything goes wrong, you don't know who to hold responsible. Even if the owners decide to invest, their aim is to secure the system and not add more chains because the level keeps increasing. The guarantee of security and no malice happening on the chain is taken by the validation rules. The rules applied for block selection ensure that enough computational resources are put in the valid blocks only. A node that can damage a network if at all is added to the network then it will need more resources in the form of monetary value. This is difficult for any investor, and hence the security of the network is maintained [11]. The mechanical design concept requires that the resources are distributed fairly. The resources allocated are as per the choice of the owner and fulfill its criteria. But because of anonymity, the selection process is mostly random.

The serial dictatorship approach makes one dictator that directs what all resources are to be allocated, the dictator is chosen in a decent manner and not random. The bitcoin structure uses commitment made before. It is expensive. The only drawback is anyone can wish to be a part of the network and can damage the internal working to an extent. Thus it is clear; the problem of Byzantine can be solved effectively by bitcoin. The need is economic value or money management. The money can be justified by equal utilization of resources, which is in the interest of the network.

The hash function is a secure quantity, and the only way to solve it by using hit and trial method or the brute force method. In this method, all the combinations which can be possible are applied to the puzzle to solve it. The ones who solve this puzzle are called miners and they need to have high computing power to solve it. This method of deciding the leader is widely accepted because it guarantees the following:

1. The puzzle is hard and finding its solution is tough.
2. The solution, once found, can be easily verified if it is right or wrong.

The miner who successfully solves the puzzle is awarded with a reward. The reward is some currency or transaction fees, this is an incentive, which is a form of motivation for the miners to continue mining. The computational power is an overhead in the blockchain technology. The power is often less and hence the miners are awarded for not cheating also. The award value is definitely less on the system in comparison to what it will have to pay off the network is compromised due to hardware failure. The advantage of PoW is it provides security, but a downside is that it consumes a lot of energy. It is believed that, as the network grows, the power of the system has to be increased and the hash function will be tougher.

5.2.3 The PoW Algorithm

Miners have the power to solve the puzzle. The accuracy of the solution depends on the load of the network and energy utilization. The security of the blockchain is strengthened as the hash of a block includes the hash code of the previously linked block. This feature stops any kind of violation. Miners can solve the mathematical puzzle, create a new block and the transaction is validated. PoW is a part of the cryptocurrency. Its most popular application is obviously bitcoin. The consensus was started for bitcoin and hence its creation. The algorithm has thc power to later the total network power based on the difficulty level of the puzzle. A block is created in average 10 minutes. Litecoin also follows the same type of consensus algorithm. Ethereum which is a platform of blockchain technology also follows this model. There are two main reasons for the application proof of work algorithm [12].

Fighting DoS attacks: the actions that take place on the network and their effect can be lessened. This is done by the PoW algorithm. Any attack would require efforts to be made for execution. The power of computation needed will be high. The possibility of attacks is very much there but the cost is very high; hence, it is curtailed.

Mining options: With the current scenario, it is mostly the person with more money who has more power. The blockchain technology used PoW to restrict the owners if they have more bitcoins in their wallets. The addition of blocks is based on the solution of the puzzle rather than the quantity of coins held.

The hardware requirement for executing a complicated algorithm like PoW is very expensive. The cost is most of the time only made for specific pools of mining. The power consumptions of such machines are also very high. This can lead to making the system centralized.

5.2.4 Proof of Stake

Proof of Stake (PoS) holds the property that a person who has more coins in his wallet can make more number of transactions, which can eventually be validated. Validation provides confirmation, thus more mining power to the person. The PoS technology is followed by Peercoin, Nxt, Blackchain, and ShadowCoin. One of the major downside to mining is that the process consumes a lot of electricity. According to a calculation made in 2015, it was observed that the computation power needed in a new block in the blockchain would require as much electricity as was needed by 1.57 houses in America per day. As a result, the miners would cover this cost by selling their bitcoins at a lower cost; this further reduced the value of the bitcoin. This disadvantage is taken care of by proof of stake (PoS), where instead of using the whole energy for solving the cryptographic puzzle, the miner can limit its task. The miner will only use as much power as is required by him, which is its contribution in the block [3]. There is no need to put 100% stake but only the percentage in the whole process, say 2%. The proof of work can be affected by a threat called "Tragedy of Commons." The threat is defined as in the future if no reward is given to the miner, then the value of a bitcoin will reduce. The miners will only be given the value that they earn by making the transactions on block and with time, this value will also lessen. Out of the many cryptocoin available like Litecoin, Nxt, Peer coin, Ethereum has changed its process from PoW to PoS. The efficient use of energy and security are the major advantages of PoS.

The decision as to who will start the mining is a random process in PoS. The power of computation is not considered but which miner has more cryptocurrency and for how long it has that currency. More is the cryptocurrency and for more time it has it, the miner wins and creates a block. The concept of randomization ensures that a miner who is richest is not always benefitted. This property makes PoS more powerful than PoW. There is another benefit; the total coins generated to give as incentive to miners will also be reduced, thereby putting fewer burdens on the system. The control of the complete network is not on just one miner as in the case PoW. The reputation of a miner with the highest cryptocurrency will be at stake, because if an attack in the network occurs, then he will be responsible for it [10].

However, there are some disadvantages associated with PoS. in case the consensus fails, then because no one has anything to lose so the conflict will remain unsolved and the blockchain will be stuck.

The total work done by PoS is more towards the upside. It saves energy, reduces power consumption, and is efficient and robust. The algorithm is modified by many companies to produce delegated proof of stake, or DpoS (Figure 5.2).

5.2.5 Proof of Burn

The proof of burn (PoB) is also an alternative to PoW. The PoB is defined as the consensus algorithm where a miner can burn the coins. The burning process helps to stop double spending of the coin. There are many versions of PoB available but this one was given to Ian Stewart. The destroyed or burned coins are sent to an unknown address. The proof is supposed to be submitted. The energy consumption in this case will be there but it will be less than the coins lying unused and adding to the currency value. The set of nodes that are working on the block of the form of read and write operations agree and confirm and it is then only that the block is added on the chain. It is because this process is autonomous and follows decentralization there the consensus algorithms play a major role. A block is written only when the blockchain nodes agree on a set of transactions that the nodes consider as valid. Due to the autonomous and decentralized nature of the working of the blockchain network, an automated mechanism is required to ensure that the participating nodes agree

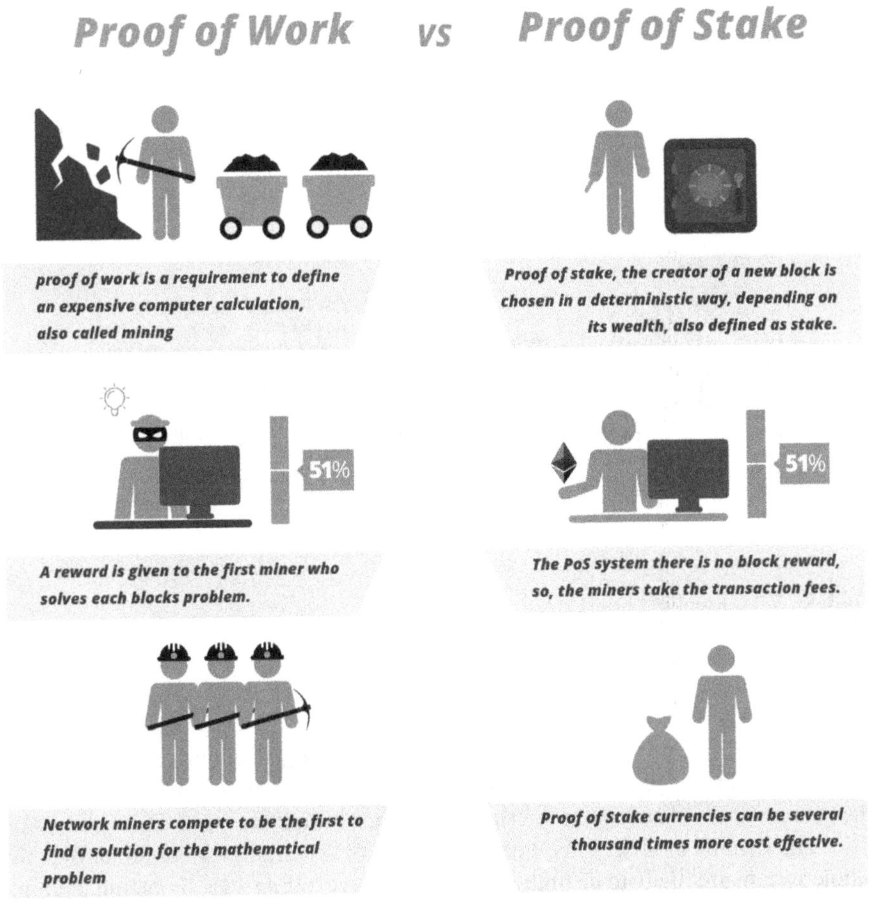

FIGURE 5.2 PoW vs. PoS [14].

on only the valid transactions. This important task is performed by the consensus mechanism algorithms [7]. As we have studied above, PoW algorithm suffers from high consumption of energy issue and this is solved to a certain extent by power of stake (PoS), which gives power to the miners based on their share of the coins.

The PoB works for the PoW drawback. It is for this reason PoB is also called PoW minus energy wastage. The virtual currency, called coins or tokens, can be burned or destroyed. The total number of coins burned by a miner gives him the power to mine that number of coins. The miners who burn the coin receive the address to send the unused coins. This address is verified and leaves no loop end for fraud. The miners who send the coins are given a reward, which incentivizes them to burn the coins. This process guarantees that the network is always working and in a state of action. The power of mining is not restricted to just a one-time activity. The miners are encouraged to do the burning process as a periodic activity so that the network is never silent. As soon as a coin is burned, the value that it had is brought back to life

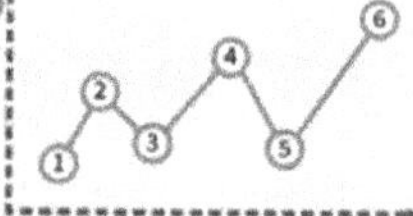

FIGURE 5.3 Proof of burn [15].

by the creation of a new block. Thus the blockchain technology continues to work effectively. The PoB algorithm can be modified according to requirements. This is followed by Slim coin, which a virtual currency. The implementation of the power of burn algorithm can be increased. Once a miner burns a coin, it can look forward to the next coin and also acquire the right to get more coins in a year. This is based on three algorithmic models: PoW, PoS, and core PoB. The rights of the miner over the whole system are likely to be higher if the miner is spending a great amount of coins. This gives balance (Figure 5.3).

5.2.6 Difficulty Level

As we have studied so far, each block in the blockchain is associated with a hash code. The difficulty is mathematically measured, which gives the level the toughness to find the hash of a given target. Bitcoin technology has set a standard difficulty level and the blocks that are created are supposed to have a difficulty level in the range lesser than the global block difference that is set. The pools of mining cater to this level by setting another pool-specific difficulty setting. Another definition of difficulty is the time taken by a miner to add a block to the blockchain along with its transactions. The level of difficulty is changed and updated in 2 weeks and it makes sure that no miner takes more than 10 minutes to add a new block in the blockchain. The aim of a consistent working of blockchains is that the blocks should be added to the blockchain regularly. The process is made no different by the addition of new miners in the network. The level of difficulty, if kept the same, will ensure less time for a new miner to enter the network and add new blocks. It therefore can be implied:

1. If the difficulty level is more than 1, then the difficulty further increase and, against expectation, the blocks will be mined faster.
2. If the difficulty level is less than 1, then the difficulty further decreases and, against expectation, the blocks will be mined slower.

The amount of time taken by a miner to add a new number in the range controls the time in which a new block is added. If the target is lower than the level will be higher to compute. The bitcoin technology works on a similar example. It sets a target first. The miners then begin to hash; the value of the hash should be lower than the target set. The numbers generated will be many because of the miners. The thousands of numbers generated by miners will make the chances of finding the winning number less likely.

5.2.7 Sybil Attack

The peer network has many nodes and it is easy to attack the system; therefore, to prevent attacks, Sybil attack is used, by making many identities on the network. All of these are fake. The outsiders will assume that these identities are real and will be treated as unique users. These many entities are in turn controlled by one single entity. This single entity that will control will give it more power. The voting power of the single entity will be more; it can easily influence the whole network. An example of a Sybil attack was reported on Facebook, which is a famous social media application. The fake accounts that were running on the website and, in hindsight, they were controlled by one entity and were reported later, once the investigation was over. These attacks are easy to hide and take time to reveal one single entity that is involved [3].

The Sybil attack is given its name after a woman named "Sybil," who suffered from an identity disorder. The first mention of Sybil attacks was reported by Microsoft in the year 2000. In the next sections we will study how these attacks damage the network and what are the effects on the blockchain network. The Sybil attacks are infamous because they are able to create multiple accounts that are fake. The accounts are pseudo are further controlled by one single entity. In one instance, the reviews that we come across on Amazon are often controlled by one but because the accounts are many, you can read more than one review. This approach is a fraud for a consumer who is looking for honest feedback. In a blockchain network, a single entity can easily take control of a large part of the network. The fake accounts can also vote for genuine accounts on the blockchain. The natural flow of the network working can be easily failed. The information traveling across the network can be compromised. The database, also referred to as ledger, can get its data altered. Following are the different ways in which the negative impact of the Sybil attack can be put under control:

1. The cost involved in identity creation: the creation of varied identities on the network is sometimes useful also, redundancy, sharing of resources anonymity and reliability are some of the advantages. The many to one

ratio can make the identities resource intensive. This way the cost of identity creation is increased. In the case of blockchains, protection from a Sybil attack is guaranteed through mining. The nodes can be added to a chain by Proof Work method. The process of creating identities will require more than one processor to increase the processing power. This will influence the network and the cost will be attached for creation. If the model of proof of stake is followed then purchasing power to compute will be required. This will put a cost on the creation of multiple accounts.

2. Trust chain: the trust chain is an approach which adds credibility to the new nodes. For example, if more than one account want to be created then the old node owners who can be trusted can take the guarantee of new nodes. If an old node is not old enough to do so, it can be in the system for a certain period of time and gain trust. This way the power of voting will not be misused. The trust can be granted by following a two-way authentication process. There are some peer networks which have set a rule to submit an identification proof before making the trust authenticated. Others are based on IP addresses associated with the account. The privileges associated with voting can only be given after crossing this verification step.
3. Unequal reputation: the merit of a node is often more than another. This is based on the reputation it has on the chain. The nodes which have been on the network for a long duration tend to enjoy a good reputation. Thus instead of democracy, power is distributed based on merit. The new accounts cannot thus be given power because they might fail to come under a good reputation scene (Figure 5.4).

For any attack to be detected, it is needed that the network shows the damage clearly. In case of Sybil attack, the damage caused is often not so obvious. The identities that are into the system are fake and are difficult to be traced. The severity of the attack thus becomes very high causing damage beyond repair. Bitcoins try to reduce the effect of such attacks by following every possible approach because the base system involves currency even though it may be virtual.

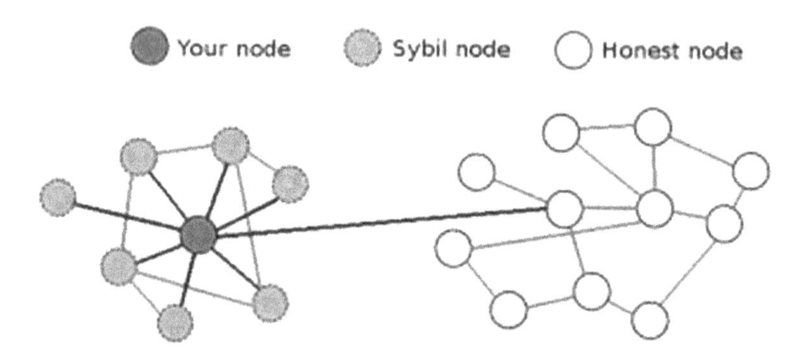

FIGURE 5.4 Sybil attack [16].

5.2.7.1 Eclipse Attack

In a recent study it was found that Sybil attack was modified and made to work as an eclipse attack in a blockchain. The attacks aim to point those nodes to which the information has to be transferred. They take the information midway and use it to damage the system. Thus instead of attacking the complete network, a single node data is compromised. The eclipse attack cans strengthen the 51%attck made on the network of a blockchain. The singling out of a node from the network helps to solve the hash in a comparatively easy way because it is de-linked from its neighbors. The power of the peer nodes is wasted and it is often referred to as selfish mining. Thus a Sybil attack is not as much of a threat as much is the eclipse attack.

Many thousands of years ago, the biggest challenge for mankind was how to generate enough electricity. Now that the technology has progressed and we have reached a phase where energy production is not so difficult anymore. The major discoveries of inventions in technology rely on the energy sector. It is but natural to question how to generate enough power to keep the advancement in the technologies running without having to compromise on the usages of the given technology. To counter the expensive nature of energy production, small batteries that are rechargeable have been popularly and extensively employed. However, the challenges still remain as to how we can reduce the gap between the advancing bigger technologies and the power sector. The blockchain technology has given a way forward in the green computing environment. Even though it aims to utilize energy effectively, the energy requirements for technology are still under consideration (Figure 5.5).

So far we are aware of some of the terms used extensively in the blockchain technology, for clarity's sake some are as follows:

A *cryptocurrency* is a form of currency to be used on a digital medium. It can be used for verification of funds and the transfer of an amount from one node to another in a blockchain.

A *node* in a networking environment is an agent that keeps the network connected. It could be a peripheral device like a printer or a computer. The blockchain technology has nodes that are added by miners.

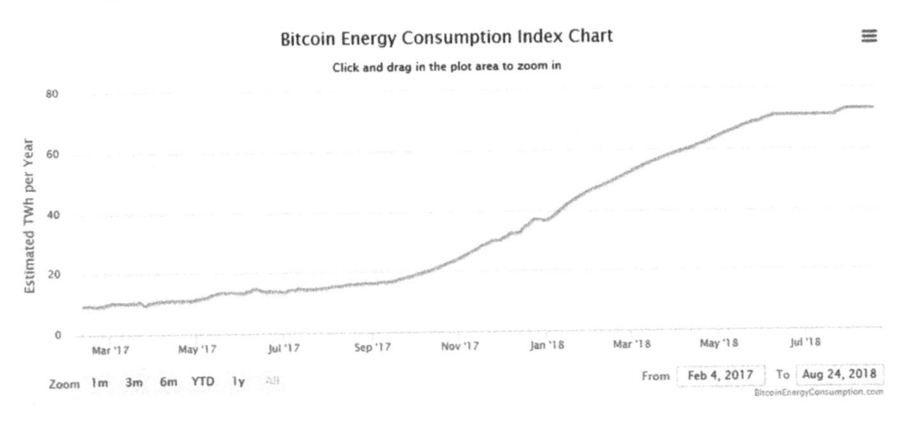

FIGURE 5.5 Bitcoin energy consumption index [17].

A *miner* is a special type of node which computes the read and write operations which are called transactions. The miners are given rewards for making a node good enough to be added to the chain.

The proof of work model, which has already been discussed, brings forth the energy consumption. Each miner is given a hash code, the hash code also has a hash code of the previous node. This hash code needs to be solved to get the power to add a block to the blockchain. As a reward, the miner can validate the block in the database. The miners are often seen wondering about the computational power which is needed to solve the puzzle. The more the difficulty level of a puzzle, the more will be the power that will be used. But the uses of bitcoins of blockchain technology are such that the cost of energy consumption could be balanced [5]. The currency assures to make a miner rich enough tom afford the cost of computation. The downside of this approach is that even though proof of work is an effective model, yet the total blockchain does not follow this consensus algorithm. There is still a demand for a more cost-effective model which does not compromise with decentralization and efficiency of the blockchain. Proof of authority (PoA) is another approach to tackle the issue. The nodes used can be chosen prior to the transactions which will add a node in the blockchain, they can then get authority using PoA. Thus, the selected nodes can have permission, which gives them special power and security. They create sidechains in the actual blockchain. It is believed that proof of stake (PoS) makes use of the energy efficiently. It does not utilize all, the nodes but only those that contribute to the blockchain more. That is higher the stake higher will be the participation.

The data which is stored on the ledger or the database also consumes energy. The data is the form of transactions that have taken place on the block. If the transaction appears quickly then the cost of energy consumption rises faster. Data storage is thus an essential; element of any cryptocurrency, even though it is cheaper storage than other functions. The future of cryptocurrency is dependent on the efficient utilization of energy without compromising on the benefits that the technology is offering.

5.2.8 Hyperledger Fabric: A Blockchain Development

The background of hyperledger fabric is related to the development of a platform for providing solutions to distributed ledger. A ledger is a storage area and because it is distributed, therefore easier to maintain security. The architecture of the fabric provides benefits of confidentiality, scalability, resiliency, and flexibility. The ease of use makes the fabric easily adaptable to any industry. The hyperledger fabric includes components of consensus and membership, which are more about plug and play. The fabric has a set of protocols for a business and they are referred to as "chaincode." The chaincode makes it possible to host smart contracts. The complete structure of the economy has varied features; the hyperledger fabric provides pluggable components which make it possible to deal with complex versions. The distributed platform of the hyperledger fabric allows running of distributed applications designed in blockchain. The hyperledger takes advantage of the idea that there is no technology that fits all domains and hence it is scalable. The use of consensus protocols allows for the implementation of use cases and give more trust to the business. The distributed applications that are run on the fabric are written in general purpose programming languages and are not dependent on a cryptocurrency. the general case is that the blockchain platform that runs smart contracts has codes written in domain-specific language or they are heavily dependent on the cryptocurrency. The industry standard of identity management can be implemented using the hyperledger fabric. The model makes use of improvisation and implements steps to reduce risks on the business. The fabric gives the functionality of creating channels where different participants can form separate ledgers where they process the transactions. The network designed is for a specific purpose where many participates would not be willing to share confidential information with all but select participants only. The participants of the channel are the only ones who get access to that channel and no outsider. Public blockchain cannot be used by the businesses because of the following [12]:

1. Privacy Issues: Bitcoin is an example of public blockchain and it is permissionless. The pseudonymous users and anonymous users can easily join it. The privacy requirements of many business models like health care and finance are strict. The confidentiality factor is critical in such sectors and they cannot allow anyone or everyone to see the updates.
2. Scalability: The public blockchains are not scalable. All the nodes that form its integral part are used in the transaction. The output is throughput is reduced.
3. Immutability of smart contracts: The smart contracts cannot be changed. The effect of the deployment and how they result on the application cannot be reversed. The caution that needs to be taken into account is the deployment without any bugs.
4. Storage Issues: A node in a public blockchain stores the complete data; therefore, a high storage facility needs to be in place. This requirement continues to grow and has to be met without causing redundancy or wastage.
5. The consensus algorithm is unsustainable: The proof of work algorithm used by bitcoin and ethereum has an immense need for computing power and energy. These requirements increase gradually and are difficult to be fulfilled
6. Lack of governance: The improvement of the projects is the sole responsibility of the developer or the community. The control in public blockchains is not restricted at all. For any business, the requirement of regulated governance is of high importance.

Every business has a specific requirement when it comes to the network of the blockchain:

- The demand is high for a validation process that can allow new users to be added to the system. The new user and its identity need to be checked well in advance.

- The execution and scalability is always the first thing on the mind of the enterprises.
- The sensitive and confidential information should be accessible to not every user but with control.
- The role definition of a requestor, approver, and the keeper of records should be done effectively by following a consensus algorithm.
- The resilience of the system to fault should be high.
- The process of troubleshooting should be automated.
- Maintenance of a blockchain in a network should be high.
- The long term processes for proactive governance is essential.

A hyperledger fabric deals with business requirements in the following ways:

- The code should be the basis for creating an open-source enterprise scalable blockchain.
- A community-driven atmosphere which can delay the business deals should be discouraged.
- The use of use cases and trails should be given more technical leverage.
- More communities should be educated to deal with blockchains.
- Promote an ecosystem of blockchain in an enterprise.

An enterprise can make use of the following features of the hyperledger fabric:

- Integration of components such as consensus algorithm and membership services by using plug and play services.
- A container technology can be implemented by making use of smart contracts, which are referred to as chaincodes.
- Confidential transactions can be used by availing the concept of channels in the fabric.
- The transactions can be delivered to their peers by using the ordering technology.
- Transaction validation can be easily implemented by endorsement policies.
- Database services like CouchDB.

The hyperledger fabric does not follow PoW or a cryptomining process. The transactions that take place in the ledger provide scalability and speed. The normal workflow is as follows:

1. Every transaction undergoes three separate phases:
    - The distributed structure to process logic and use of chaincodes
    - Ordering transactions
    - Validation and commit of the transaction
2. The various types of nodes offer reduced overhead and work smoothly.
3. A transaction lifecycle follows this path:

- A requester is the one that forwards a proposal of the transaction to the endorser
- The transaction requirement of the number of endorsers and the endorsement policy is stated
- Simulation of chaincode for the proposal using read/write set
- The endorsements, which are the signed proposed responses, are sent back
- The orderer receives these transactions with digital signature by the client
- Block of transactions is made by the orderer and forwarded to the peers
- The peers crosscheck for any conflict with the endorsement policy in the transaction
- The peer commits the block in the ledger

5.3 CONCLUSIONS AND DISCUSSIONS

The distributed hash tables use the concept of consistent hashing. The reason for this is the leader agent requires that data is uniformly distributed over the network. A key goes through a randomization a function made with the help of the hash algorithm. The idea for this approach is equal opportunity to nodes for being chosen to store the key and value pair. The complete data is not stored on the leader agent, therefore to access a node that has the key/pair value; a node needs a routing layer. The routing table and its construction, how the nodes join and leave the network is different from the algorithm of leader agent. The routing layer will store less information about the node on the network. Therefore, the routing information has to be spread across a small number of peers. Since less information is stored on the routing table, therefore, the multiple nodes have to check to find the key. The common complexity of lookup is O (log n) where n is the total nodes in the network. This means if there are a million nodes then 20 would need to be found.

The lookup operations in different initial conditions with leader agent and PoS are classified as recursive and iterative. In case of iterative, the requesting node will query another node for a key/value pair, in case that node does not have the pair, it will return one or more nodes that are closer. This process continues until the key is found or else an error message is generated stating that the key cannot be found. In the case of recursive lookups, the query is made to the next node, which further queries the next node, and the result is returned through the chain of nodes that is created. Sharing the actual input values infringes the agents' confidentiality and this has encouraged researchers to establish privacy measures that prevent the feedback of an agent from being exposed to some other agent in the network or an eavesdropper monitoring the agents' communications throughout the consensus protocol. Some of the typical knowledge consensus applications with important confidentiality concerns include online voting by social networks (peer-to-peer communication), total power consumption, medical surveys, online product sales monitoring, multiagent rendezvous systems, and online census.

REFERENCES

1. H. Liu et al., "Distributed identification of the most critical node for average consensus," *IEEE Trans. Signal Process*, vol. 63, no. 16, pp. 4315–4328, 2015.
2. A. Bertrand and M. Moonen, "Distributed computation of the Fiedler vector with application to topology inference in ad hoc networks," *Signal Process*, vol. 93, no. 5, pp. 1106–1117, 2013.
3. L. Xiao and S. Boyd, "Fast linear iterations for distributed averaging," *Syst. Control Lett.*, vol. 53, no. 1, pp. 65–78, 2004.
4. A. Bertrand and M. Moonen, "Topology-aware distributed adaptation of Laplacian weights for in-network averaging," in *Proc. Eur. Signal Process. Conf.*, Marrakech, Morocco, September 2013, pp. 1–5.
5. A. Ghosh and S. Boyd, "Growing well-connected graphs," in *Proc. IEEE Conf. Decision Control*, December 2006, pp. 6605–6611. http://citeseerx.ist.psu.edu/viewdoc/download?doi=10.1.1.85.320&rep=rep1&type=pdf
6. P.-Y. Chen and A. Hero, "Local Fiedler vector centrality for detection of deep and overlapping communities in networks," in *Proc. IEEE Int. Conf. Acoust., Speech Signal Process* (ICASSP), Florence, Italy, May 2014, pp. 1120–1124. doi: 10.1109/ICASSP.2014.6853771.
7. L. Xiao and S. Boyd, "Fast linear iterations for distributed averaging," in *Proc. IEEE Conf. Decision Control*, Maui, Hawaii, December 2003.
8. S. Boyd, A. Ghosh, B. Prabhakar, and D. Shah, "Randomized gossip algorithms," *IEEE Trans. Inform. Theory*, vol. 52, no. 6, pp. 2506–2530, 2006.
9. P. Gupta and P. R. Kumar, "The capacity of wireless networks," *IEEE Trans. Inform. Theory*, vol. 46, no. 2, pp. 388–404, 2000.
10. www.slideshare.net/philippecamacho/analyzing-bitcoin-security.
11. H. E. Bell, "Gerschgorin's theorem and the zeros of polynomials," *Am. Math. Mon.*, vol. 72, pp. 292–295, 1965.
12. D. Kempe, A. Dobra and J. Gehrke, "Gossip-based computation of aggregate information," in *Proc. IEEE Symp. on Foundations of Computer.* Cambridge, MA, 2003, pp. 482–491. doi: 10.1109/SFCS.2003.1238221.
13. https://lisk.io/academy/blockchain-basics/how-does-blockchain-work/proof-of-work
14. https://blockgeeks.com/guides/proof-of-work-vs-proof-of-stake/
15. https://coinweez.com/consensus-algorithms-pob-poi/
16. https://coincentral.com/sybil-attack-blockchain/
17. https://www.quora.com/Will-Blockchain-power-consumption-the-high-electricity-billeventually-kill-Bitcoin

6 DAI for Information Retrieval

Annu Mishra and Satya Prakash Yadav

CONTENTS

6.1 INTRODUCTION

The first thing that comes to mind when we see the topic "DAI for Information Retrieval" is what is information retrieval means. As we know, the amount of data available today is increasing at a very high rate—for example, on the Internet and the World Wide Web. Data can be in many forms, including audio, video, image, text, etc. Information retrieval is a way to obtain relevant data from that available. Information retrieval deals with search processes in which we need to identify a suitable subset of information. Distributed artificial intelligence (DAI) is responsible for managing communication between smart and keen agents. DAI endeavors to develop smart operators that settle on choices that permit them to accomplish their objectives in a world populated by other smart operators having their individual objectives. For the past two decades, semantic networks have played an important role in knowledge representation and artificial intelligence, i.e. basically in cognitive science-related research fields. It has a wide area of application, ranging from medical science, cognitive behavioral therapy, data mining, games, and politics. It has also been used in search, discovery, matching, content delivery, and synchronization of activity and information. The use of semantic network analysis has also been observed in thematic analysis of Twitter for the #Metoo hashtag. "Semantic networks" is a keyword that has been used to represent the concept. A semantic network or net is a graph structure for representing knowledge in patterns of interconnected nodes and arcs.

In this chapter, we present the recent work done by researchers in the field of semantic network representation and organizations of information and knowledge and basic concepts involved in semantic network.

According to Ponomarev and Voronkov (2017), a DAI can be characterized on the basis of three principle qualities:

1. It is a technique for the dispersion of jobs between operators.
2. It is a technique for dispersion of forces.
3. It is a technique for communicating among the participating agents.

The main aim of a distributed artificial intelligence framework is to inspect the collaboration of different agents to expound an exceptionally troublesome issue. This sort of framework is commonly portrayed level-wise, primarily on a miniaturized scale level. An expert is perceived as an autonomous object with information and that is capable of distinguishing some restricted errands. The commitment of each object might be basic: the strength of the framework is the after-effect of the cooperation between the various experts. The large-scale level, which is the second level, depicts the expert's environment that finds a solution for the given issue. A multiexpert design permits the framework to examine the participation of the experts. The advantages include critical thinking. The design of such a framework is modular in nature, prompting a superior working. Additionally, the advancement of the framework and the reutilization of modules inside different frameworks will be simpler.

Distributed intelligent multiagent systems offer secluded, adaptable, versatile, and generalizable calculations and frameworks answers for data recovery, extraction, combination, and information-driven information disclosure utilizing heterogeneous, appropriated information, and information sources in data-rich, open conditions. Such frameworks comprise of numerous cooperating shrewd programming specialists. Distributed intelligent multiagent systems offer modular, flexible, scalable, and generalizable algorithms and systems solutions for information retrieval, extraction, fusion, and data-driven knowledge discovery using heterogeneous, distributed data and knowledge sources in information rich, open environments. These systems comprise of number of interacting smart software agents.

The scope or region of DAI can be divided into two subfields: distributed problem-solving (DPS) and multispecialist systems (MAS). The vision of DPS is on the data the executive parts of frameworks where a few agents are cooperating in a conveyed way towards finding solutions for a given issue and in this manner attempting to accomplish the common objective. Multiagent systems are identified with DPS [1].

Smart software operators are those objects that play out a number of errands for the benefit of a client with some level of self-rule. Such receptive, dynamic, expectant, objective-driven, versatile, intelligent, information chasing, self-governing, intuitive, open, collective agents and multioperator frameworks discover applications in Internet-based data frameworks, versatile (adjustable) programming frameworks, enormous scope information-driven information revelation from heterogeneous, dispersed information and sources, cooperative logical disclosure (e.g., in bioinformatics), and wise DSS networks (e.g., observing and control of intricate, disseminated, dynamic frameworks).

6.2 DISTRIBUTED PROBLEM-SOLVING

In DPS, various operators cooperate to find solutions for a given issue. The central matter in these frameworks is that collaboration is required in light of the fact that no individual operator has adequate data, information, and the capacity to take care of the total issue.

6.3 MULTIAGENTS

Multiagent systems are fundamentally a class of algorithms in which agents cooperate and communicate among themselves on the basis of preestablished rules/constraints and, as a consequence, a collective behavior that proves to be "good enough" comes out of those communications. These interactions that happen to be in the systems are both in between the agents and/or in between the agents and the environment [2]. Multiple-agent systems are used to solve problems that are difficult to solve by an individual agent. One of the applications of multiagent systems is problem-solving.

Distributed Problem Solving: It is a substitute to centralized problem solving. This is for two reasons: first, the problems are distributed by themselves and, second, because of the distribution of problem among various agents, the system failure can be handled easily and thus provides the fault tolerant property to the system [3] (Figure 6.1).

6.4 A MULTIAGENT APPROACH FOR PEER-TO-PEER-BASED INFORMATION RECOUPMENT SYSTEMS

Recently, peer-to-peer (P2P) systems have changed the manner in which we successfully utilize and offer disseminated assets. Other than the conventional customer server design, P2P frameworks are community-oriented frameworks in which agents cooperate to play out specific undertakings; in this manner, multioperator framework innovation would appear to be pertinent for actualizing these sorts of frameworks.

Compared to record sharing frameworks which depend on a precise match, information recovery frameworks are designed to discover semantically pertinent reports which probably won't contain the entire keywords in the inquiries. This inquiry includes finding and recovering important archives conveyed to at least one database. This kind of framework manages to make us expect that there is a main mediator that

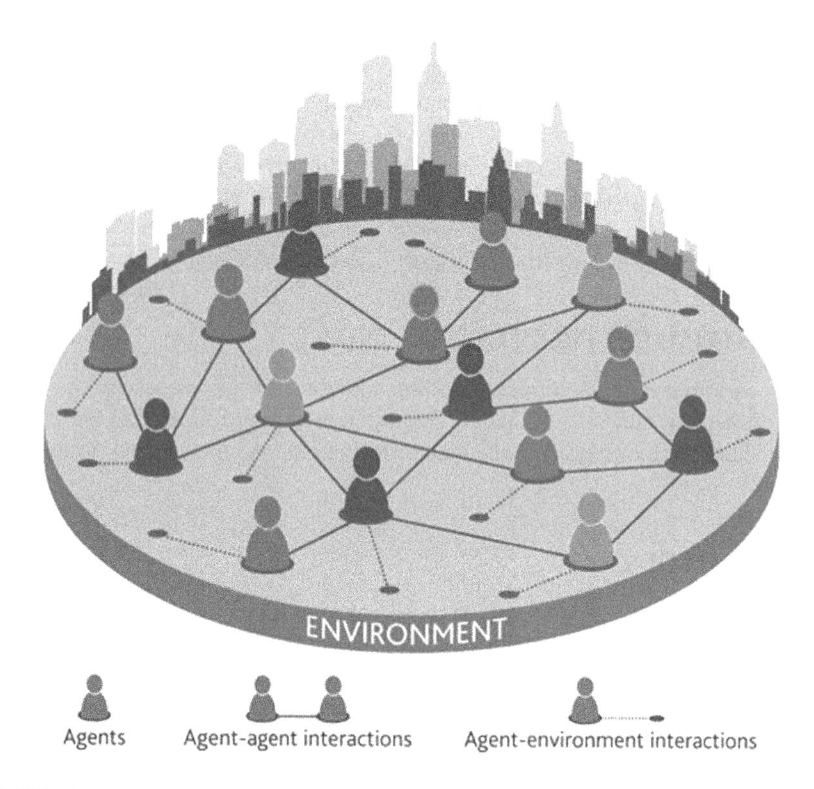

FIGURE 6.1 Multiagent interaction scenario [4,5].

can get to the asset depictions of all of the customer subassortments. Nevertheless, this unified framework has various limitations that inspire progressively decentralized methodology. To start with, the main mediator in these frameworks will in general be overpowered by countless approaching inoculations, hence turning into a bottleneck and a solitary purpose of disappointment. Restricted by the limit of the main mediator, the unified framework does not fit enough. To add to this, short meeting lengths make a weblike arrangement (crawlers, indexing, recovering) unfeasible because of the regular operator high and low points that influence the main mediator capacity to follow collection updates. In this chapter, we will examine an alternate methodology to eliminate the aforementioned weaknesses, a mediator-less P2P design that has no agent in the framework to possess overall perspective on the collection. In cases in which the fragmented idea of the nonglobal perspectives on singular agents regarding the collections retained by different operators are given, the data recovery issue in P2P systems is usually organized as a disseminated search process [6].

6.4.1 A Mediator-Free Framework

Because of the nonaccessibility of a mediator, the agents collaborate to pass the inquiries among themselves to discover proper operators, rank the collections, and, last, to restore and consolide the outcomes so as to satisfy the data recovery errand in a disseminated situation. Every agent is in four parts: a collection descriptor, a web crawler, an operator view framework, and a control center. The collection is a group of reports that impart information to different peers. The collection descriptor can be regarded as the "signature" of the collection. By appropriating collection descriptors, agents can possess superior information on how information is dispersed in the operator society.

6.4.2 Agent-View Algorithm

A conventional way for forming an early agent-view by the agents is, when they join the system for the first time, to initialize the protocol by sending packets containing their IP addresses. In this system, the aforementioned procedure is vaguely modified by adding the document collection model exhibited by the agent during transmission. This agent discovery approach ensues in an arbitrary-graph like arrangement of networks being built, wherein every agent belonging to the network creates an agent-view framework along with the collection models and IP addresses of its neighbors. Nevertheless, one of the main limitations of these kinds of arrangement of networks is the absence of search efficiency as there is no bridge between the agent-view and how to proactively look for agents that possesses pertinent documents. Therefore, this technique uses an agent-view reorganization algorithm (AVRA) hinged on the earliest agent-view. The objective of AVRA is to construct an agent-view that is comprised of agents whose possession is equivalent, creating semantically matched agent clusters. For instance, we can have a "sports" cluster, an "economics" cluster, and so forth. Clusters mentioned are not dissociated, which signifies that an agent can be a part of various clusters available. For instance, an agent can be a part of both a "basketball" cluster and a "college" cluster on the basis of its content. Clusters are linked among themselves; this means that if a query is forwarded to an agent, in the absence of relevant documents it can be forwarded in a timely manner to an appropriate cluster. Therefore, agents can share their local agent-views to stretch the extent of view so that every agent is adequately enlightened in the matter of content distribution over the whole network. Every agent decides locally about the agent belonging to its agent-view with which they would interact. This helps in building expanded views in a directed manner. This decision leads to the sending of an Expand? Message to the relevant agent. The Expand? Message incorporates the IP address of the sending agent so that target agents can return an answer. After receiving the Expand? Message, every agent forwards its agent view to the inquiring agent. This kind of communication informs every agent's local view with more intelligence on the content dissemination in the agent society, and permits the agent to take informed decisions regarding which agent to pick to forward the queries [6] (Figure 6.2).

The algorithm works as follows: we evaluate similarity Wcc(Ai, Aj) with the neighbors of Ai(every agent of the system) where Aj is the neighbor.

Following the ranking process, agent Ai inquests its maximum comparable K neighbor operators using "Expand?" Messages. Here, $K = 1$ is applied to reduce the

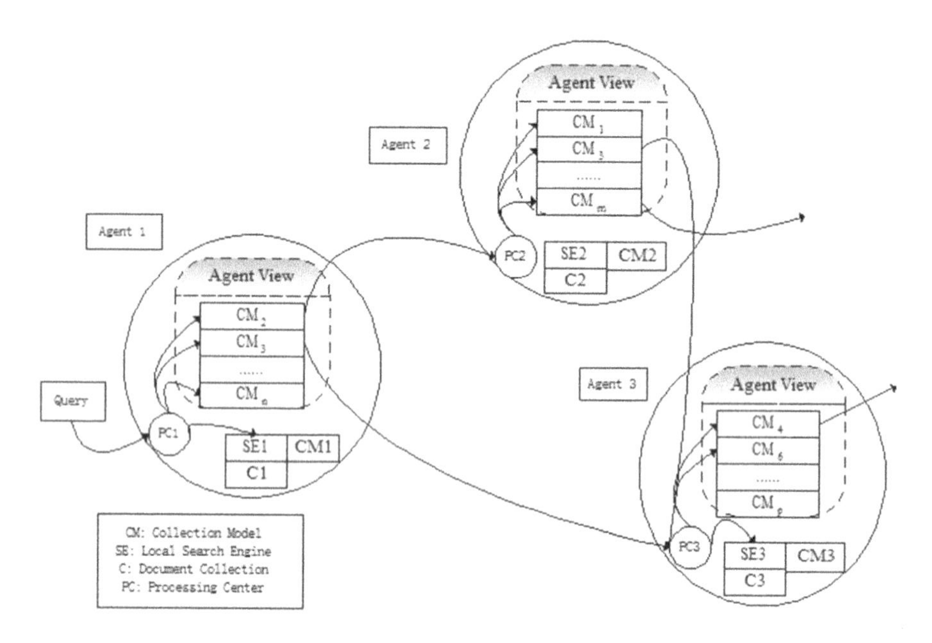

FIGURE 6.2 A mediator-free multiagent system [6].

communication. After accepting Expand? Messages, every agent acknowledges with its current agent-view. To avoid the unvaried agents from being nominated again and again, we set out that agents cannot be chosen more than two times in three consecutive cycles. This heuristic enables an agent to build a greater bounded view in order for the restructuring process to continue persuasively.

After incrementing the view by communicating with the neighboring agents, the policies mentioned are followed:

(1) M% of the total degree of Ai are classified as its maximum similar neighbors, while the remaining (1-M%) neighbors are arbitrarily selected from the agent-view collected. Discussed randomization results in prolonging existence in the agent society. It can be observed that on the condition that all of the neighbors are selected out of the maximum similar neighbors (M is a 100%), the consequent network suffers because of the association. In other words, this includes many distinguishable "clusters," although those clusters are semantically close. After checking out various values, we assign M to 80.

(2) If there is a scenario that the wide variety of incoming connections (in-degree) of Agent Ai lies beneath 2, an empirical threshold that stipulates if the agent is effortlessly reached through the global elements, then the agent approaches its neighbors to request for including it among their neighbors in the local agent-views.

6.4.3 Distributed Search Algorithms

The distributed search procedure starts as soon as the agent receives a query. The agent thereafter makes numerous local decisions, for instance if it must carry out a local search to find out whether it has the caliber to fulfill the query locally and whether it needs to forward this query to different neighbors (if yes, then to whom), or else the query must be dropped. In the course of this technique, if any of the agents receive a query that it formerly handled, it will bypass it immediately. The query then continues within the community unless all of the agents receiving the queries drop the message. Without this message, it is impossible to recognize that the query is now not being handled via any agent inside the network. [6]

Two search algorithms, k Nearest Neighbors (kNN) collection model-based method and Gradient Search Scheme, are utilized.

6.5 BLACKBOARD MODEL

The necessity of cooperation and communication between knowledge-based systems (KBSs) has given rise to research in the field of DAI. Various models have been propounded—comprising the blackboard model. The blackboard was originally designed as a method to manage complicated, inexplicit problems. The first well-known example is the Hearsay II speech recognition system.

A blackboard system has three components:

1) blackboard
2) knowledge source
3) control component (Figure 6.3)
   - Knowledge sources (KSs) are unbiased modules that include the knowledge required to resolve the issue.
   - The blackboard is a nonlocal database comprising input records, partial solutions, and other information, which can be in numerous states.
   - A control component facilitates runtime selections about the direction of problem-fixing and the cost involved in problem-solving amenities. This component is different from the individual KSs as it is dynamic in nature in terms of path selection and run time changes to be made, unlike the KSs where the knowledge is fixed for the particular problem.

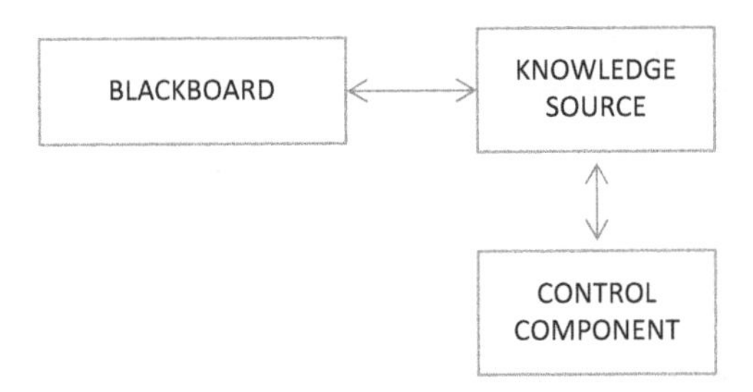

FIGURE 6.3 Basic components of the blackboard model.

6.6 DIALECT 2: AN INFORMATION RECOUPMENT SYSTEM

DIALECT 2 is a data retrieval system with the usage of DAI equipment that is composed of two modules: the first module is called the linguistic parser and the second module is known as the reformulation module. First is the linguistic parser, which finds the syntactic shape of a request so as to create a chain of templates that incorporate the essential data factors of the request; and second is the reformulation module, which utilizes the network so as to extract the required data for problem-solving. On each facet, the structure of each module is formed on the blackboard version [7]. Task of the system calls for several tiers, which are shown in Figures 6.4 and 6.5.

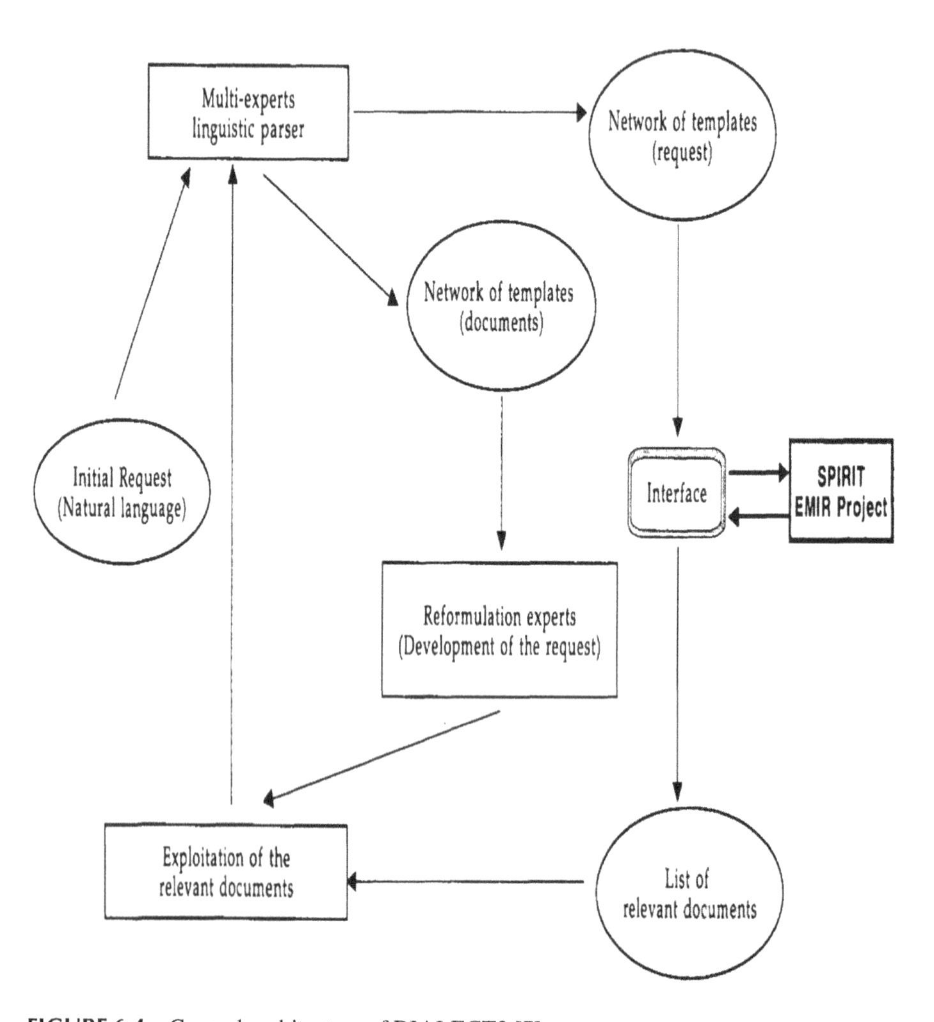

FIGURE 6.4 General architecture of DIALECT2 [7].

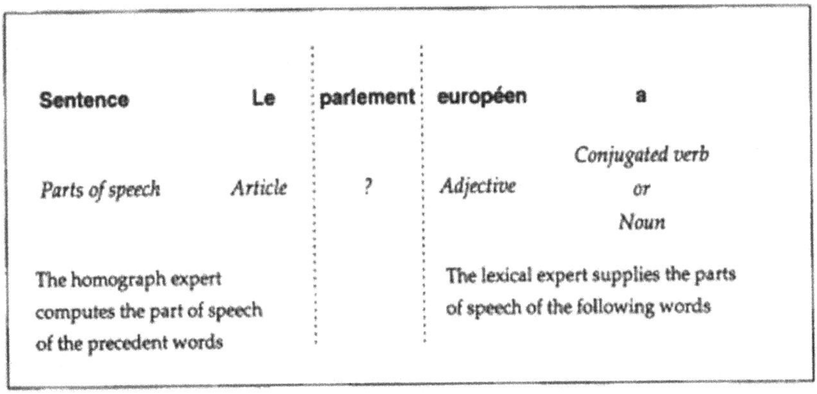

Sentence	Le	parlement	européen	a
Parts of speech	*Article*	*?*	*Adjective*	*Conjugated verb or Noun*

The homograph expert computes the part of speech of the precedent words

The lexical expert supplies the parts of speech of the following words

FIGURE 6.5 Process of a misspelt word [7].

The multiexpert linguistic parser recognizes the syntactic arrangement of the request. Thereafter, it fabricates a system of formats. A format depicts expedient linguistic pattern and the system of layouts comprises the principle components of data present in the parsed content. This errand is finished by a few operators [7]:

- The lexical entries master examines each expression of the sentence utilizing Gross' electronic dictionary [8]. Thereafter, grammatical forms of each word are given.
- The expressions of a sentence frequently have a few grammatical forms: the job of the homographs master is to illuminate vagueness. This master utilizes a number of condition-activity protocols. The condition part of the protocol is a grouping of grammatical forms and the activity side shows the credibility related to the succession of grammatical features.
- If the scenario arises when a word is obscure, the incorrect spelling master, the lexical entries expert, and the homographs master collaborate so as to supplant it: the homographs master prefers the potential grammatical features of the obscure word. The lexical entries expert gives a rundown of words that may supplant it. The incorrect spelling expert utilizes this data to address the incorrect spelling. On the off chance that the remedy comes up short, the homographs master is capable of deducing the potential grammatical forms of the obscure words.
- The templates expert along with the grammatical expert and word expert construct the templates. Grammatical expert completes the templates built by the templates expert if "special" words as conjunctions, relative pronouns, or commas occur in the sentence. The two specialists utilize a number of condition-activity rules. The condition side is an arrangement of

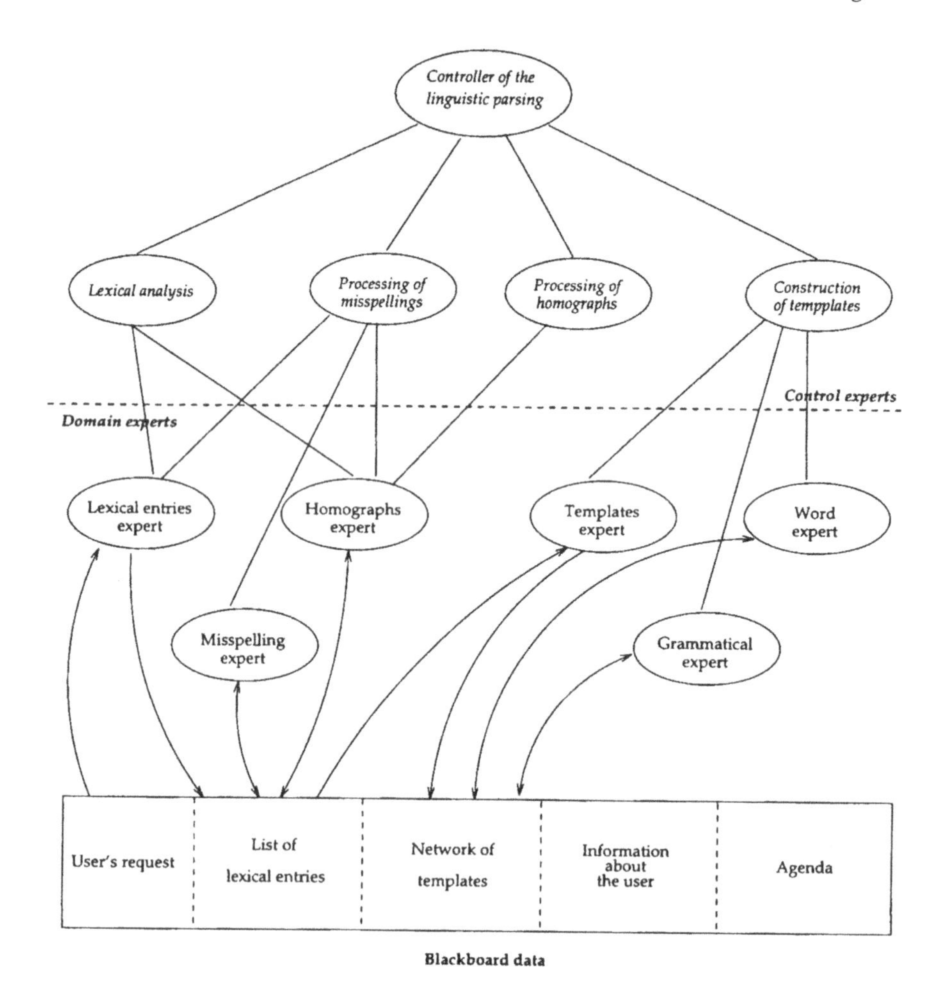

FIGURE 6.6 The control of the linguistic parser [7].

grammatical forms. The activity side demonstrates how these specialists construct the formats. The word master checks the formats that connect an action word to its supplements.

Let us consider the following sentence: "Le parlment européen a créé une commission d'enquête." (in English: "The European parlment (misspelt word) created a committee of inquiry."). The second word of the sentence is unknown. Several experts work together in order to correct the mistake. First, the homographs expert creates a set of the prospective parts of speech of the unspecified word. The homographs expert identifies the part of speech of each precedent word and the lexical entries expert supplies the determined parts of speech of the following words (see Figure 6.6). Then, the homographs expert can determine the section of speech of the unspecified word by comparing the sequence of section of speech with the condition sides of its rules. For example, the prospective section of speech of "parlment" is an adjective or noun:

Sequence of parts of speech	*Condition sides of rules*
article ? adjective	article *adjective* adjective article *noun* adjective
? adjective noun	*article* adjective noun *adjective* adjective noun *preposition* adjective noun *preposition* adjective noun
? adjective conj. verb	*noun* adjective conj. verb

The lexical entries expert nominates those words which could replace the misspelt word. This nomination is based on three retrieval criteria:

- the nominated word comprises of any prospective section of speech
- the nominated word starts with initial alphabet of the wrongly spelled word
- the length of the nominated word and the misspelled word are nearly equal

In conclusion, the incorrect spelling master figures the interval between every word and incorrectly spelled word. The word that contains the most minimal interval is chosen to address the misstep. It uses the interval characterized by C. Fouquere [9], which is known in as the French difference estimation distance. In this model, "parlement" (in English: "Parliament") is chosen to supplant "parlment." The best solutions are shown below:

parlement	palmette	parlant	parlements	payement
parement	paravent	parlante	parmesan	purulent
paiement	parements	parlants	pavement	pédiment

At the point, when each expressions of the sentence are examined, the linguistic parser of DIALECT 2 processes the sentence. The result is depicted in the following tables:

Word	*Parts of speech*
le	article or pronoun
parlement	noun
européen	adjective
a	conjugated verb or noun
créé	past participle

Word	*Parts of speech*
une	article, adjective or noun
commission	noun
de (d')	preposition or article
enquête	noun or conjugated verb
.	punctuation

The homographs expert selects the first part of speech of each word.

(européen, est, parlement)	*(European, is, Parliament)*
(création, par, parlement)	*(creation, by, Parliament)*
(création, de, commission)	*(creation, of, committee)*
(commission, de, enquête)	*(commission, of, inquiry)*

These templates are built by DIALECT 2 from the sentence.

DIALECT 2 recovers a few parts of applicable documents.

After this, the system carries out a mechanized reformulation of the request so that it can propound to the user various pertinent documents. These documents are in natural language. Therefore, every sentence could be parsed with the aid of the linguistic parser of DIALECT 2 to create any other network of formats. The newly formed network of formats is utilized by the agents of reformulation to broaden the preliminary request. Thereafter, the reformulation procedure stops while all of the sentences of every parts of applicable documents are analyzed.

From the ameliorated request, DIALECT 2 gains different components of pertinent documents that were never propounded. These parts of documents are also fixed by the reformulation operators as a way to ameliorate the request. The automated reformulation is concluded while not even a single applicable document is gained within the database. The blackboard architecture is a preferred version for solving any problem.

6.6.1 The Control in Blackboard Systems

In 1975, HEARSAY II [10] appeared to be the earliest blackboard framework. Subsequently, various analyses and research on the related topic were conducted and three fundamental kinds of control were propounded: the procedural control, multilevelled control and the blackboard control. The similarity measurement between these types of control was done based on the following:

- Efficiency of the system: the system can quickly select the succeeding activity that will be carried out
- flexibility of the arrangement: the control strategies evolve following the current solution (opportunistic resolution).
- Moreover, the system can easily be completed by new modules.
- Uniformity of the system: the control problem and the domain problem are solved in the same way.
- Simplicity of the architecture: simple representation of the strategies and structured organization of the knowledge.

Next, we see the primary kind of control that were propounded in different blackboard systems [7]:

1. The procedural control: the system makes use of a bundle of previously available heuristics in an effort to pick the expert having maximum priority (in blackboard terminology, this is known as a knowledge source (KS)). This form of control is recognized with the aid of its high performance and its proper flexibility when the scope of the problem is widely recognized and structured. However, its performance and its maintenance emerge as quite complex. The procedural control was architected by way of HEARSAY II.
2. The hierarchical control: the system comprises two sorts of knowledge source (KS), which might be the domain KS and the control KS. These sources are arranged level-wise: the domain KSs commonly hold the lowest level of the arrangement and they are capable of changing blackboard facts. The higher tiers are held with the aid of the control KS. At the highest level, a control KS invokes one of the KSs, which is present in the bottom level, in line with the existing state of the decision. The KS additionally invokes one more KS belonging to the lower level and this procedure is going on till the level of domain KSs is attained. The hierarchical control possesses numerous superiorities: understanding of each module, better enactment of the issue, and the proficiency of the control.
   - The blackboard control: the KSs belonging to the system can be categorized in two categories: the domain KSs and the control KSs. Each category utilizes a specific blackboard. The former KSs construct the interpretive of the issue on the domain blackboard and the control KSs manages the collaboration of the domain KSs thanks to the control blackboard.

6.6.2 Control in DIALECT 2

We will now explain control in DIALECT 2. The authority of the linguistic parser is a multilevelled control. We will then explain management of the reformulation module on the basis of blackboard control.

6.6.2.1 The Linguistic Parser

The authority of the linguistic parser is a multilevelled control. In DIALECT 2 the parsing of a sentence is accomplished by a few levels: the morphological stage and the lexical stage, the handling of the homographs, the processing of incorrect spellings, and the development of formats. However, these levels are not all that autonomous.

The hierarchical control permits a specification of the issue wherein each and every control information and every scope of expertise are arranged level-wise.

In the multilevel control, at the highest level, a solitary KS watches the parsing. The KS points out the plan of the blackboard so as to supervize the advancement of critical thinking and to launch one KS at the bottom level. Every KS belonging to the level possesses individual area of ability: the lexical investigation, the handling of incorrect spellings, the processing of homographs and the development of layouts and they actuate the domain KSs of the top most level. Many KSs may be summoned via various control KSs on the grounds that the insight is required so as to accomplish various errands. The procedure is rehashed until the production of the considerable number of formats is completed. Toward the start of parsing, the blackboard delays calling the client. In the course of parsing, domain KSs develop a rundown of lexical sections and the system of layouts.

6.6.2.2 The Reformation Module

A reformulation phase deals in substituting some words belonging to request or some words belonging to text so as to ameliorate the purpose of the propounded texts. Two basic concepts have been defined: *recall* and *precision*.

Recall is the segment of pertinent information gained in response to a search, whereas the precision is the segment of gained items which are actually relevant. The purpose of the reformulation is to improve the recall and the precision. In contrast to the linguistic parser, the multilevel control cannot adjust with a reformulation module in light of the fact that the various domain KSs does not follow the leveled organization. At present we tend to learn an answer to the rein of the reformulation module that relies on blackboard control. The layouts belonging to the arrangement will be portrayed. The reformulation module possessed by the DIALECT 2 comprises of a few domain KSs which propound various kinds of reformulation [7]:

- *Normalization of the lexical entries*: normalization allows the system to have only one representative for the lexical entries which have a lot of inflexions. This processing is applied to verbs, nouns, and adjectives and their respective representatives are the nominal form (or the infinitive one if the nominal one does not exist), the singular, and the plural. This processing is performed on the lexical entries of the blackboard.
- *Word families*: the expert replaces a word of the request with some other word which belongs to the same word family. The expert extracts the stem of the word of the request and uses a list of possible suffixes. It searches the words of the dictionary that have the same stem and keeps the ones which have a possible suffix.

  In linguistics, a *stem* is a part of a word used with slightly different meanings and would depend on the morphology of the language in question.
- *Construction of semantic classes*: a semantic class is made up by words which have the same part of speech. These words come from two sentences which are very close. One sentence is a part of a relevant document and the other is a part of the request. The system compares the networks of templates which are built from these sentences: it looks for common templates or common parts of templates and it produces semantic classes according to the result of the comparison. The semantic classes which are built by this expert are stored on the blackboard.
- *Thesaurus*: a thesaurus allows replacing a word by another one which has no morphologic link. So, the user can formulate the request without using the words of the texts.

  Furthermore, when the reformulation is finished, the semantic classes which are not present in the first thesaurus (called *the general thesaurus*), are stored in another thesaurus (called *the user's thesaurus*). The responsibilities of control KSs are as mentioned:

- Activating the domain KS of reformulation, which improves the recall and the precision.
- Management of development of the semantic classes: two kinds of client are recognized by DIALECT 2: infrequent client and experienced client so as to guide the development of the semantic classes. The principal case has all the fabricated semantic classes are retained. However, the second case involves a process in which confirmation is sent to the client prior to retaining every semantic class.
- Refurbishment of the client's thesaurus toward the accomplishment of reformulation.
- Regularizing ellipses and pronouns so as to upgrade reformulation as it gives the flexibility to view the entire range of phrases (Figure 6.6).

6.7 ANALYSIS AND DISCUSSION

The reason for this examination is to sum up the current psychological positioning model assessment, break down existing procedures, and add some understanding for the future. Through model engineering and configuration learning, we built up a solitary structure over the subjective arrangement models and assessed existing models dependent on this definition from various measurements. Specialists broke down existing models for model engineering investigation to comprehend their fundamental ideas and central plan standards, including how to oversee inputs, how to discover explicit highlights, and how to break them down. Scientists audited famous learning targets and prepared techniques embraced for psychological client-situated administrations for evaluating learning examination. Specialists additionally tended to many developing subjects that will be significant in the future. Similarly, as there was a flood of new ideas for other profound learning-based methodologies, we see that neural network positioning models developed quickly and extended in application terms. We trust this review can help scientists through new looks at past techniques. We anticipate that noteworthy discoveries in this space should be cultivated sooner rather than later through the network's endeavors.

6.8 CONCLUSION

The research reported revealed in this chapter tends to a significant number of the issues that emerge when full content data recovery is applied in situations containing numerous content databases constrained by numerous free gatherings. Agents of DAI require provisions for keeping up and spreading and projecting their point of view for real-world problems. But inefficient communication due to slow speed, distinct nature of the agents, and—sometimes—a lack of resources results in the poor working of the DAI system. To overcome these limitations, agents are required to keep up with the system and reexamine or update the network to which it is connected. This refreshing of the network must be done every time agents receive any information. Structures and tasks of DAI frameworks, which proceed true to form, present noteworthy difficulties. Due to the fact that the framework control is distributed in nature, every agent—also known as the operator—has restricted capacities and data, information is concentrated, and calculations are simultaneous. Therefore, the best way to ensure proper execution of the DAI framework is to permit suitable communication among operators. Subsequently, the problem of which operators ought to collaborate to solve a problem and how they ought to interface, what ought to be their methodologies, and how to determine any contention that may emerge are significant for DAI frameworks. The arrangements incorporate strategies for procuring depictions of assets constrained by uncooperative gatherings, utilizing asset portrayals to rank content databases by their probability of fulfilling a question, and combining the report rankings returned by various content databases. Altogether, these methods speak to a start-to-finish answer for the issues that emerge in disseminated data recovery. The circulated IR arrangements created in this chapter are powerful under a wide arrangement of conditions. The exploration revealed in this chapter speaks to a huge initial move toward making a total multidatabase model of full content data recovery. A basic circulated IR framework can be assembled today dependent on the calculations presented here. Be that as it may, a considerable lot of the customary IR apparatuses, for example, significance input presently cannot be applied to multidatabase situations. This outcome is only of academic interest for now until there is a general technique for making inquiry extension databases that precisely speak to numerous different databases..

REFERENCES

1. G. Weiss (ed.), *Multiagent Systems: A Modern Approach to Distributed Artificial Intelligence*, MIT Press.
2. S. Ponomarev, A. E. Voronkov, (2017). "Multi-agent systems and decentralized artificial superintelligence", arXiv preprint arXiv:1702.08529.
3. J. Ferber, (1999). *Multi-Agent System: An Introduction to Distributed Artificial Intelligence*. Harlow: Addison Wesley Longman, Paper: ISBN 0-201-36048-9.
4. F. Corea, (2019). "Distributed artificial intelligence, a primer on multi-agent systems, agent-based modeling, and swarm intelligence".
5. A. Turrell, (2016). "Agent-based models: Understanding the economy from the bottom up", *Bank of England Quarterly Bulletin 2016 Q4*.
6. H. Zhang, W. B. Croft, B. Levine, V. Lesser, (July 2004). "A multi-agent approach for peer-to-peer-based information retrieval systems", *AAMAS* (Vol. 4, pp. 456–463).
7. M. Braunwarth, A. Mekaouche, J.-C. Bassano, (1994). "Information retrieval system using distributed artificial intelligence tools", *Intelligent Multimedia Information Retrieval Systems and Management* (Vol. 1, pp. 449–460).
8. M. Gross, "Lexicon grammar and syntactic analysis of French", *Proceedings Coling* 84, France.
9. C. Fouquere, "Système d'analyse tolérante du langage naturel," These de Doctorat d'Universite, Universite Paris-Nord, 1988.
10. L. D. Erman, F. Hayes-Roth, V. R. Lesser, D. R. Reddy, (1988). "The HEARSAY II speech-understanding system: Integrating knowledge to resolve uncertainty," in *Blackboardsystems*, eds. R. Engel, T. Morgan, chap. 3, pp. 31–86. Addison Wesley, New York.

7 Decision Procedures

Aishwarya Gupta

CONTENTS

7.1 MOTIVATION

The main question that arises in this chapter is how individual systems should solve their subtasks. Nowadays, parallel distributed architectures are being used for implementation of these systems and their subsystems for performing the subtasks. The basic points to be discussed here are:

1. A global task has been replaced by several local ones, each handled by an individual agent or process. The main reason behind discussing this point is that just to let others know that various subtasks may interact or communicate with each other if required. The problem of dividing these subtasks is not discussed here.
2. The actions of the single agents may interact, in that success or failure for any one agent may be partially or wholly contingent upon an action taken by another. It simply says that single action, success or failure of any agent can affect the other agents too. If in any case agents do not have any form of interaction with each other then it is by default understood that they have enough communication for the evaluation of results or conclusion while integrating. It is assumed that every agent knows about other agents' payoff function.

7.2 INTRODUCTION

AI is defined as a thing in the computer world that can sense or study the environment, learn based on experience, and react in response to what they have been trained for or what they are sensing. This is what makes AI a powerful tool, which if used in the right way can completely transform the way we do business and radicalize decision-making process. Decision-making processes are increasing rapidly in customer-driven market complexity where it is very important to understand customer's needs and desires and then aligning products according to it. In order to make the best market decisions, it is important to handle changing customer behavior. To help businesses make insightful decisions, AI uses modeling and simulation techniques that help in predicting consumers' behavior as well as providing insights into the consumer's persona.

7.3 DISTRIBUTED ARTIFICIAL INTELLIGENCE

Distributed Artificial Intelligence (DAI) is a domain that deals with various classes of technologies and techniques which range from swarm intelligence to multi-agent technologies. It is the study of a group of intelligent systems which unite together to solve a problem. All subsystems are interrelated in some or the other way not completely independent. The main focus is to handle the development of distributed solutions for a target problem.

Basically, it is used for reasoning, planning, learning, and applying for some specific problems. Being a subset of artificial intelligence, simulation plays a very important role in prediction or recommendation. This requires automated learning processes so that agents may reach conclusions with the help of communication and interaction between various agents and systems. The benefit of neural networks and deep neural networks is that they do not work with the same amount of data every time.

There are some basic characteristics of distributed artificial intelligence (DAI) that will make a system considered to be Distributed AI. They are:

1. It generally distributes various tasks between the agents.
2. It distributes powers between agents.
3. It allows agents to interact with them.
4. After the completion of tasks these agents integrate to produce the result.

In the context of distributed AI, multi-agent systems work far better; to achieve a common task, they break down that task into several subtasks while looking for an emerging solution from the communication of various agents. In a DAI system, the designer takes complete control over all the parts of every agent environment which involves negotiations with many other agents simultaneously. So a way to save time is to constrain the negotiation problem to make the optimal decisions.

Recently, DAI has been recognized as a subdomain of the field of Artificial Intelligence. Artificial Intelligence is the field of study to produce intelligent behavior in machines. Central focus areas of AI are heuristics and symbolic processing. A heuristic may not produce a solution to some problem or may not produce a solution at all. But, in scenarios where formal approaches do not exist or are inefficient, heuristics provide a way to solve a problem, whereas symbolic processing involves the concept of manipulating the object structures that represent concepts. It can either be algorithmic or heuristic in nature and is usually more understandable by humans rather than numerical processing.

For getting an outcome as high-level performance, AI requires a knowledge base about a particular problem. The concerned areas of AI that are most optimal for decision-making are:

1. Knowledge representation: It is the art of selecting methods by which heuristics (static information) and dynamic information are represented in the AI systems in order to produce an efficient solution.
2. Problem-solving and inference: Standard AI Knowledge Representations are semantic nets, predicate calculus, prediction rules, computer programs, and a combination of problem-solving techniques involves the inference of new knowledge. Inference can be inductive (learning by example) or deductive (backward chaining/forward chaining).
3. Hypothesis formation: Some AI problems might be divided into a static analysis of the environment, response planning, environment monitoring while executing the plan. Hypothesis Formation involves the interpretation or understanding of real-world situations.

4. Planning: Artificial Intelligence planning consists of present situation reasoning, specific goal, a set of operators to generate an ordered set of operators; when applied to a present state will produce the goal state.
5. Control of environment: When a plan is finally executed, then evaluating the uncertainty of usefulness of the plan is what motivates the AI control.

7.4 APPLYING ARTIFICIAL INTELLIGENCE TO DECISION-MAKING

Let us walk through a real-time example of the applicability of AI to decision-making is in the field of Air Force Command and Control situations. In this, all the decision-making process is distributed, solely based on real-time data as well as on large sets of data. Keeping these constraints in mind, it is a challenging activity to make decisions, involving solution assessment, making decisions on the basis of the assessments, implementing the decision, and then monitoring its execution.

7.5 AUTOMATED DECISION-MAKING BY AI

In simpler terms, automated decision-making can be defined as "making a decision directly with the machine without any human intervention." No human involvement or interference is there in any of the processes in between. In other words, it can be described as a machine gathering large amounts of data at a single place, getting it processed with the help of some algorithms, and then using those experiences and data knowledge in making its own decision.

Those automated decisions will be taken by keeping in mind the various parameters in the surroundings. This further leads to manual decision processes becoming automated as well as serving as a support for human decision-makers in many contexts. But in general, it is observed that there are some limitations and risks even with these automated decision-making systems.

The importance of understanding the core drivers of automated decision-making perception becomes even more pressing considering that what algorithms could do or are capable of doing not only in terms of their performance even in a manner that how they evaluate and analyze automated decision-making as a whole. The current study has contributed a lot toward the attributes of automated decision-making.

7.5.1 Impact of Automated Decision System

There is a big difference between the different levels of impact of decisions made by automated decision-making, which can be as low as generating a list of recommended news articles, or as high as sentencing in the judicial sector or various decisions in the healthcare domain. By doing so it is being explicitly compared to perception of similar decisions made by human experts or automated machines. From that, one can make clear perceptions of the critical level of automated decisions made using algorithms, and those decisions made by humans.

The three sectors with the greatest advent of automated decision-making at the present time are public health, media, and justice. The significant impact will be on individual rights, wellbeing, and function in society. The feedback from these automated decisions is used for improving the automation for future use. An algorithm is nothing but "a set of encoded procedures for transforming the input data into the output based on some specific calculation and computations." But how we conceptualize them further through time depends on societal, institutional, and individual contexts between humans and machines. Mainly, automation here will focus on ongoing production of the process without any human indulgence.

At times, the decisions can be biased as the area is getting broader and enlarged day by day. According to a survey on health, media, and justice, it has been analyzed that usage and usefulness of algorithms in decision-making may adversely impact humans.

7.5.2 Forms of Automated Decision System

It can take any form, ranging from a recommendation system (to help human decision-makers), to various other decision support systems which are completely automated and work on the behalf of an institution, company, or organization without human involvement. In these cases, humans rely on automated decision-making systems up to a great extent which may be either related to them or someone else. The level of involvement of any human in these automated systems varies according to the type of system made. If it is a recommender system then what to suggest to a user will depend on user only if that suggestion has to be accepted or not. So, it completely depends on the type of automated system you are dealing with and what form of problem it is solving.

Another form is a fully automated decision-making system, which keeps the recipient in the dark without letting them know what data were used to make the decisions. There is no scope of any involvement from humans in any form. The main reason of interest in this domain is its way to process and communicate with the user.

These systems are strategically articulated very carefully so that the objective of socio-technical actors in the discourse surrounding their implementation and usage in different aspects of daily life. This may further lead to the expectation of fairness and objectivity of a machine. Mostly, these expert systems show rational and objective behavior. It gave rise to an algorithmic era where people prefer suggestions and recommendations by an automated system over human recommendation. Automated systems can be inscrutable as they might impact the user's choice of accepting the decision or recommendation. Intelligent machines or automated decision-making systems are more trusted than human non-experts, but less than human experts for their subjective decisions or for managerial decisions which require human or physical (mechanical) skill.

7.5.3 Application of Automated Decision System

The main applications of automated decision-making are:

- Communication
- Recommendation

- Personalizing various online behaviors
- Regulating user activities on social media
- Identifying suspicious profiles
- Fake news stories

7.5.4 Cyber Privacy Concerns

Automated decision-making is concerned with data-driven decisions. Much earlier research indicated that a high level of privacy is required so as to deal with slapdash behavior toward automated personalization of news according to user behavior. It has been assumed that a more concerned user will expect a more critical evaluation of automated decision-making systems. If they trust the system completely and are confident about it then a user will try to protect their own privacy using these automated systems.

This domain is not only limited to the computer science field but also to social sciences. As research continues, there is analysis of what extent individual and contextual characteristics are affecting an automated decision-making system. The result of the survey referred to above shows that there is a split in responses received, but in all cases of health, justice, and media, automated decision systems remain on the upper part.

It also shows that knowledge of both domain-specific and in general also. Both are associated with increased expectations about the usefulness of automated decision-making systems. This scenario works easily with new recommendation systems. When a person deals with a highly knowledgeable individual, then they come to know that knowledge has no connections to risk or fairness scenarios, and it is concerned only with the content of knowledge of that person.

The outcomes that we achieve in a specific scenario of automated decision-making are aligned with the machine heuristic and algorithmic appreciation. This will help in providing a new level of nuance to the theory of algorithmic decision-making.

Firstly, we try to showcase that automated decision-making in artificial intelligence is better than human experts, and then reinforce the machine heuristic in taking roles in automated decision-making in artificial intelligence.

Secondly, it has been analyzed that algorithmic decisions depend completely on whether the subject is objective or subjective or requires any human or mechanical skill or not.

Thirdly, the earlier researchers also said that human experts are better than artificial intelligence. In the case of a recommendation system there is a difference in attitude toward automated decision-making in artificial intelligence, as it is the perspective of an individual that they understand the subject of the decision and decide what input–output will be received. This may be due to differences in societal attitudes about automated decision-making. Still it is a topic of research for the future.

Another reason might be the lack of trust in human decision-making due to which automated decision-making is considered as the better option by making fairer decisions. Finally, the contexts of media, healthcare, and judiciary are showing similar but different trends.

7.5.5 Discussion and Future Impact

By the means of surveying various experiments with high quality samples, one study brings many views regarding the perception of fairness, risks, and usefulness of automated decision-making by artificial intelligence. While evaluating various societal scenarios of automated decision-making in artificial intelligence in general, the points that come up concern the risks and mixed views about usefulness and fairness. When respondents had to decide on any one decision-maker for some specific decisions on the basis of potential risk, fairness, and usefulness then they opt for automated decision-making by artificial intelligence even better than human experts for their high impact decisions. The domain-specific knowledge and online self-efficacy were both associated with positive attitudes regarding usefulness, fairness, and risks of automated decisions made by artificial intelligence. The people who are more concerned about online privacy trusted the decisions made by automated decision systems and find them fair and useful for various purposes.

After analyzing these important contributions toward the understanding of algorithmic appreciation in context with all domains, altogether it has been noted that there are some limitations too. So, comparison between automated decision-making and human experts as decision-makers were done between groups.

- In the present scenario all future researchers should take care that their evaulation is unbiased.
- All the respondents were presented the evaluators' decisions and asked to imagine that they were given any real-life situation scenario to check the consistency and comparability.
- It can also be extended by exploring the perception of people who act as the subject of these decisions and try to implement a research design that is realistic while manipulating context-specific decisions.
- The findings could also be extended by comparing with other countries' different privacy concerns and expectations.

7.6 COOPERATION IN MULTI-AGENT ENVIRONMENTS

In their paper, Ginsberg et al. discussed that intelligent agents should be able to interact even without any communication. The task is difficult but not impossible. The leading factor is basic rationality in this context. Many assumptions are being made about independence but exactly what comes into focus is a shift in the definition of rationality from individual actions to decision procedures.

According to the previous work in the field of Distributed AI (DAI), these agents are mutually cooperative because of their inbuilt design and nature. But the question arises that with no conflict of interest, how are these agents to achieve their goal? How are they able to communicate smoothly in sync without any interference?

There are many application areas like resource management applications and military applications where autonomous independently motivated agents may interact with each other only because of designers' creativity. The agents interact with each other, finding other agents with potentially conflicting goals. So, by allowing interaction due to conflict of interest, we see why these agents choose to cooperate with one another and how they communicate.

The basic inputs for an environment class E where an agent is to operate and performance measure U will evaluate the sequence of various states by which an agent goes through the actual environment. Let V (f,E,U) denote the expected value according to U obtained by agent f in environment class E. Then, the agent function is defined by the following equation (Figure 7.1):

$$F_{\text{opt}} = \text{argmax}_f V \left(f, E, U\right) \tag{7.1}$$

7.6.1 Notations and Workflow

For an agent "i," consider a payoff function "p," which assigns to any alternative "m" for further evaluation. Here, "M" represents the set of all possible courses of action, "p" using the function

$$p: M \rightarrow R \tag{7.2}$$

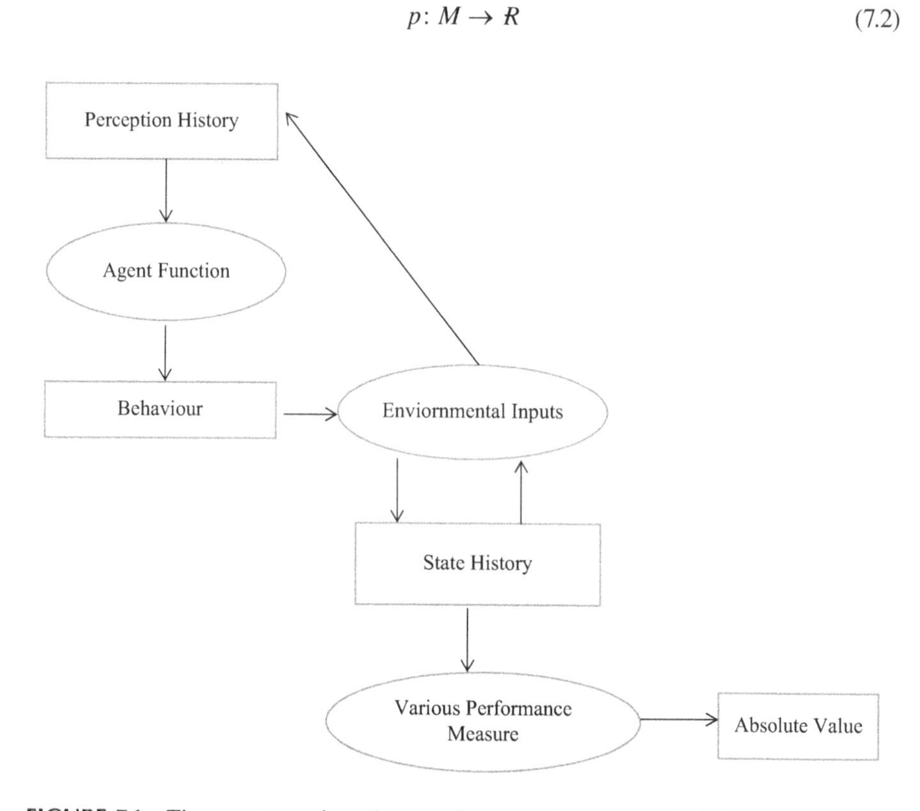

FIGURE 7.1 The agent perceives the experiences and generates the behavior that allows the environment to produce a history of states and evaluate various performance measures to arrive at an absolute value of an agent.

That the range of "p" is R as opposed to {0, 1} reflects our allowing for a varying degree of success or failure [1]. Determining the function "p" is a task that lies squarely within the province of non-distributed AI research. Given such a function, selecting the alternative m which maximizes p{m) is straightforward.

In the absence of interaction, the problems of the individual agents would now be solved. If the success or failure of some agent depends on actions taken by another, however, this is not the case: the function p in Equation 1.1 will have as its domain the set of actions available to all of the agents as opposed to a single one. The basic interaction can be noticed through the dependence of an agent's utility on the action of another. This dependence can be characterized by defining the payoff for each agent "i" in an interaction "s" as a function p_i^s that maps every joint action into a real number designating the resulting utility for "i." Assume that M and N are the sets of possible moves of the two agents respectively. Then we have

$$`p_i^{s}' : M \times N \rightarrow R. \tag{7.3}$$

The values of this function can be represented in the form of matrices for each interaction as shown in Figure 7.2. p denotes the payoff to agent J if the agents perform the corresponding actions, and the digit in the upper right-hand corner denotes the payoff to K. For example, if agent J performs action "a" in this situation and agent K performs action "c," the result will be four units of utility for J and one unit for K. Each agent is interested in maximizing its utility.

While dealing with other agents it is generally reasonable to assume that the agent is basically rational. Both the basic rationality and mutual rationality are analogous to each other. This basic rationality helps in deriving the iterated dominance analysis. It handles the column dominance problem as explained above. Using the basic rationality of K, it can be action "d" is irrational for K. Due to which neither "ad" nor "bd" is the possible outcome. Out of the remaining outcomes "ac" dominated "bc," so "b" is irrational for J.

7.6.2 Action Independence

There are various situations that cannot be handled by the basic rationality on their own. As there is no way to account for the dependencies between the interacting

		K	
		c	d
J	a	3 4	1 2
	b	4 1	2 3

FIGURE 7.2 Column dominance problem.

		K	
		c	d
J	a	2 4	4 2
	b	3 3	1 1

FIGURE 7.3 Case analysis problem.

agents, there are several other approaches to deal with this weakness. Complete independence is the one way to resolve this. The assumption states that each agent's choice of action is independent of another agent. Each agent's reaction value is same for every other agent's action. For all m, m', n, n', the values are:

$$A_K^s(m) = A_K^s(m') \text{ and } A_J^s(n) = A_J^s(n') \tag{7.4}$$

The consequence of independence is "case analysis." For every fixed move of K, J's action is superior to the others and the latter one is skipped. The difference between case and dominance analysis is that it allows J to compare two possible actions for each action of K without taking the cross term.

The payoff matrix given in Figure 7.3 deals with utility maximizing agent J, which will perform action "a" and if K performs action "c" then J gets four units of utility and if K performs action "d" then J gets two units of utility. Dominance Analysis cannot play any role in this case as the payoff of "ad" is less than "bc."

After analyzing both basic rationality and independence together, they imply the case analysis problem. Thereafter combining mutual rationality and independence, it shows the correctness of an iterated case analysis. So in Figure 7.4 one cannot apply dominance analysis, iterated dominance analysis, or case analysis to select an action. So, if two decision procedures are not independent, then the independence assumption can lead to non-optimal results.

Now, as the final payoff matrix given in Figure 7.5 is ready, the utilities present in a payoff matrix can take on any value. Only the ordering of outcomes is required

		K	
		c	d
J	a	3 3	4 2
	b	1 1	2 4

FIGURE 7.4 Iterated case analysis problem.

		K	
		c	d
J	a	3 3	4 2
	b	1 1	2 4

FIGURE 7.5 A payoff matrix.

for analysis. Therefore, only the numbers from 1 to 4 will be used for denoting the outcomes. An agent's main task in this scenario is to decide which action to perform. The decision procedure will be characterized for agent "i" as a function W_i from situation to action. If S is a set of possible interaction, then we have

$$W_i \ : S \rightarrow M \tag{7.5}$$

7.7 GAME THEORY SCENARIO

Game theory has been considered as an important tool used by intelligent artificial agents for interaction between them and with the environment also. It is difficult to implement in the real world practically.

For example, game theorists have never modeled the reason behind how players play to find a solution. They never spoke about or wrote about the reasoning process that leads players to come up with a solution. It always remains a challenge to predict an outcome. After that, one has to explain how rational agents would proceed to infer that outcome from the given information

7.8 DATA-DRIVEN OR AI-DRIVEN

AI has allowed many companies to adopt a data-driven approach for operational decision-making. It will help in making decision processes better and efficient. For that they will be requiring the right processor to get the best outcome from it. So, in the end we need to move toward an AI-driven approach and workflows.

Both data-driven and AI-driven processes describe their own importance.

1. The former targets data as a resulting metric, whereas the latter targets processing ability as a metric.
2. Data will help in taking the decisions better knowing about the insights whereas processing will lead that insight in a proper format to take suitable actions.
3. Data-driven methods focus on enhancing and improving the internal arrangement of a system while AI-driven deals with the implementation

capabilities and tries to improve the decision process for external arrangement.

7.8.1 Human Judgment

Earlier the decision-making system relied on human judgment, which was known as the central processor of business decision-making. Leaders used to make decisions based on their intuition and sixth sense, which they think is the best and only way to make any decision. According to them, their intuition can never go wrong as it comes from years of experience. So, this gut instinct always works for them in all scenarios.

In reality, their intuitions work far from ideal decision-making processes. As we all know, our mind works in a very predictable way with cognitive biases. This format does not work for real life problems now. The enhancement in basic structure is mandatory. A quick and unconscious decision may lead to bad results at times. Like in early times, a group of hunter-gatherers make decisions on the basis of past experiences, which in today's scenario is called learning. They know that they are sacrificing accuracy in some or other way but are left with no other option at that time.

But nowadays, systems have immense capacity, better processing capabilities, and better decision-making procedures, which allow them to give outstanding results. At this point in time, it is better to have impulsive decision-making procedures and less information processing. Our preloaded brains with cognitive biases influence our judgment and decision-making behavior in a way that departs from rational objectivity (Figure 7.6).

7.8.2 Data-Driven Decision-Making

With the inflow of immense volumes of data, every microsecond we need to adapt to this data-rich environment. Every transaction, every customer gesture, every micro and macroeconomic indicator can provide us with better information and help in better decision-making processes. By using various machines such as distributed file systems (DFSs) and various tools like dashboards, spreadsheets, documents, etc., it is easy to summarize all the data in a proper manageable form. Even highly processed data and manageable one can be used to present for decision-making. This complete workflow is known as a data-driven workflow. Human judgment now also acts as a central processor but with summarized data as input.

Humans playing as a central processor can create some limitations too. The human mind can handle only some average selling price, numeric values, tabulated data, or other information, but can't occupy infinite information:

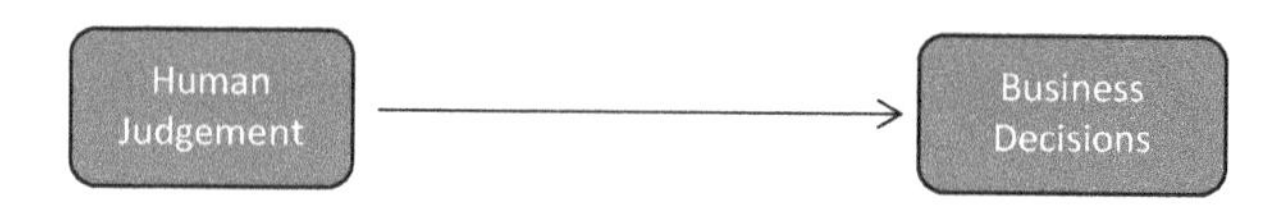

FIGURE 7.6 Human judgment-based decision-making model.

1. It can't leverage the whole data and datasets.
2. Data reduction is a necessary process to accommodate original data in less space because summarized data do not represent the complete picture.
3. It may obscure various patterns, insights, and other information.
4. The human mind can be overloaded with these thoughts and gets shut down or goes blank which may cause loss of information. After a period of time the relationship between data elements and information lost gets aggregated and leads to formation of incomplete new information. Good decision-making is not possible with that data.
5. There are many confounding factors which give the appearance of a positive relationship even when it is not and once the data is aggregated it is very difficult to recover the information in the original format. Even AI may not be able to properly control these factors.
6. By default data summaries prepared by humans are prone to cognitive biases.
7. It is easier to think of a linear relationship between elements because the human mind can process it easily.

7.8.3 Working of Data-Driven Decisions

The complete workflow of any data-driven decision-making process can be analyzed from the given framework. Figure 7.7 explains how a data-driven model will be communication and work with all components (Figure 7.8).

7.8.4 AI-Driven Decision-Making

This approach overcomes the issue of cognitive bias faced by the data-driven approach. In general routine decisions, only structured data can be relied upon. AI

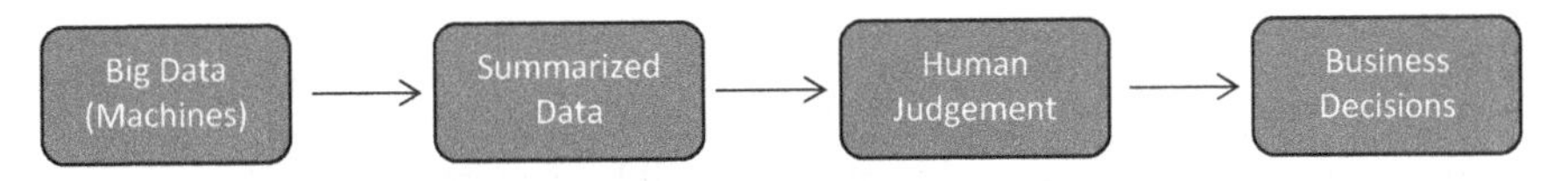

FIGURE 7.7 Data-driven approach using summarized data.

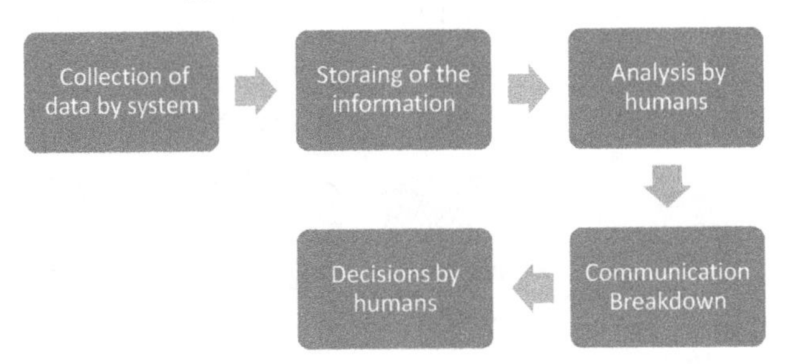

FIGURE 7.8 Workflow of a data-driven decision in an automated decision-making process or model.

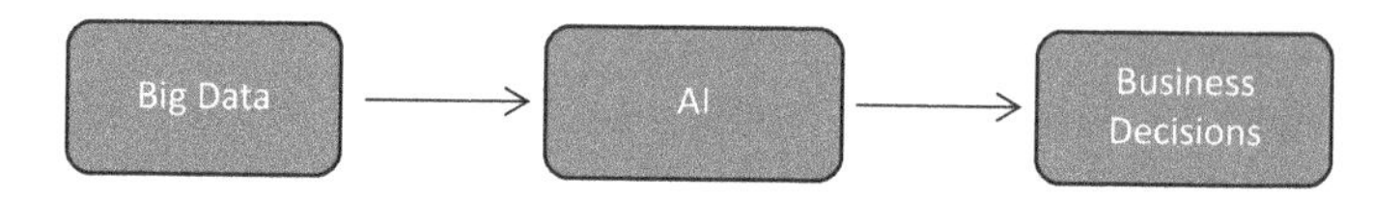

FIGURE 7.9 AI-driven decision-making model.

has the capability to find the small segments in the population with more variance at fine-grain levels even if they not in human perceptions. AI can deal with millions of groupings in a single chance. AI can deal with non-linear relationships like power laws, binomial distribution, geometric series, and exponentials, etc., even better.

As we can see in Figure 7.9, there is no human judgment interference, and only the automation is left. This will help in reducing the cost and make performance better. This AI-driven approach works better and leverages the information present in the data. It is more consistent and objective in the final decision. It can properly identify the basic features and requirements to be taken care of.

7.8.5 Leveraging Human and AI-Driven Workflows Together

Removal of humans from any process or workflow will lead toward structured data only. It doesn't mean that humans are obsolete. Many businesses are dependent on human judgment only, such as for company strategy, vision statements, and corporate values, etc. All of these require mind and non-digital forms of communication. The main difference is that shown in Figure 7.10. Human intervention is not directly from input rather than after processing the data through an AI system. So, rationality is the key feature required for decision-making procedures. This is done explicitly and necessitates the party being fully informed. Thus, it is better to leverage both approaches in combination and make good decisions.

7.9 CALCULATIVE RATIONALITY

Calculative rationality plays an important role in the decision-theoretic and logical methods. In the case of the former, only the performance measure of those behaviors that achieve the specific goal in all cases will be evaluated. Newell defined that rational actions are the ones which guarantee to achieve the agent's goal. Logical systems like theorem provers use situation calculus for satisfying the condition of

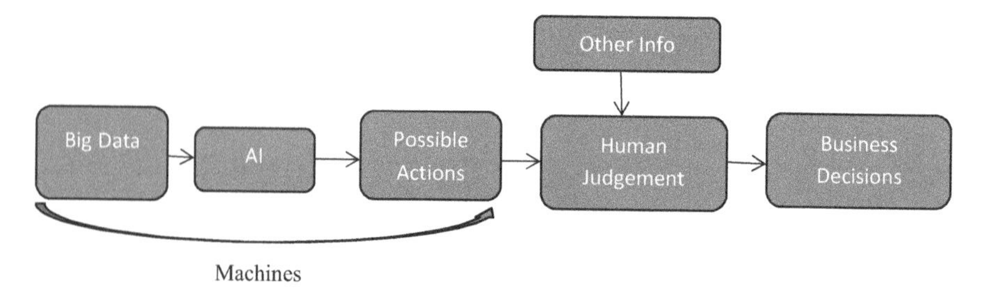

FIGURE 7.10 Combination of power of AI and human judgment.

calculative rationality. In the latter, design of calculative rational agents has moved outside artificial intelligence. For example, stochastic optimal control theory shows how calculative and computable the problems are at times, which can take exponential amounts of memory and time.

The representation format was always state-based instead of sentential-based and also the solvable problems were very small and specific. Artificial intelligence has opened up new possibilities in the field of probabilistic or belief networks. Various systems based on the influence diagrams satisfy the decision part of calculative rationality. The influence diagram is the probabilistic network with action and value nodes added to it. In any case, neither the logical nor decision-theoretic methods can ignore the intractability of the decision problem given by the calculative rationality. One solution is to cut down various sources of exponential complexity in various reasoning and representation tasks so that both the perfect and calculative rationality coincide. On the tractable sublanguages, the research results can be seen as an indication of where complexity may be a problem rather than a solution.

In distributed artificial intelligence (DAI), the system designer controls all the part of each agent's environment which involves negotiations with another agent:

- One possible solution can be controlling the complexity to constrain the negotiation problem in a way that the optimal decision can be taken.
- The other possible response for complexity is to take shortcuts and approximations in the hope of achieving reasonable behavior.

Artificial intelligence has developed a powerful memory for reducing the complexity using methods, which include state representation and sparse representation into sentential form and environmental models respectively.

Many representation models should be modeled to offer a varying mix of cost and quality that should be taken in account of the cost and benefits as perceived by the system's user. This means that they suggest a solution based on meta-reasoning. Some methods like approximate parameterized representation, partial order planning, and abstraction can retain the guarantee of effective and moderately large problems. It is inevitable that intelligent agents will be unable to act rationally in all circumstances or any situation. This observation was made in the very beginning, yet the system that selects suboptimal actions falls outside calculative rationality and it requires good theory and concept to understand it.

7.10 META-LEVEL RATIONALITY AND META-REASONING

There is a balance to be maintained between computational cost and decision procedures. The main aim was very clear that to take advantage of meta-level architecture by implementing the trade-off. Meta-level architecture is defined as a philosophy used for designing intelligent agents who categorize the agent into many notional parts. The computations with the application domain are carried out by object level computations; some of these domains are projecting the results of physical actions, computing the utility of certain states, and so on. Object-level computations are actions with costs and benefits. As per the expected utility, various computations are being selected by rational meta-level.

Meta-level is another decision-making process that consists of computational objects and object-level computations. Various meta-level strategies can be embodied by the desirability of object-level search operations. Two of them are mentioned below:

- Selective search methods
- Pruning strategies

Meta-level can do the right thinking as formalized by the theory of meta-reasoning. Rational meta-reasoning has an idea of the information value concept. The value of decision theory can be computed by taking the extra information by simulating the given process which will follow each possible outcome of the information request, which helps in estimating the expected improvement in the quality of decisions.

According to work done by Eric Wefald, a search algorithm in which the projections result in a course of time contain object level computations. Taking an example of a chess program, in which every object-level computation extends a leaf node of the game tree. The main task of a meta-level problem is to choose the nodes for expansion and to stop the search at an appropriate point.

- The main issue with meta-reasoning is the local effects in the computations don't directly translate into an improved decision because there are various difficult processes of propagating the local effects back to the root. It is used as a normal formula for computing the values that can be found in terms of local effect and propagating function such as minimax propagation as the formula can be instantiated for any specific object-level system compiled and efficiently executed at runtime. This basically works for two-player games, one with chance nodes and also single-agent search.
- The other class of meta-reasoning problem gets expanded with time or flexible algorithm which returns the result with varying quality. The increase in the decision quality for a specific algorithm is measured against the cost of time. This is the simplest form of meta-reasoning trade-off. A termination condition is said to be optimal if its second derivative is negative.
- Various difficult problems rises when anyone wants to build a complex real-time system from various unknown or old components. First, we need to look at the interruptibility of the formed system to ensure the robustness of the system.
- The only solution to this problem is to allocate the time to each component. It will double the system's overall improvement. Following this method it will be easy to build a system that can handle random real-time demands in an exact manner that time is available with a small factor of speed's penalty. Also, the available computations should be allocated optimally between the various components to increase the overall output quality.

Might this problem be an NP Hard problem, but can be solved in linear time with tree structured components. As these results are being derived for the algorithm having well-defined performance profiles but there is a need to check the capability of real-time decision-making of the system in the real-world scenario in a complex system so that it can be handled as a normal case. The robustness of the complex system can be checked only if any real-time problem can be solved efficiently.

7.11 THE ROLE OF DECISION PROCEDURES IN DISTRIBUTED DECISION-MAKING

A decision-making problem contains the following components:

- An individual decision-maker or a group of decision-makers along with an environment. Whenever any decision made by the decision-makers is implemented, it influences the environment. This is because; all the decisions are made either on the available online or real-time information or on the basis of prior information.
- A decision process consists of several steps:
  - Problem recognition
  - Problem structuring
  - Alternatives generation
  - Decision selection
  - Implementation

From Figure 7.11, it can be easily inferred that before making or implementing any decision on a problem, it is important to iterate through these steps a number of times. So, we can say that decision-making is nothing but a choice of an appropriate action from a set of possible alternatives from the available information.

Decision-making is a crucial process which is applied in all domains ranging from market analysis to dealing with Air Force mission planning. A number of technologies, like control theory, operations research, system theory, or artificial intelligence, can be applied to a decision-making process. Rather than using a centralized design, it is preferable to divide the decision-making tasks into various automated as well as human decision-makers. When we talk about a distributed decision-making (DDM) environment, its prime objective is to develop DDM techniques and to

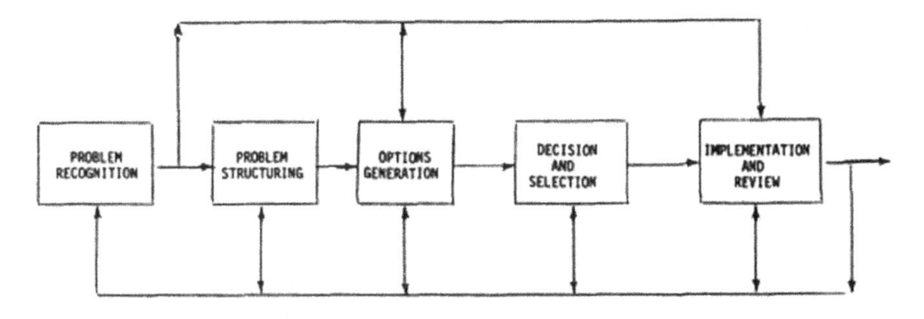

FIGURE 7.11 Process flow of distributed decision-making.

develop a design methodology that would be used for decision-making in a distributed environment.

7.12 ADVANTAGES OF DISTRIBUTED DECISION-MAKING

There are many factors available for adopting distributed decision-making over a centralized structure of decision-making. Some of the factors are:

1. Better performance
2. Efficient utilization of communication as well as computational resources
3. Easy design as one central design is replaced with several smaller design units
4. Better growth potential
5. Increased reliability as failure to a central decision-maker would be havoc to the system

In a distributed decision-making process, either the computation involved in the process is distributed, or the information is distributed, or both are distributed at the same time. There are several ways of distributing information as well as computational results depending upon the algorithm being used for the decision-making process.

7.13 OPTIMIZATION DECISION THEORY

Optimization is a static as well as a dynamic problem. It includes both optimal control as well as decision theory. Apart from straightforward optimization, these areas have their own special features integrated with them. For instance, in statistical decision theory, the presence of uncertainty is important whereas control theory deals primarily with the dynamic systems. Optimization is used to select the optimal decisions from a set of alternatives when applied to decision-making. For that, it is required to assume the existence of a set of feasible alternatives and a well-defined objective function.

Most commonly used optimization techniques are linear/non-linear programming, dynamic programming, and network flow techniques. In scenarios where a decision problem is large, sometimes it is essential and desirable to distribute the large problem or its computation into smaller processing units (decision-makers), each handling a smaller problem.

In decision theory, there are two major classes of algorithms where:

1. Decision-makers are arranged in a hierarchy.
2. Decision-makers are distributed with no particular center.

7.13.1 Multi-Level (Hierarchical) Algorithms

These algorithms are based on the concept of coordination and decomposition. In this, a large-scale problem is transformed or manipulated into a form which can further be ready for decomposition. These decomposed problems are then handled

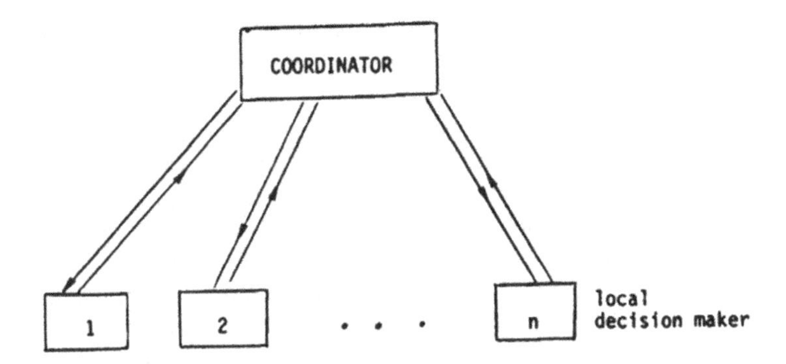

FIGURE 7.12 Hierarchical process flow.

and solved by the local decision-makers individually. However, the original problem cannot be decomposed completely. So, to influence the problem of local decision-makers, a coordinator is needed in order to ensure a global solution.

As shown in Figure 7.12, the lower level decision-makers and the coordinator form a hierarchy. In this, based on the received coordination parameters, local decision-makers solve their problems independently. Once a solution is obtained, then information about the solutions to local problems is reported to the coordinator who then adjusts the coordination parameters. This process continues until no further improvements can be made.

The prime reason to opt for a hierarchical approach is to distribute computation involved in a large problem. On using the distributed approach, computation time rises faster than at a linear rate with problem size, and then the solution of many smaller problems will be more efficient as compared to the solution of the original problem.

7.14 DYNAMIC PROGRAMMING

It is a type of decomposition method that is used for handling optimization problems over large-scale. Dynamic programming application works on an optimization problem by distributing the bigger problem into a sequence of smaller optimization problems which are easy to solve. Decomposition is with respect to stage and time. But, this is best suited for the optimization of dynamic systems whose temporal structure is already known.

- In *Spatial Dynamic Programming*, the state corresponds to the interaction variables, and, a stage variable now corresponds to each subsystem. For each subsystem, on each possible interaction variable, the optimal solution is parametrized. These partially generated optimal solutions are then combined to form a global optimal solution.
  - With respect to subsystem failure, spatial dynamic programming is reliable. The algorithm stays in its modular structure for as long as there is a local decision-maker to deal with the optimization. The overall

system can be re-optimized whenever there is addition or removal on the subsystems irrespective of the need to solve the optimization of the subproblem again.

- In *Distributed Dynamic Programming*, multiple schemes are available to distribute all the mathematical calculations involved in the recursion of the dynamic programming algorithm amongst several processes. In such schemes, it is required for all the processors to complete their assigned sub-problem computation so that a new recursion stage can begin.
  - These types of distributed dynamic programming algorithms involve asynchronous distribution of computation for the problems including shortest-path problems as well as finite and infinite time horizon-based optimal control problems.

7.15 NETWORK FLOW

To build a better use of distributed algorithms, mathematical structure of the network problems would be helpful. To control the problem of maximum network flow, a decentralized algorithm was designed. For this, a set of spatially separated oscillators were considered to achieve a common steady-state frequency.

Store-and-forward packet switching routing algorithm in communication networks has been the most successful distributed algorithm application. Except the algorithms with static and hierarchical nature, all other algorithms deal with quasi-static routing. This type of routing algorithm is used in cases where because of changing patterns like termination/establishment of a communication session, or addition/removal of a link, changes in the routes are needed while the network is in operation.

It is not practical to use a centralized approach to some problem collecting all the information from all other nodes in regard to traffic information and then using it to solve ongoing routing problems. Following are the reasons:

- Protocols should be followed in order to communicate the information between the central node as well as all other nodes.
- Central routing will suffer from a chicken-and-egg problem if the network fails. This would then require new routes to be established in order to communicate all the network information.

Routing algorithms are of two types:

- Single-path algorithms
- Multi-path algorithms
  - In single-path algorithms, a packet is sent over a route in order to minimize the delay to its destination irrespective of other packet delays
  - In multi-path algorithms, each node calculates the fraction of the total traffic being sent along various paths from that node in order to minimize the overall delay in the network for all the messages

7.16 LARGE-SCALE DECISION-MAKING (LSDM)

The basic paradigm of any decision-making algorithm is the number of decision-makers take part in any decision-making process or system. Nowadays, the scope of large-scale decision-making is growing very rapidly and it has established itself by attracting concepts. When people's participation in any decision-making process gets increased then these types of phenomena come to existence, as their participation as affected parties will add more in the decision process rather than as a stakeholder.

To make your decision process accepted by all, a dynamic and iterative consensus reaching process is important. The main aim of any large-scale decision-making process is to make most or all stakeholders agree on the final decision. This will further reduce complaints, dissatisfaction, and unnecessary arguments at the end. Many large-scale decision-making processes require artificial intelligence-based tools to handle the process, for example clustering, visualization, self-organizing maps, and agent-based theory.

7.16.1 Key Elements in an LSDM Model

Various key elements in any large-scale decision-making process are as follows:

- Subgroup clustering: Various clustering methods can be approached for large-scale decision-making problems, which covers some main aspects such as interest preferences, relationship of decision-makers with each other, and distance to the opinion collected from all decision-makers.
- Consensus measurement: It helps in finding a consensual decision made by all decision-makers altogether. It is based on consensual degree and supporting rate to the defined consensus. The most frequently used approach is identifying the degree of similarity among decisions made by all decision-makers, which is based on a distance function and can be interpreted as supporting rate of the decision-makers for a random alternative.
- Behavior management: At times when decision-makers are the experts or representative for various stakeholders, they can be assigned as organizers of an LSDM event.
- Feedback and preference modification: Here some of the decision-makers are asked to make modifications in the decision for a given assessment so as to improve the consensus level.

7.17 CONCLUSION

In the context of distributed artificial intelligence (DAI), the decision procedure paradigm is not complete but yet to be explored in depth. Examining multi-agent interaction and cooperation without any communication is still one of the research challenges. It may either be game theory, rationality concept, or AI applications. After going through all the concepts of rationality and independence, it can be concluded that no assumptions are required for various decision methods of several agents. They all work independently with any dependency.

The agent's utility maximization part ensures no glitch in the process of analyzing the independence between the agents. These decision procedures work in a dynamic format and mostly cover up the artificial intelligence domain completely. Without better decision-making processes, all AI systems and models will be lacking in providing the best outcome.

REFERENCE

1. T.B. Araujo, N. Helberger, S. Kruikemeier, & C.H. de Vreese (2020). In AI we trust? Perceptions about automated decision-making by artificial intelligence. *AI & Society*. https://doi.org/10.1007/s00146-019-00931-w.

8 Cooperation through Communication in a Distributed Problem-Solving Network

Anisha Singh, Akarshita Jain, and Bipin Kumar Rai

CONTENTS

8.1 INTRODUCTION

Analysis in AI reasoning has been generally situated towards individual agent conditions. In the current methodology, an agent develops in a static domain, and its primary exercises are: gathering data, arranging, and executing some intend to accomplish its objectives. This methodology has been demonstrated deficient because of the inescapable nearness of various agents in reality. As per the survey described, we should design agents' exercises while remembering the other agents' exercises to either help or baffle him. Subsequently, established researchers are keen on DAI to concentrate on such kinds of communication. DAI, a generally new yet developing group of research in AI, depends on an unexpected model compared to customary man-made reasoning. As per the present scenario, the last specified that an intelligent framework mimics a specific type of human thinking, information, and ability for a given undertaking, though DAI frameworks were imagined as a gathering of intelligent substances, called agents, that associated by collaboration, by concurrence, or by rivalry (Figure 8.1).

8.2 DISTRIBUTED CONTROL SYSTEM

The disseminated control framework comprises of a control application and a PC framework. The last gives the equipment and programming stage that empowers the execution of the previous, for example, the control calculations that cooperate with the persistent and discrete-occasion portions of the controlled procedure. The computer system comprises of nodes which are interconnected by a sequential, communicate based, communication organize. Every node is controlled by various modules actualizing an interactive interface to the broadcast bus, an I/O interface to the robot application with at least one joint and a chip module. The nodes' seclusion permits changes of I/O and interaction framework, just as various asset structures execute. A host node deals with the man-machine interface (MMI) and can likewise be customized to carry out portions of the control task.

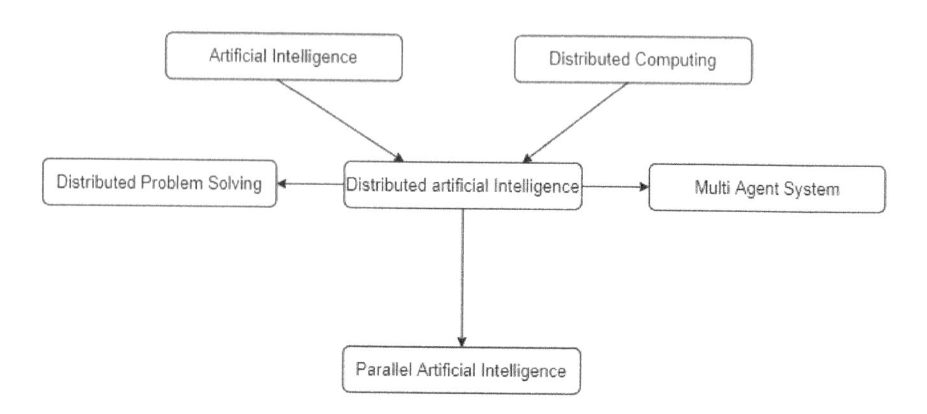

FIGURE 8.1 Distributed artificial intelligence.

8.2.1 Design Decisions

Nowadays, people like to do bird watching but to identify bird species [1], they need the help of zoological science. Bird watching provides health benefits that we get by enjoying nature. Species identification is a challenging task that may result in many different labels. Sometimes even experts may disagree on species. It is both difficult for humans and computers which hit the limit of visual abilities. To help bird watchers, we created a deep learning platform that helps detect species of birds to help users recognize species of birds using software based on image processing. This software would recognize the input image by comparing the model with a trained model and thereby predict the bird species. The details would be displayed as an output. Also, it will help us to build the dataset. If any image captured or uploaded by the user is unavailable in the dataset, the user can add it to the dataset.

8.2.2 Host Node Software Communication

Software Node Network [1] model has three layers, which are, input layer, hidden layer, and output layer. A software Node Network works exactly like a human brain that identifies relations between datasets with some algorithms' help. The system of software Node Network may be organic or artificial.

INPUT LAYER—It fetches the data into the system, and then the other layers of software Node Network conduct further processing on it.

HIDDEN LAYER—It is present between the input layer and the output layer. It computes the weighted inputs and produces net input, which further produces actual output with the activation function.

OUTPUT LAYER—Last layer of software Node Network. It produces output for the system (Figure 8.2).

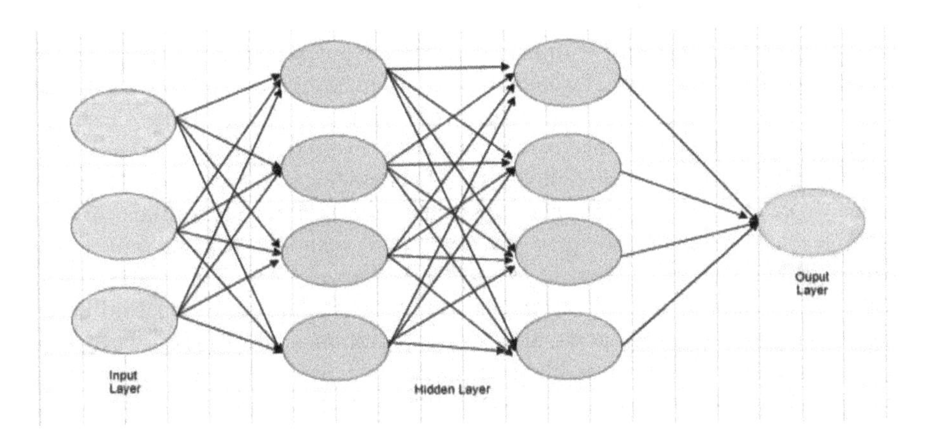

FIGURE 8.2 Software node network.

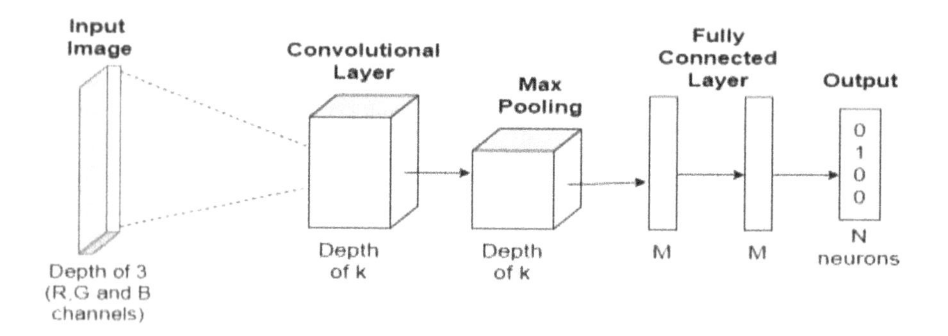

FIGURE 8.3 Schematic representation of architecture of CNN.

8.2.3 Convolutional Software Node Network

Convolutional software Node Network (CNN) [1] are feed-forward networks in which information flows unidirectional, i.e., from inputs to outputs.

CNN is used to identify images by transforming input image through different convolutional layers to a score sheet (Figure 8.3).

CNN consists of convolutional and pooling layers, which are used for image classification. An input image is passed through the convolution and pooling layer. After that, the input image of many symbolized parts of a bird were gathered. The feature extraction takes place in which each generic part is identified and differentiated based on shape, size, and color. After that, the CNN model training takes place with some pictures in a graphics processing unit to extract the features with the above-mentioned features, and the trained dataset was stored in a server to target instance.

Lastly, we get information as an output result from an image the end-user uploaded, captured using a camera. So, by the image, we can obtain information and predict the species from the trained model.

8.2.4 Assessment of Distributed Situation

An intelligent agent engaged with DAI may have, at any second, a wide assortment of decisions for what to accept, what information can be taken as pertinent, and what activities to seek after. As such, the agents do what, and when, can be viewed as DAI's essential inquiry. To take care of this issue, every individual agent (or just a single agent if the control is brought together) must evaluate the appropriated circumstance where develop numerous agents. For the most part, the circumstance evaluation includes obtaining, arranging, and abstracting data about nature, which may relate well with the agents' desires or serve to make new ones. This inquiry might be replied to partly by utilizing arranging that includes numerous agents working in unique universes. Up to this point, all work in programmed arranging has been worried about speaking to and unraveling what may be known as the traditional arranging issues. Issues of assessments, the world is considered as one of the conceivably interminable numbers of states. Performing an activity makes the world go through from one state then onto the next. In determining an issue, the article is given many objectives, a lot of moderate activity, and a depiction of the world's underlying class. The arranging framework is then solicited to discover a succession from activities that will change the world from any state fulfilling the underlying state portrayal to one that fulfills the objective depiction. This structure has ordinarily been utilized to demonstrate certifiable issues that include a solitary agent working in a situation that changes just as the consequence of the agent's activity, and in any case, stays static.

8.2.5 Computer-Aided Control Engineering (CACE)

The utilization of PCs [2] in the structure of control frameworks has a long and genuinely recognized history. It starts before the beginning of the cutting-edge data age with simple processing gadgets that were utilized to make tables of ballistic information for big guns and hostile to airplane heavy armament specialists, and proceeds to the current day in which present-day work area machines have processing power previously undreamed of. Present-day control speculations were set down in the twentieth century. Present-day control system design (CACSD) has been made conceivable by combining a few key advancements in figuring. The development and proceeding strength of high-level procedural languages, for example, FORTRAN, empowered the development and dissemination of standard numerical programming.

For example, the rise of completely intuitive working frameworks, such as UNIX and its client "shells," impacted the development of CACSD packages along similar lines. The prepared accessibility and affordability of raster-graphic displays have given the on-screen show of information from control frameworks examination, the creation of instruments for modeling control frameworks utilizing recognizable block diagrams and can make order-of-magnitude enhancements in the usability, simplicity of-control, and effectiveness of the connection between the control designer, his model, investigation devices, and the final result—programming for installed controllers. The main thrust of every one of these improvements is the nonstop increment in programming power year-on-year. The outcome has been to make PCs available to enormous quantities of individuals.

The term PC supported control framework configuration might be characterized as:

The utilization of advanced PCs as a virtual device during the displaying, recognizable proof, examination, and structure periods of control designing.

CACSD devices and bundles ordinarily give all-around incorporated help to the examination and structure of natural plant and controllers, albeit numerous cutting edge bundles additionally offer help for the displaying, recreation, and linearization of non-linear frameworks, and some can actualize a control law in programming.

Figure delineates the advancement of CACSD bundles in the course of the most recent four decades. Other keys affecting components, essential equipment, and programming advancements are likewise appeared to place occasions into an appropriate setting. In this area, we portray the foundation to the rise of CACSD devices in more detail, beginning with innovative turns of events and afterward proceeding onward to UI perspectives. The point is to comprehend the present cutting edge by inspecting the authentic setting in which these instruments have been created (Figure 8.4).

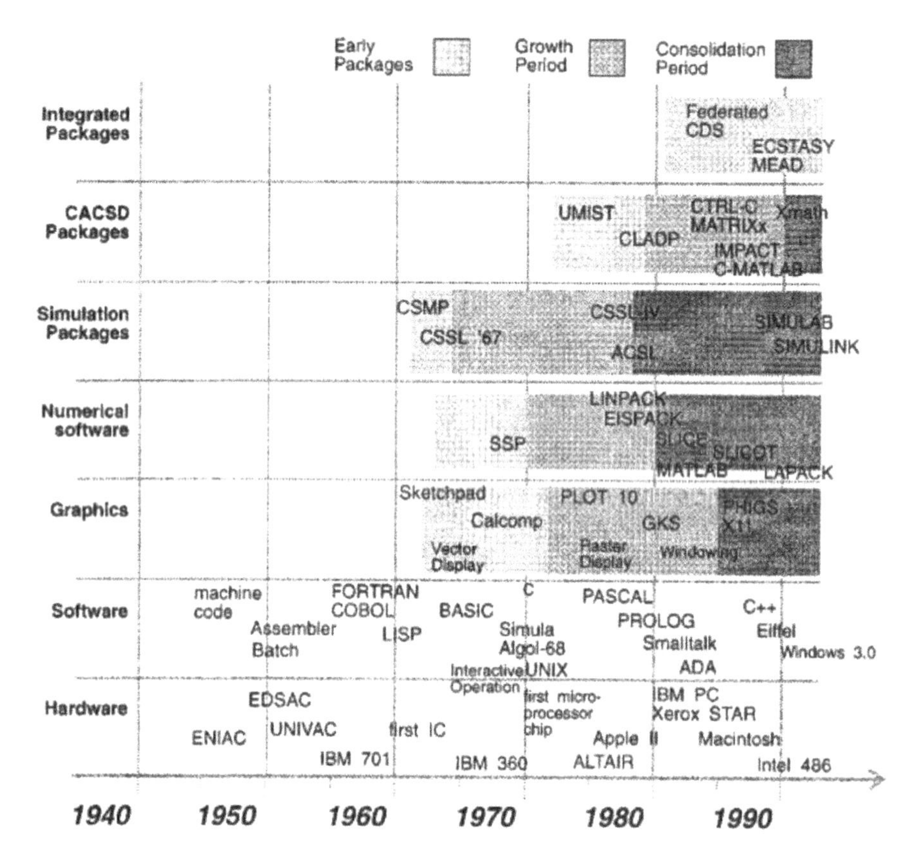

FIGURE 8.4 The recorded improvement of intuitive CACSD devices demonstrating the accessibility of related hardware and software. Some genuine items are incorporated to show the best in class [2].

8.2.6 Knowledge Base

A knowledge base is a place where the minute information of the image is stored. The knowledge base can be convoluted as a dataset comprising a locale's resolution satellite images adding change-detection applications (Figure 8.5).

Initially, software node networks start with random weights and biases, but it repeatedly trains until it reaches its best performance. It is done by calculating the amount of error they have at that moment, known as the cost of software Node Network. This is premeditated by finding the difference between the network's prediction and the desired result and finding the sum of those error values' squares. Training software Node Network for picture acknowledgment from a set of information is created. TensorFlow, an open-source software for numerical calculation utilizing data flow charts created by analysts and designers on the Google Brain Team inside Google's Machine Intelligence, examines machine learning associations. Deep software Node Networks explore, assumes a significant job. The data flow graph consists of nodes that show mathematical operations, and the edges of the

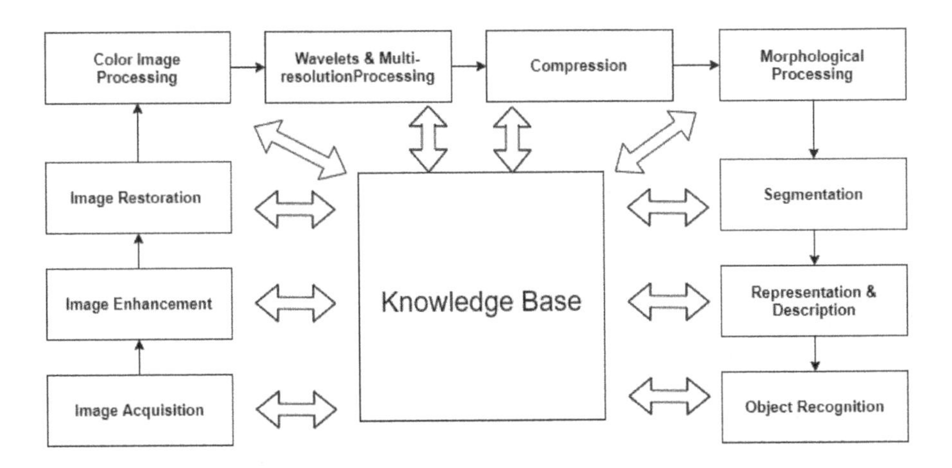

FIGURE 8.5 Flow chart showing different phases in image processing.

graph show multi-dimensional data arrays share between them. In machine learning, CNN is a feed-forward network in which neurons are connected with each other similar to the animal visual cortex setup.

Receptive field is a restricted area where cortical neurons respond to stimuli. CNNs have multiple layers of receptive fields. Collection of small neurons process some part of the input image. This is then piled up so that their input regions overlap to achieve a higher-resolution portrayal of the original image. This process is repeated with every layer. Piling up of results may give the ability to CNNs to handle the conversion of the input image. CNN may have local or global pooling layers, which joins the outputs of neuron clusters. A convolution operation on small parts of input is done to minimize the number of free parameters. A software node network consists of neurons connected. All software Node Network connection has some weights with it that tells the dominance of the relationships within the neuron when multiplied with the input value. Every neuron is associated with an activation function used to present non-linearity in the network's modeling capabilities that tell the neuron's output. Preparing of Software Node Network is a finding out about the estimations of parameters and this learning procedure in a Software Node Network as a monotonous procedure of "proceeding to return" through all the layers of neurons. Starting from the last layer, for example, the yield layer, misfortune data goes to all the concealed layer neurons that add to the output. The hidden layer neurons only receive some parts of the loss's total signal, based on the relative contribution that every neuron has contributed to the actual output. This process is done, again and again, layer by layer, till all the neurons in the network have received a loss signal that tells their relative contribution to the total loss.

8.2.7 Training Dataset

The execution of bird species' learning with the assistance of CNNs was performed on a GPU workstation with a 12 Intel Xeon CPU, 32 GB of memory, and an Nvidia

GeForce 2 11 GB GTX 1080 Ti illustrations card on a TensorFlow stage. The input images of fixed size 112 × 112 pixels were provided to the convolutional software Node Network for the feature extraction and recognition process. A dataset of bird species is used for training, validation, and testing in this study. Different input image features are extracted like colors, shapes, body, legs, angle, head, etc. while we pass it through the pile of convolutional layers. The input image's transformation into pixels occurs in the first convolutional layer, moving forward to the next layer. Then, feature extraction takes place until the image classification is completed with a probability distribution. To record all the input image elements, the convolutional filters have a kernel and a high feature map that can glide over the input volume. The stride must be set at solidarity as we move the kernel by one unit to assume responsibility for the filter looping around the information the other pixel with the goal that the yield would not lessen and give out a whole number worth as opposed to a division. By utilizing this condition, $(I - k + 2q)/(s + 1)$, where I is the information tallness or length, k is the filter, q is the padding, and s is the stride. The padding was obliged to one round the info picture to maintain the output activation map's spatial goals after convolution, which is finished by utilizing this condition, $q = (k - 1)/2$. The spatial pooling implementation was done to localize and detach the hunks of images sized 2×2 pixels, carpooling, and two strides, the pixel rate of maximum value was taken under consideration. To maintain a constant volume through the network, the pile of convolutional layers was chased by an activation function, the ReLU, also known as Rectified Linear Unit (ReLU). The skip connection is implemented to perform downsampling by conv3 and conv4 with a stride of 2. Skip connections are directly used when both the input and output have the same dimensions.

8.3 MOTIVATION AND DEVELOPMENT OF THE ICE ARCHITECTURE

8.3.1 History of ICE Model

The ICE design [3] was grown freely in the mid to late seventies to get correspondence and participation among numerous specialists. The hypothesis behind ICE was created in three significant stages.

8.3.1.1 Operators on Information States

To start with, in my work on the worldly modular rationale of games, I had built up a special conventional semantics that included data states as a component of the semantics. Research describes that one ought to have the option to utilize the formalization of data inside semantics to build up a hypothesis of how data is communicated by language. All the more especially, research provides how the data condition of operators changes as they convey. Methodology built a hypothesis of how correspondence changes data states. This is the thing that I called a down to business hypothesis of the significance of informational messages. The subject or specialist had a psychological express. His data state and this state were changed by the informative messages he got and deciphered.

8.3.1.2 Relations to Observable Quantum Mechanics

This prompted a view that importance resembles an administrator on the subject's data condition, changing the data state by working on it to shape another data state. Autonomously, methodology perusing in quantum mechanics, specifically the mathematical foundation of quantum mechanics by von Neumann, and the hypothesis of how observations go as operators on the state (as a vector in a Hilbert space) the wave function. By surveying the eyewitness' state itself as a wave work, we can see the perception as changing the spectator's condition and the watched. Physicists had contemplated such administrators' intriguing properties, and I understood that they had analogs in ordinary correspondence. Moreover, the picking up of data about a stage space in factual mechanics, with its decrease of entropy, as the stage space decreased, was like how correspondence administrators diminish the data set characterizing the operator's data condition. These similitudes with other unique hypotheses recommended that there was some basic, all-inclusive part of data that had been caught by the hypothesis of data states and their administrators.

8.3.1.3 The Influence of Sociology and Intentional States

This authentic administrator model of correspondence was broadened when animated by the compositions of the social scientist Ju¨rgen Habermas. It became evident that the most significant parts of human correspondence, to be specific correspondence in agreeable social action, couldn't be clarified by correspondence about state data alone. Some other authentic structure was required. Something that controlled the agent's activities: Control data. In this way, a methodology built up the hypothesis of purposeful states with the specific point that such states must be powerfully transformable (worked on) by open communications. The thought was that correspondence between agents powerfully structure the goals of agents setting up their purposeful states, so planned movement and collaboration were conceivable. Research commented as the establishment for a hypothesis of society. I saw the deliberate state that controls the agents' actions based on the agents' assessments and state data about the world.

8.3.2 Requirements of a Theory of Animal and Robotics Communication

We attempted to build up a general hypothesis of the requirements of a theory of animal and robotics communication on human correspondence and collaboration and creature and automated correspondence and participation. Creatures display very advanced social coordination and flagging. A hypothesis of correspondence and coordination ought to have the option to clarify such wonder too. Since they are presumably more straightforward than human correspondence and social coordination, the article shows them to be an intriguing region to test a correspondence and collaboration hypothesis.

If one needed to manufacture robots with different degrees of capacities, informative and social ability, a hypothesis of authentic states (for example, the ICE engineering), correspondence, and collaboration ought to be sufficiently conceptual to allow the plan of basic imparting participating robots. Such robots may not reason emblematically and may need advanced aims and convictions; however, they need key control states and data expresses that are manipulatable by some type of crude correspondence despite everything. By and large receptive operators may have extremely basic, non-symbolic, non-anthropomorphic vital control states. However, these might, in any case, react to a crude, or even mind-boggling, practically deciphered language.

8.4 A BRIEF CONCEPTUAL HISTORY OF FORMAL SEMANTICS

We currently [3] take a gander at the work in formal semantics that has affected the proper investigation of multi-agent frameworks and formal correspondence hypothesis.

8.4.1 Tarski Semantics

In the first place, there is the convention of present-day numerical rationale, which started with Boole and proceeded in this century with its accentuation on proverbial formal frameworks. Crafted by Tarski is huge because it was the main formalization of the first request rationale's semantics with quantifiers over people. Tarski saw semantics as setting up a connection between the language of rationale he called the object language and the world he called a model. Tarski recognized the item language from the meta-language. The meta-language is the language the philosopher uses to explore the article language. The meta-language contains the item language; however, it also comprises customary English and other numerical languages that the philosopher decides to utilize. Along these lines, the semantics of the object language are portrayed in a semi-formal meta-language. The meta-language need have no exact syntax since it is generally a casual blend of common language and formal numerical language.

To be named a conventional semantics, there must be an exact scientific portrayal of the world, called the model, comprising objects, properties, and connections. There must likewise be an exact mapping between the item language's terms and predicates, from one perspective, and the objects, properties, and relations in the model, then again. What's more, there must be an exact meaning of the conditions under which any given sentence of the article language is valid or bogus. The last is known as reality states of the semantics. All the more officially, a Tarski model for a language L comprises where O is a lot of objects, Rn is a lot of properties and relations on O, and Φ is an assessment work that doles out truth and deception to sentences of the item language L.

With Tarski's work, one begins to have an inclination or instinct that it is the beginnings of a reasonable hypothesis of importance for a language. Indeed, crafted by Tarski has affected conventional rationale, phonetics, and software engineering. Lewis first explored the language structure and a portion of the proverbial frameworks of modular rationale. Hintikka, Prior, and Kripke made some historic examinations concerning the semantics of modular rationale. Kripke's work is maybe the most conceptual and by and large material.

8.4.2 Possible World Semantics

Kripke semantics adds possible universes to Tarski models. Rather than having a model speak to a world, a model currently contains numerous potential universes. An openness relationship R between possible universes shows what different universes are conceivable from a given world. For instance, assume our model M contains many possible universes W with wl and w2 part universes in W. Besides, our model will contain an availability relationship R that holds if one world can be reached from a different universe. Let Φ be a capacity that relegates truth T and bogus F to sentences contingent upon the conditions that hold in the model M. At that point, the reality condition for the sentence 3α, can be written as follows:

$$\Phi\ (3\alpha,\ w_1) = T \text{ iff } \exists w_2 \in W, \text{ if } w_1 R w_2 \text{ then } \Phi\ \left(\alpha,\ w_2\right) = T$$

8.4.3 Semantics of Temporal Logic

ANN (artificial neural network) earlier examined the semantics tense administrators, similar to "The reality of the situation will prove that α," Fα, before Kripke. In any case, it worked out that one can rework the availability relationship in Kripke semantics from numerous points of view. One route is to decipher the availability relationship as a worldly requesting relationship. R is simply <, i.e, less than, the connection between time and focuses on account of straight time. This would then be utilized to give a proper semantics to the Prior tense administrators just as others. To summarize, in Kripke's argument to Tarski semantics, a model is a tuple where W is a lot of possible universes, and R is an openness connection between possible universes. The subportion < W,R,O,Pn > is known as a model structure as it is, to some degree, autonomous of the specific truth conditions work Φ of the total model.

8.4.4 Limitations of Kripke Possible World Semantics

Note, there are no operators in a Kripke model. A further nuance is that any modern object language contains both tense administrators and modular administrators. For instance, one can talk about what was conceivable or what will be essential. Such blended modular frameworks that contain a few kinds of modular administrators are getting progressively normal and being utilized in the proper determination and confirmation of dispersed and multi-specialist frameworks. They were first researched in. The actual outcomes were republished and are accessible in. Both modular administrators with shortlists, just as rationales with the two tenses and modular administrators, were researched.

8.5 RELATED WORK

Chaib-draaet al. [4] tells most work done in DAI had focused on tangible systems, including airport regulation, urban traffic control, and mechanical frameworks. The fundamental explanation is that these applications require disseminated understanding and appropriated arranging by methods for intelligent sensors. Arranging incorporates the exercises to be attempted, yet additionally, the utilization of material and personal assets to achieve translation assignments and arranging undertakings. These application territories are likewise portrayed by a characteristic appropriation of sensors and beneficiaries in space. As such, the tangible information translation assignments and activity arranging are between subordinate in existence. For instance, in aviation authority, an arrangement for directing an airplane must be composed of the plans of others close by the airplane to maintain a strategic distance from impacts.

Shekhawat et al. [5] explained the expanding advancement of the present data period represents certain difficulties in customary data innovation frameworks. Intelligent Agents and agent-based programming innovation are quickly advancing to fulfill the needs of this new data period. Be that as it may, before agent-based arrangements can be routinely and effectively abused in real issues, certain essential research and programming designing issues must be considered. Apart from this, for actualizing multi-operators, many difficulties incorporate understanding the significance, validation, mystery, and security. Taking a gander at the pace of improvements and progress, it isn't impossible that in this century just, we will see practical multi-agent organizations. Robocup competitions have just made a decent beginning. The pursuit of the web is being improved ever by sending delicate yet savvy agents on the net. Many complex computational issues are now being understood on the system named framework registering, though with less insight. The time is not far when such endeavors will approach the human feet.

Van Dyke Parunaket al. [6] says in numerous mechanical applications, huge concentrated programming frameworks are not as compelling as circulated systems of moderately easier modernized specialists. For instance, to contend successfully in the present markets, producers must have the configuration option, actualize, reconfigure, resize, and keep up assembling offices quickly and reasonably. Since current assembling relies intensely upon PC frameworks, these equivalent prerequisites apply to assemble control programming, and are more effortlessly fulfilled by little modules than by huge monolithic systems' frameworks. This paper surveys mechanical requirements for Distributed Artificial Intelligence (DAI), focusing on frameworks for assembling planning and control. It portrays a scientific categorization of such frameworks, gives contextual analyses of a few propelled look into applications and genuine mechanical establishments, and recognizes steps that should be taken to send these innovations all the more comprehensively.

Durfeeet al. [7] tells DAI and BTNN dier in numerous regards One regard is obviously the granularity of calculation comparative with correspondence DAI for the most part accept that correspondence is time, devouring, error, inclined, and exorbitant Thus, correspondence choices are made sensibly, what's more, messages among elements in DAI frameworks are at the image, as opposed to flag, level, encoding a lot more extravagant semantic substance because correspondence is at such a premium, and since imagining the effect of a message requires a model of the listener to that message, agents in DAI frameworks normally have unequivocal models of different agents, including their inclinations, capacities, and desires Decision making in DAI agents is in this way a mind boggling procedure of mapping potential activities counting correspondence acts, into express models of the envisioned exercises of others, prompting a wide scope of practices Often, to foresee the activities of another, a DAI agents will execute its inferences forms on its model of that other agents, reaching inferences on what the other agents could be thinking or doing by,putting itself from different point of view.

Brazdilet al. [8] tells how the investigation of multi-agent learning offers new conversation starters that should be investigated. For instance, when should the individual frameworks collaborate and how. The motivation behind this paper is to examine a few distinct methodologies that have been taken. This conversation won't endeavor to be comprehensive but instead focus for the most part on the work done around there by this paper's creators. In any case, an endeavor will be made to introduce this work in a bound together point of view and propose bearings for additional work. In this, we will talk about independent operator learning. Next will be devoted to multi-agent learning. It will portray certain models that we can utilize when looking at changed frameworks. This segment depicts a few existing frameworks and approaches and is mostly situated towards the creators' prior work. The last segment will talk about new skylines and future work.

Verbraeken et al. [9] explains the quick advancement of innovations lately has prompted an exceptional development of information assortment. AI (ML) calculations are progressively being utilized to break down datasets and manufacture dynamic frameworks for which an algorithmic arrangement isn't possible due to the issue's multi-faceted nature. Models incorporate controlling self-driving vehicles, speech recognition, or prediction of customer conduct. Sometimes, the long runtime of preparing the models steers arrangement originators towards utilizing distributed frameworks for an expansion of parallelization and the aggregate sum of I/O data transmission, as the training of information required for advanced applications can without much of a stretch be in the request for terabytes. In different cases, a brought together arrangement isn't even an alternative when information is innately distributed or too enormous to store on single machines. Models incorporate exchange, preparing bigger undertakings on information put away in various areas or cosmic information that is too enormous to even think about moving and unify. To make these kinds of datasets available as preparing information for AI issues, calculations must be picked and executed that empower equal calculation, information dispersion, and strength to disappointments. A rich and different biological system of research has been led in this field, which we order and examine in this article.

8.6 DYNAMIC POSSIBLE WORLD SEMANTICS

In Kripke semantics, the openness relationship is static. It doesn't change with time. However, in reality, prospects change with time, and instinctively universes open even from a parallel world can change time. In chess, you have the exemplary model that you can't castle the king if the king has been moved previously, regardless of whether the king has been moved back to its original position. In this way, what is conceivable after the king has moved is not, at this point, conceivable later on regardless of whether the condition of the chess game is indistinguishable from one where one could somehow or another castle. All the more, for the most part, it is useful to consider connections that change with time when exploring the dynamic circumstances that agents end up in, particularly when the agents no longer have ideal data about the condition of the world.

A world history, much like the world line in material science, can be seen as a world that adjustments in time. In this way, comparable to world narratives, a changing relationship can be characterized as a capacity from times to possible static connections. The outcome is a unique connection that changes with time. Specifically, Dynamic possible world semantics can sum up the Kripke availability relationship to one that changes with time. All the more officially, a unique openness connection R is a capacity from times T to relations on Ω. Let R^t be the estimation of the dynamic openness connection R at time t. Two chronicles, H and K, are connected by R at time t if HR^tK. Review an ordinary static connection is only a lot of requested sets. In this view, two articles a, b remain in a static relationship R if the arranged pair (a,b) is an individual from R. A powerful connection R is a changing arrangement of requested sets. So, the estimation of a unique relationship at a time t, in images R^t, is a static arrangement of requested sets. Subsequently, H is identified with K by R^t if and only if the arranged pair (H, K) is an individual from R^t, which is what HR^tK says.

8.7 SITUATION SEMANTICS AND PRAGMATICS

While examining [3] a huge albeit moderately level section of English, in which refutation, disjunctions, quantifiers, and so on were going to cause troubles on the off-chance that one attempted to give a direct social structure of semantics to all sentences of English. Since, be that as it may, language refers to and connects with the world, it is expected to formalize the propositional or referential substance of sentences. To take care of the difficult semantic and pragmatics we partition the hypothesis of importance into two layers: the main layer, called the profound semantics, is a semantics alluding to circumstances like social space-time structures; a subsequent layer, the pragmatics, permits the translation of progressively complex sentences by portraying their impact on the specialist's data and deliberate conditions. The profound semantics was evolved out of the need to give a full compositional pragmatics for English-like parts. The compositional pragmatics emerged out of the endeavor to get correspondence. Furthermore, the hypothesis of correspondence emerged out of the endeavor to comprehend collaboration between gatherings of operators. This strategy of having one control compel another is very helpful for the improvement of thoughts.

8.8 MODELING DISTRIBUTED AI SYSTEMS AS A DISTRIBUTED GOAL SEARCH PROBLEM

A few creators [10] have, as of late, described DAI as a type of appropriated objective hunt with numerous loci of control. Receiving Lesser's fundamental formalism, the activities of Agent1 and Agent2 in illuminating goals G10 and G 0 separately can be communicated as an old-style AND/OR objective structure search2. The traditional diagram structure has been expanded to portray the interdependencies between the goals and show the assets expected to explain the simple goals. Interdependencies can exist between elevated level kin goals, for example, G^1_1 and G^1_2, or progressively inaccessible in the objective structure (e.g., somewhere in the range of $G^1_{1,1}$ and $G^2_{p,2}$). In the last case, G^1_1 and G^2_p become connecting goals if $G^1_{1,1}$ is utilized to illuminate G^1_1. Circuitous conditions exist between goals through shared assets (e.g., $G^1_{m,1,2}$ and $G^2_{p,2,2}$ through asset d^1_j). Asset conditions can be expelled basically by giving a greater amount of the asset being referred to; conditions between objectives, then again, can't be bypassed as they are a consistent outcome of the network's condition. In every other perspective, the two sorts of reliance are indistinguishable (Figure 8.6).

Interdependencies can be ordered along two symmetrical measurements: whether they are powerless or solid, and whether they are one-directional or two-directional. Solid conditions must be fulfilled if the dependent goal is to succeed; unstable conditions encourage or compel critical thinking yet need not be satisfied for the dependent goal to succeed. A case of a solid reliance is the place the yield of a goal (G) is required information (I) for the dependent goal (DG) and where G is the main wellspring of I in the network. A frail reliance exists if there is more than one hotspot for I or a discretionary contribution for DG. A one-directional reliance (composed of

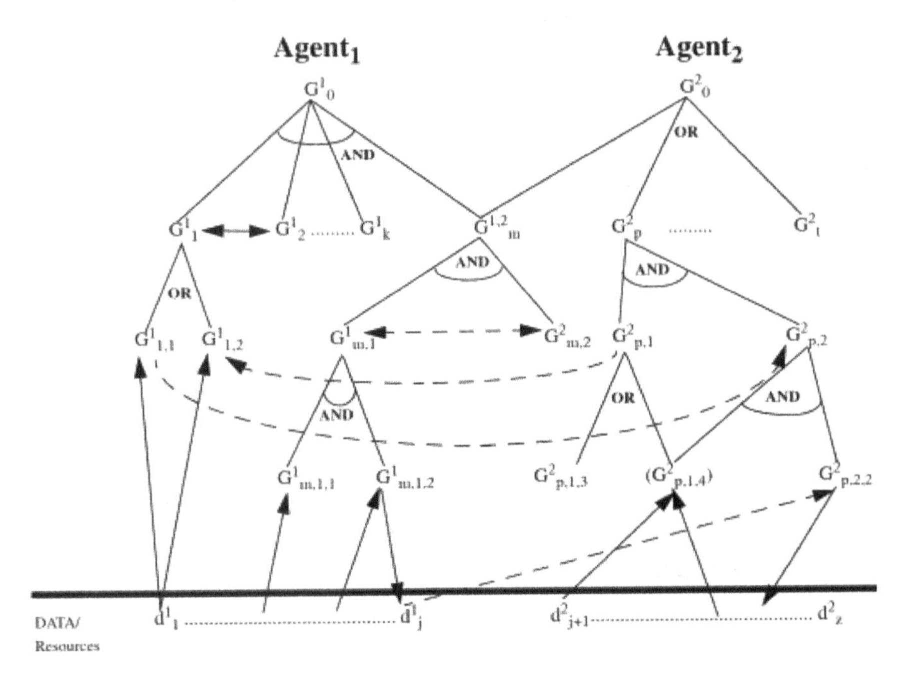

FIGURE 8.6 A distributed goal search tree including Agent1 and Agent2. The specked bolts demonstrate interdependencies among goals and information in various agents, strong bolts conditions inside an agent. The superscripts related to goals and information show the agent which contains them [10].

$G^1_{1,1} \rightarrow G^2_{p,2}$) implies that agent2's goal $G_{p,2}$ is dependent (either emphatically or pitifully) on agent1's goal $G_{1,1}$; however, $G^1_{1,1}$ is unaffected by $G^2_{p,2}$; with two-directional conditions (composed of $G^1_{m,1} \leftrightarrow G^2_{m,2}$) the goals of the two agents are influenced. The arrangement of data I by goal G for DG is a case of a one-directional dependent ($G \rightarrow DG$); a two-directional reliance happens, for instance, when two goals should be performed at the same time.

The concept of inter-agent dependencies is the basic determinant of the type of coordination which will occur. For instance, if Agent1 realizes that $G^2_{p,2,2}$ requires asset d^1_j before it can begin (solid dependency, one-directional), at that point, it might choose to implement $G^1_{m,1,2}$ (to create the important asset) before $G^1_{m,1,1}$ if no other data is recognizing these two other options. Besides, the connection somewhere in the range of $G^1_{m,1}$ and $G^2_{m,2}$ may specify that the two activities should be performed simultaneously (solid dependency, two-directional), in which case the two agents need to agree about the separate implementation times. At last, if Agent1 picked $G^1_{1,1}$ as a method for fulfilling G^1_1, the after-effect of this errand may give important data (frail dependency, one-directional) which Agent2 could utilize when explaining $G^2_{p,2}$ (e.g., it might give an incomplete outcome which empowers $G^2_{p,2}$ to be essentially smaller). Knowing this, Agent1 will conjure a data-sharing type of collaboration to flexibly Agent2 with the essential outcome when it opens up.

It was important to stretch out Lesser's diagram formalism to permit common goals (e.g., $G^{1,2}_m$) because joint goals are the premise of joint activity (for example, there can be no joint activity except if there is an initial a joint goal). Joint activities are an advanced type of participation where a group of agents choose to seek a shared goal helpfully (this stands out from less difficult types of collaboration, for example, requesting that a specialist play out a solitary undertaking or suddenly chipping in pertinent data to intrigued associates). This type of connection can be described as having the accompanying properties: (i) the colleagues are commonly receptive to each other, (ii) the colleagues have a joint responsibility to the joint action, and (iii) the colleagues are resolved to be commonly steady of each other during the quest for their joint goal. Common goals contrast from singular goals in that they are not legitimately connected with activities—consequently, they should be mapped onto singular goals as just individual agents can act. Anyway, joint goals can be in the brain of every person going about as a feature, suggesting that singular agents can control all things required for group conduct, although the point references the system. In this manner, the common goal $G^{1,2}_m$ is disguised inside Agent1 and Agent2 and results in Agent1 performing $G^1_{m,1}$ and Agent2 performing $G^2_{m,2}$.

8.9 DISCUSSION

New roads of research are developing, filling holes in DAI writing and practice. Especially, new speculations, both practical and theoretical, are becoming visible. In this manner, a few analysts start to create theories closer to human science. However, different specialists are growing new techniques dependent on old-style AI, which is generally gone to brain research, phonetic and psychological science. Likewise, there are new critical thinking models for DAI that have started to supplement before

DAI structures, for example, writing boards, multi-operator arranging, and undertaking portions. Notice additionally that there is currently a wide scope of points of view on arranging in multi-specialist universes, which can be composed along a range from those managing basic numerically characterizable operators, working in dynamic obliged universes, to tentatively helpful yet hypothetically uncharacterized multi-operator organizers, and on to progressively subjective models or 'arranged' hypotheses of helpful arranging. Works in DAI likewise show that specialists are starting to characterize their objects of investigation painstakingly and research strategies utilizing similar clear examinations to reinforce DAI's logical premise.

At last, it appears that it is currently a lot simpler to utilize cautious exploratory strategies in DAI. This is because of numerous long stretches of exertion in building critical thinking models, distributed object-oriented languages, and programming conditions.

8.10 CONCLUSION

This chapter has researched inclines in DAI, which is a branch of AI. The use of AI methods and the assessment of human cooperation and social association are undoubtedly key to the DAI region. From a perspective, AI methods permit an agent to have a nearby modern control to reason about its own critical thinking and how this fits in with critical thinking by different agents. Then again, assessing the procedure of human association and social association permits DAI planners to consider a progressively versatile association of agents. The chapter likewise audits some ongoing work done in DAI and shows how DAI looks into uncovering the intricacy of gathering coordination and rationality. It likewise reflects how DAI looks into the cutting-edge issues in zones, such as thoughtfulness, arranging, language, and thinking about conviction. As Nilsson anticipated in his initial ramifications in DAI, inquiry about distributed artificial intelligence forces researchers to show a significant number of AI's fundamental issues.

REFERENCES

1. Singh, A., Jain, A., & Rai, B.K. Image based bird species identification.
2. Rimvall, C.M., & Jobling, C.P. (1993). Computer-aided control system design. *IEEE Control Systems Magazine*, 13(2), 14–16.
3. Werner, E. (1996). Logical foundations of distributed artificial intelligence. *Foundations of Distributed Artificial Intelligence*, 2, 57–117.
4. Chaib-draa, B. (1995). Industrial applications of distributed AI. *Communications of the ACM*, 38(11), 49–53.
5. Shekhawat, R. (2000). Distributed artificial intelligence and multi-agents. *Journal of International Institute of Information and Management.* https://www.researchgate.net/publication/277009140_Distributed_Artificial_Intelligence_and_Multi-Agents.
6. Van Dyke Parunak, H. (1996). Applications of distributed artificial intelligence in industry. *Foundations of Distributed Artificial Intelligence*, 2, 139–164.
7. Durfee, E.H., Lesser, V.R., & Corkill, D.D. (1987). Distributed Artificial Intelligence. Cooperation Through Communication in a Distributed Problem Solving-Network. Edited by M.N. Huhns. Morgan Kauffman, Los Altos, CA.

8. Brazdil, P., Gams, M., Sian, S., Torgo, L., & Van de Velde, W. (1991, March). Learning in distributed systems and multiagent environments. In Ana L.C. Bazzan (ed.), *European Working Session on Learning*. Springer, Berlin, 412–423.
9. Verbraeken, J., Wolting, M., Katzy, J., Kloppenburg, J., Verbelen, T., & Rellermeyer, J.S. (2019). A survey on distributed machine learning. https://arxiv.org/abs/1912.09789.
10. Jennings, N.R. (1996). Coordination techniques for distributed artificial intelligence. *Agents, Interactions & Complexity*.

9 Instantiating Descriptions of Organizational Structures

Niharika Dhingra, Mahima Gupta, Neha Bhati, Pallavi Kumari, and Rijwan Khan

CONTENTS

9.1 INTRODUCTION

The organizational structure is an instrument used to split, organize, and coordinate multiagent system (MAS) organizational activities. It represents authority relations and responsibility for goals, providing a typical way to assign tasks to agents. An explicit organizational structure helps agents know where they are relative to others and their responsibilities.

Various analysts have identified the need to depict huge and complex procedure structures to launch them on explicit processor designs and give data to the working framework for asset designation choices and correspondence directing. They have created dialects for this reason. These dialects incorporate TASK, ODL, PCL-14, HISDL, PRONET, and DPL-82 [1]. These dialects are exceptionally frail in their capacity to show the intricate preparing structures important for the up-and-coming age of system models and dispersed applications. This is valid for applications with intently communicating undertakings executed on systems, including heterogeneous organizations of databases, effectors, sensors, and processors with different preparing velocities and memory sizes. For instance, the preparing structure of a disseminated handling system that per structures signal understanding requires a mind-boggling, area explicit, correspondence connection between deciphering hubs and detecting hubs. This correspondence connection requires every deciphering hub to discuss just with the littlest gathering of detecting hubs that can furnish it with data about the district for which it is mindful. Simultaneously, each detecting hub is required to speak with a set number of incorporating hubs to limit the time it must dispense for correspondence.

The determination of such complex procedure structures includes distinguishing useful segments (for example, deciphering and detecting hubs), their obligations (giving understandings of the signs recognized in a specific area), and asset prerequisites (processor speed and memory size, information about deciphering signals, and so forth), and the relations among them (correspondence and authority). Together, this data is a detail of the framework's authoritative structure. We consider determination to be a hierarchical structure as parameter replacement and full-scale development, but the issue of authoritative arranging under clashing launch limitations [2]. These requirements show up from the need to determine complex relations among the segments of an association. Relations incorporate correspondence relations, authority relations that show the significance given to orders from different hubs, and closeness relations that determine spatial situating among objects. These relations might be confused by associating limitations. This was valid for the correspondence connection among detecting and deciphering hubs given above and valid for different relations. For instance, a maker of an item whose worth declines with time may necessitate that it be situated close to the shopper utilizing the item or that both be situated close to hubs of dependable transportation.

Existing dialects have executed a couple of direct relations, yet their methodology is constrained. A correspondence connection, for example, is depicted by expressly expressing that procedure U is to speak with process V. On the off chance that the procedures might be reproduced, this announcement becomes U, speaks with V., where I distinguish a particular duplicate of each procedure. This type of depiction isn't sufficiently general. On the off chance that Vs, for instance, is lost because of hub disappointment, U, should be lost. Any data it was to have gotten from V won't be prospective and it will be inactive; the creation of any data it was to have sent Vs will devour framework assets futile. Since the portrayal determines as it were that U, is to speak with Vs, it is impossible to locate a substitute. One can't be made because V's qualities correspond with U, and the significance is obscure [1,3,4].

The capacity to determine increasingly complex relations and permit organize fashioners to indicate area explicit relations (for example, the correspondence connection given above) is required. Rather than expecting fashioners to indicate correspondence relations as point-to-point associations, they ought to be approached to flexibly the standards by which such pairings can be resolved. The rules that an individual from one space of a connection uses to perceive an adequate part from another area are called requirements. Imperatives determine a connection since they show which pairings of an individual from one area with the individuals from another are allowable. All the more decisively, a connection characterizes a subset of the arranged sets (all in all, n-tuples) that is the Cartesian cross-result of every space of the connection, where every requirement in the connection is a predicate that chooses a portion of the pairings as more critical than others. We will all the more freely depict constants a characterizing another, progressively limited connection, by refining the meaning of an increasingly broad connection.

In this chapter, we portray a language, called EFIGE [1], for indicating the intricate relations expected to depict a circulated critical thinking frameworks authoritative structure. We likewise depict a mediator for EFIGE, that can start up a specific association by consolidating a portrayal (in EFIGE) of a nonexclusive class of authoritative structures and a lot of launch limitations that indicate the specific launch. The presentation of relations characterized with requirements to association depictions altogether upgrades the portrayal as an emblematic portrayal of the association. It permits the portrayal of hierarchical classes, instead of depictions of explicit examples of some class. Imperatives, in any case, convolute association launch. To start up a connection, arrangements must be discovered that fulfill every one of the requirements in the connection. This requires looking through enormous spaces of potential arrangements trying to discover esteems that will all the while fulfilling the entirety of the limitations. As a between time approach, we have adjusted a calculation from the Artificial Intelligence writing that is utilized to kill conflicting assignments of qualities to imperatives. This methodology is restricted, be that as it may, because it attempts to pick arrangements that fulfill one requirement without first playing out some investigation that will guarantee that the arrangement will be satisfactory to the rest of the imperatives. The utilization of an increasingly complex methodology anticipates further research.

9.1.1 Example of Organizational Structure

In our situation for dispersed sign translation, various types of signs are discharged by different vehicles as they travel through a district. The framework's errand is to make a background marked by vehicular movement inside the district dependent on the signs it identifies. One processor authoritative structure for performing signal understanding is the various leveled association. It has three sorts of segments: detecting hubs, which perform signal discovery and grouping; blending hubs, which cause nearby translations of the sign data they to get from the detecting hubs and incorporating modes, which consolidate understandings got from different hubs to make understandings over bigger parts of the detected area. Figure 9.1 represents an

FIGURE 9.1 Organizational structure.

example of the progressive authoritative structure that makes them coordinate hub, four orchestrating hubs, and four detecting hubs [1]. The figure likewise shows the lines of correspondence between the hubs, even though the directionality of these correspondence joins and the data transmitted isn't the equivalent between all sets of hubs. At long last, the figure shows the covering locales checked by every sensor. Figure 9.2 shows another occurrence of the various leveled authoritative structure. It has five coordinating hubs, sixteen orchestrating hubs, and sixteen detecting hubs.

Figures 9.3 and 9.4 show two examples of the equivalent authoritative class. The objective of this work is to build up a method of depicting authoritative classes, instead of portraying explicit associations that are launches of some class. The key highlights of any hierarchical class are the various kinds of parts in the disseminated signal translation application, detecting, blending, and incorporating hubs) and the relations between them (correspondence among detecting and combining hubs, orchestrating and coordinating hubs, and low-level and elevated level incorporating hubs'). Each kind of segment has a specific arrangement of obligations to complete (signal identification, translation, incorporation) and a lot of prerequisites for assets to be used in meeting its duties (preparing equipment, information about sign understanding, and so on).

9.1.2 Purpose

An association is a gathering of at least one person whose reason for existing is to play out some arrangement of undertakings trying to accomplish a lot of objectives while watching a lot of imperatives. Limitations on how the objectives are to be accomplished decide the pace of preparing required and, thus, influence the size and multifaceted nature of the association. For instance, the objective of the

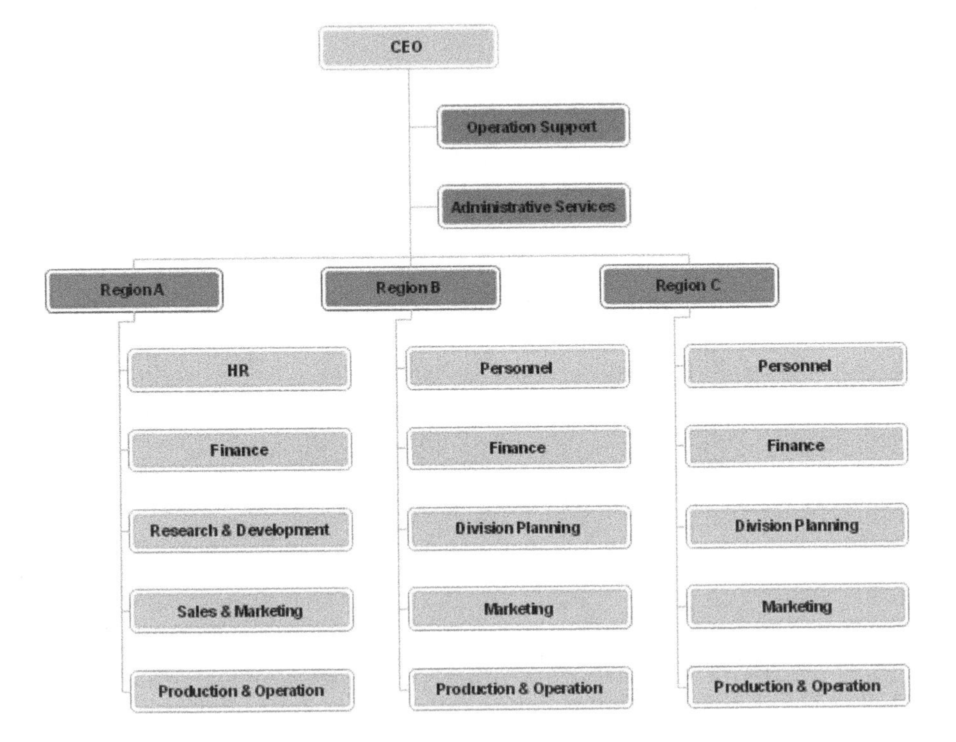

FIGURE 9.2 Hierarchical organizational chart.

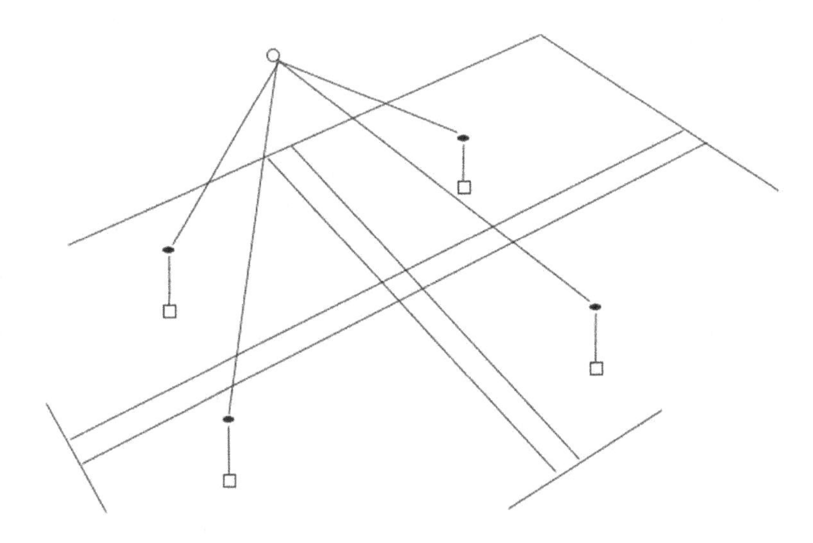

FIGURE 9.3 An instance of hierarchical organization structure with one integrating node (circle), four synthesizing nodes (dots), and four sensing nodes (squares).

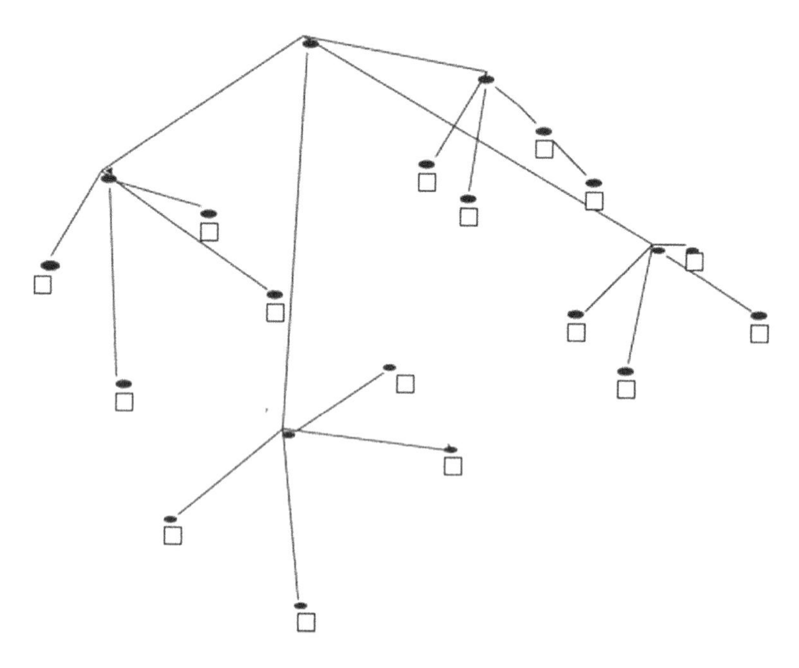

FIGURE 9.4 The hierarchical organizational structure with 5 integrating nodes, 16 synthesizing nodes, and 16 sensing nodes.

progressive association is to make a significant level history of vehicular action over an area. The assignments required to accomplish the objective incorporate the recognition and characterization of acoustic signs produced by the vehicles, the weighing of proof for the nearness of specific kind of vehicle dependent on the sign sorts distinguished, and deciding the ways of vehicles through the district and recording them.

Imperatives to accomplishing the association's objective stress preparing exchange off between such highlights in topicality, creation costs, vigor, fulfillment, and quality For instance, in the sign understanding assignment, we may demand that the framework produce profoundly appraised translations of the information as fast as could be expected under the circumstances, in this way accentuating maximal qualities for topicality (short reaction time) and quality right understandings), to the detriment, maybe, of creation costs (the pace of handling expected to determine the appropriate response). Moreover, appropriated frameworks are ordinarily expected to be powerful, ready to acclimate to hub disappointments, and to have execution corrupt effortlessly as a blunder in the framework increments.

9.1.3 Components

Organizations are made out of parts. The progressive association, for example, has three segments: detecting, combining, and coordinating hubs. What these parts share practically speaking are sets of duties and assets to be utilized in meeting them.

9.1.3.1 Obligations

Parts perform errands. These include: a subset of the assignments vital for achieving the association's motivation, the board undertakings acquired as hierarchical overhead; and-particularly in human frameworks errands that counter or don't contribute towards the association's motivation, however, are, for peculiar reasons, imperative to the segment. One method of determining obligations is by allotting parts subregions of the critical thinking space characterized by the hierarchical undertaking [2]. For the sign understanding assignment, the elements of the critical thinking space may be the physical area observed by the framework, critical thinking occasions, (for example, the location of a sign of a specific sort, the choice that a gathering of signs was delivered by a specific kind of vehicle, and so forth.), deliberation levels (signs of various kinds, gatherings of signs, vehicle types, examples of vehicles), and time. Out of the entirety of the errands that an association for signal translation needs to perform to meet its objectives, detecting hubs perform just the sign identification task. Different parts are liable for playing out the rest of the undertakings.

9.1.3.2 Assets

Components have certain assets with which they are relied upon to play out their assignments, therefore the assets required by a part will rely upon the jobs it plays in the association. We depict three "flavors" of assets: programming assets (information), equipment assets (apparatuses), and different parts (specialists). Access to a part asset is access to another arrangement of programming and equipment assets and another rundown of segment contacts.

9.1.3.3 Information

We likewise depict three sorts of information: calculations, information bases, and aptitude. Calculations indicate how to process information, information bases are storehouses of data, and ability alludes to the kind of heuristic information normal for master frameworks. The issue solvers situated at every hub may fuse any or these types of information [4]. Calculations and mastery, for instance, advise a hub on how to decipher signal information as proof for the nearness of vehicles and how to follow those vehicles. Some information might be meta-level information used to decide when it is suitable to apply space explicit information.

9.1.3.4 Apparatuses

Notwithstanding information about how to play out an assignment, a specialist may require specific actualizes with which to execute the errand. These can be effectors (a robot arm or the sled or wrench that the arm may employ during a sensor (the gadgets that a detecting hub uses to distinguish signals). Utilization of an instrument necessitates that the specialist had extra information: how to utilize it.

9.1.3.5 Experts and Subcontractors

On the off chance that surprising issues emerge that are outside the scope of ability of a part, it is helpful to know about somebody who has the skill. Given this data, the part could request critical thinking guidance or agreement on the issue's answer for the master. Essentially, a segment may think that it is helpful to realize who can utilize its information, who can give it missing information, or who is accessible to share its preparing load. Smith has explored a technique for appropriated critical thinking, called the agreement net methodology, in which a hub, given an issue that it can't fathom alone, contracts for the arrangement of the issue or of its subproblems [2,5]. This strategy doesn't depend on knowing ahead of time who is fit for tackling the issues or subproblems, since they can be communicated to the system, yet this data is utilized if accessible. This is known as engaged tending to. We can envision a situation wherein a detecting hub starts sending a blending hub data about signs of a kind for which the hub has no information. If the orchestrating hub knows of another hub that has the information, it could request help. If not, it could communicate a solicitation for the information it needs.

9.1.4 Relation between Components

Segments in an association don't exist, nor do they work, freely of each other. Parts associate. Orders, data, and subassemblies (counting fractional arrangements) are passed among them, and they may work agreeably at performing a procedure on some article. These connections are communicated as relations between the parts in question.

Relations between segments can be subjectively intricate. It will rarely be the situation that lone a solitary connection will exist between just two parts. When all is said in done, a combination of relations between gatherings of parts will be required. These gatherings may, thus, be shaped from different relations,

9.1.4.1 Correspondence

The most significant connection between at least two segments is who converses with whom. This is the connection indicated most conspicuously in Figures 9.1 and 9.2, where every internode line speaks to an occasion of correspondence connection. The correspondence connection is utilized to distinguish a segment's wellsprings of an especially significant asset, data, and to recognize the purchasers of the data it produces.

Similarly, significant are the subtleties of what is traded during correspondence. The need to relate a message structure with a correspondence connection confuses its launch. It necessitates that articles fulfilling the connection should furthermore, fulfill the requirement that their message structures be good. That is, on the off chance that one article hopes to send messages comprising of certain data in a particular organization, the other item in the connection (expecting the double case) must be set up to get that data in a similar arrangement.

At last, it might be important to connect a particular correspondence procedure with a correspondence connection. Durfee, Lesser, and Corkill have examined the impacts of a few correspondence methodologies on the worldwide conduct of a dispersed critical thinking system.

9.1.4.2 Authority

Expert in a connection that shows how much accentuation ought to be given to messages from various sources or conceivably, to various messages from a similar source. On the off chance that the message has authority, the part may permit it to have a more noteworthy effect on its exercises. In the five hubs association, the coordinating hub might be given the power to guide orchestrating hubs to search for proof of vehicles in areas it assigns. Endless supply of such a message, a blending hub may stop whatever handling it had decided to do (because of the nearby data accessible to it) and take up the mentioned work.

What amount of consideration ought to be paid to a position? The part may understand that the earth has changed and the position's directions are not, at this point fitting Should they be followed, disregarded, or questioned? An incorporating hub may have exceptionally solid proof that a vehicle's way lies a specific way when it gets the coordinating hub to look somewhere else. The hub must choose if it is progressively imperative to keep handling the solid information or to adhere to the integrator's directions. Truth be told, it might be attractive to have singular variety between hubs, weighing some combining hubs with a more prominent predisposition toward the incorporating hub's power than others. Hubs with a minimal inclination toward power are called self-coordinating or suspicious. Reed and Lesser have talked about the significance of self-course in the individuals from bumble bee provinces. Corkill examines the utilization of incredulous hubs in conveyed critical thinking associations performing signal understanding. As a rule, hierarchical relations can be depicted on two levels, at a (moderately) worldwide level sketching out the connection and its members, and at the nearby level, where subtleties and individual change are expounded.

9.1.4.3 Area, Proximity, and so on

Many other significant relations may exist between the parts of an association For example if one segment is a maker of an item whose worth abatements with time, the segment utilizing that item may be found close by, or the two of them may be set close to terminals of dependable transportation arrange, Sales workplaces for a producer may be situated the nation over, rather than across the board city Sensing hubs in the associations for signal translation should be conveyed over the whole district. Integrating hubs need to speak with a detecting hub (all the more, for the most part, gathering of detecting hubs) that checks the hubs' locale of duty.

9.1.5 Description of the Organizational Structures with EFIGE

This area presents EFIGE, a language for portraying hierarchical structures. Depictions of associations in EFIGE are various leveled [1]. That is, they are made out of either individual or composite structures, and a composite structure's parts might be individual or composite. Segments are given neighborhood names, are restrictively launched, might be recreated, and data as qualities for parameters-might be parceled among them. Parameterized portrayals and restrictive launch of segments permit depictions to be characterized recursively, this is the situation with the various leveled association.

Fields are accommodated determining the person's obligations inside the association, posting the assets the individual will require to meet its obligations, and for extra data about the person that might be gathered during launch or may give data to be utilized to gauge the person's preparing attributes. Qualities for these fields are fundamentally application subordinate. The various leveled approach we have introduced for depicting associations is like the determination structure of different dialects. What gives our methodology extra agent power is the presentation of relations and limitations into this various leveled unmistakable system. EFIGE permits relations of any sort to be set up among parts and permits extra data to be related to the connection. For example, practically all dialects for portraying authoritative structures give their individual and composite structure.

9.1.6 The Constraint Solution Algorithm

The calculation we use for discovering answers for the collaborating requirements related to A connection initially applies every part's inclination and gathering imperatives, at that point picks part with the most modest number of gatherings. In this manner, an integrating hub whose bunch limitation delivered just a single arrangement will be prepared before any hub with at least two gatherings to browse. This procedure limits expanding in the hunting tree, which is significant because we have no worldwide information to apply while picking a branch [5,6]. The part's inclination limitation is utilized to choose one of its gatherings if there are multiple choices. The chose bunch is a nearby arrangement. Neighborhood arrangements are then used to manufacture the worldwide arrangement. The nearby arrangement records the detecting hubs with which this orchestrating hub will impart, the worldwide arrangement contains the entirety of the detecting integrating hub pairings. The gatherings of different individuals from the connection that don't yet have a nearby arrangement must be made steady with the arrangement just picked. For the other individuals' gatherings to be predictable with the arrangement they should either:

1. contain the name of the part simply prepared, if the arrangement contains their names.
2. not contain the name of the part simply prepared, if the arrangement doesn't contain their name.

Conflicting gatherings are erased and the natural part that currently has the littlest gathering is chosen for preparing Thus the decision of a nearby arrangement may prune the pursuit tree and influence the request wherein hubs in the tree are visited.

EFIGE shares have been actualized for all intents and purpose Lisp. Depictions have been composed of authoritative structures for use in the Distributed Vehicle Monitoring Testbed (DVMT) [4,7]. (The various leveled association utilized for instance in this part is one of these.) The DVMT reproduces the execution of a conveyed issue solver that performs signal understanding. Depictions of associations in EFIGE are deciphered and added to a record of parameters that indicate the investigation that will be done on the DVMT. One portrayal of an association can be utilized to create numerous launches of the association by shifting the qualities provided to the depiction's parameters. These outcomes in reserve funds must be produced by hand—a tedious and mistake-inclined strategy.

Base Level Constraints Currently, the EFIGE translator is allowed to truly find hubs in any place directed by the portrayal and to expect that correspondence channels exist any place required. Essentially, the framework is permitted to design the preparing system as is helpful. This is helpful for at first structuring an association outside the requirements of a current design. The subsequent stage for upgrading the EFIGE translator is to permit it to start up the "best" association of a predetermined class, given a specific system engineering. This includes determining base-level requirements speaking to the specific system engineering and joining those fixed limitations into the launch of an association [7]. Bottom-level imperatives indicate that the start-up association incorporates segments with given qualities for a few or the entirety of their properties or those specific relations be actualized. Such requirements could determine a whole preparing system, making it the activity of the translator to start up, like most ideal as, an association's useful parts and their relations over a physical system that gives not exactly ideal help To model, if base-level imperatives indicate that there are just twelve handling hubs however they launched association needs thirty-seven, the mediator should dole out various hierarchical hubs to a similar processor.

9.1.6.1 Requirement Propagation

In light of the combinatorics, it might be preposterous to apply limitation or gathering requirements to the entirety of the individuals from a space. For example, the number of ways, n orchestrating hubs can speak with detecting hubs, where any given integrating hub might be allotted from zero to of the detecting hubs, is 2". The current calculation endeavors to abstain from looking at all of the objects of this set by wiping out addendums of items based on nearby data. Accordingly, if the limitation requirement for incorporating hub A chooses to detect hub P, P is verified whether its limitation imperative chose A. If not, P's disposed of as a possibility for A [1]. This dispenses with from further thought those setups wherein P and An are combined, subsequently slicing the pursuit space down the middle Unfortunately, assessment of A's limitation imperative requires applying it to the entirety of the detecting hubs in the area and this is rehashed for the entirety of the orchestrating and detecting hubs in the connection

Another way to deal with improving effectiveness is imperative engendering. In this technique, a depiction of the accomplice required by a part in a connection is step by step developed as imperatives are assessed. Imperative engendering, it is trusted, would permit the discharge of an increasingly explicit limitation that would recognize, after just one pass say, the detecting hubs that both require and are required by an integrating mode. Engendering of limitation requirements has been acted in frameworks, for example, MOLGEN. Engendering of the more intricate limitations utilized in EFIGE, nonetheless, makes it difficult for the remaining parts to be examined.

9.1.6.2 Imperative Utility

At the point when a lot of imperatives ends up being over-compelled, it is valuable to have the option to insightfully change them so an answer can be acquired or to figure out which ones must be fulfilled and which ones can be securely disregarded or loose [1,2]. This requires information about the motivation behind the limitation (because of the authoritative objectives), no that decisions about its significance can be made, and it requires the capacity to find the contention, to figure out which imperatives to change. This may not generally be conceivable. Fox allocates requirements utility appraisals which would then be able to be utilized to decide the value of a given limitation's fulfillment, or absence of fulfillment, in a circumstance. The least helpful imperatives are more averse to antagonistically influence results on the off chance that they are not met. Utility appraisals are additionally utilized during backtracking to discover choice focuses where almost an inappropriate decision was made. Another decision is looked for and made at these focuses and the pursuit is restarted.

9.2 COMPARATIVE STUDY OF ORGANIZATION STRUCTURE

Functional: The functional structure is established on structuring and is divided up into small –scale grouping or sets with a specific piece of work or duty. For example, a business could have a mass functionality in computer technology, artificial intelligence, another in marketing (buying/selling), and also in finance and accounts. Each branch has a supervisor or managing director who responds to an executive-level up in the ranking who can supervise various sectors [6].

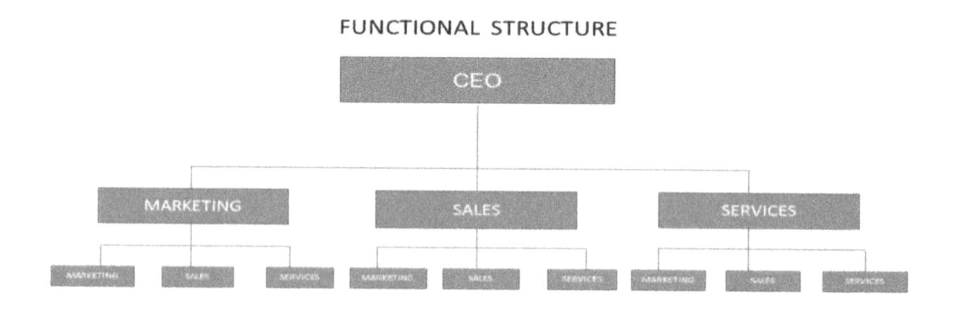

Divisional: This organization structure agrees for much more freedom in the group within the corporation. One example of this is an organization such as General Electric. GE has numerous divisions as well as aeronautics, transport, thermal, digital and renewable energy, and so on.

The following structure, each part mainly performs an operation as its organization, manages its resources, and also manages or controlling how much funds it utilizes on certain projects or aspects of the disunion.

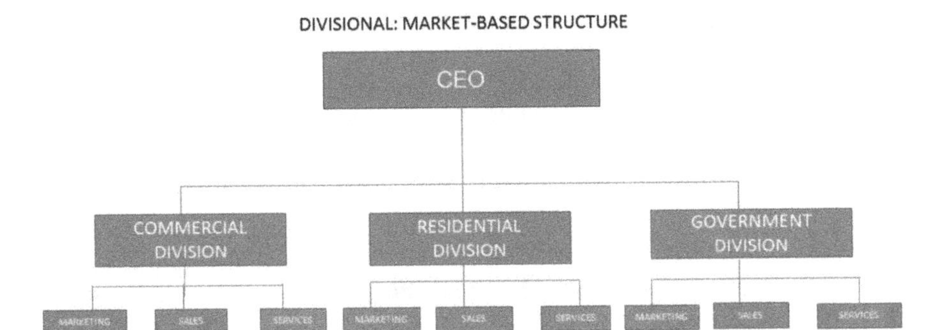

This type of structure provides large elasticity to a greater firm with much more separation, permit each one to handle as its own business with one or more individuals broadcast to the holding company's chief executive officer (CEO) or higher authority staff. Despite having each program approved at the very higher levels, those queries can respond at the divisional level.

A drawback to this kind of company structure is that by the focus on divisions or disunion, employees or hired man working in the identical task in individual divisions may be not able to communicate well between divisions. This structure also increases default with accounting practices and may have tax implications.

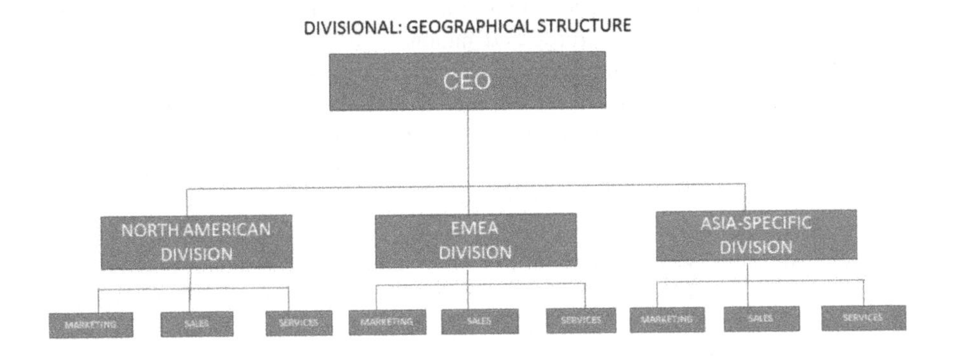

Matrix: In the given organization structure, employees or workers report more than one bosses based on situation or project. For example, under normal working circumstances, an employee at a large engineering company could job for one manager, but a new project may arise where special engineers are needed for the company. For the time duration of that project, the employee would also report to that project's supervisor, as well as his or her manager for all other daily tasks or works.

This structure is difficult because it can be tough to describe to several managers and aware of what to speak to them. That's why the engineers need to know their roles, principals, responsibilities, and priorities for the company.

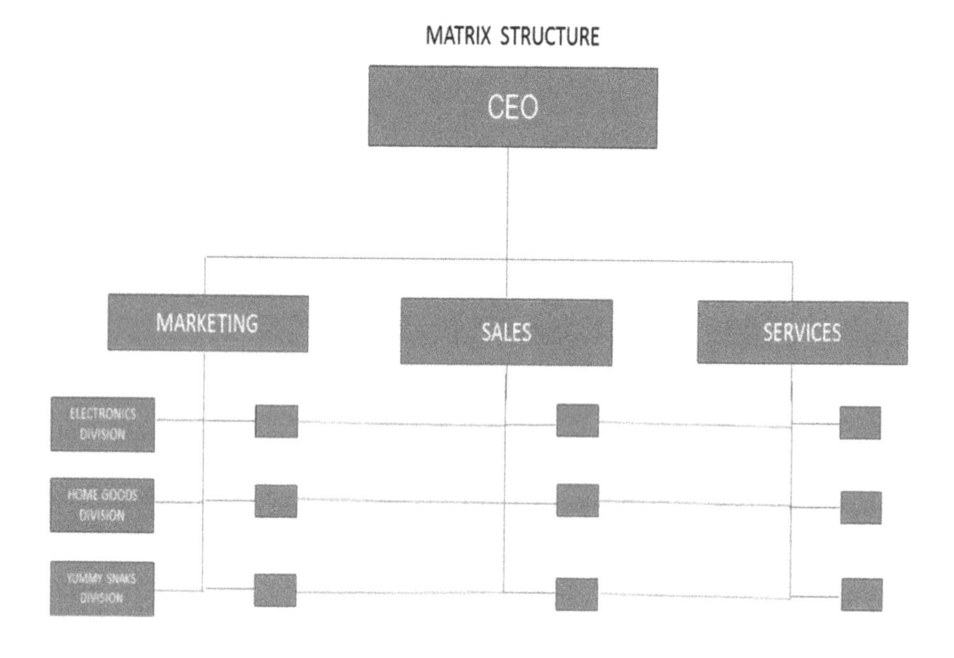

The comparative study of organizational structures highlights similarities and differences between different components influencing a company that how it operates with the internal and external environment. Compared management also contains analyses of different cultures and nations, comparisons between different organizations helping the managers identify similarities and differences between them, to increase the organizations' capability, and efficiency. To increase the performances of a company, be it on the low or high level, the development of its organizational structure is necessary. Before analyzing the organizational structure in the companies of the different countries, it is necessary to understand its definition and contents. We need to emphasize the fact that these countries differ from one another by their economic development, cultural level, and History. With a relatively well-built organizational structure, a society may not be successful, but with an improper organizational structure, no company of the developing countries can survive in the long run in the market. The government factors rely upon country tradition and previous facts. British companies are managed by a board constituting the main decisional unit of a company and the source of power in an organizational structure [1,8]. A public limited company (PIC) has to have a managing board made up of at least two managers who were appointed by shareholders. One of them is the chairman, who may be the general manager (GM) of the company, as well. Or else, the general manager shall often be appointed administrative manager. Besides the chairman, there will be a secretary of the company, responsible for the observance of the law and responsible for discipline in the administration. The other members of the board are not subjected to restrictions.

9.3 CONCLUSION

This study reconfirms that the culture needs to be oriented with a crucial objective. Organizations dragnet strategies that are needed to ponder about whether their traditions are commending to or need changes for the outcoming results. Organizations are expanding, the executives amend the need of the system to be reformed and the composition to be inducted and strengthened should be effectively implemented. Benchmarking can be used to choose the plan of action and to progress the presiding organizational approach.

We have proposed that portrayals of the hierarchical structure are significant for the instantiation and support of dispersed frameworks over huge heterogeneous systems Current dialects for depicting authoritative structure don't permit depictions of discretionary relations and are unequipped for depicting higher request relations. We have identified three kinds of limitations (limitation, gathering, and inclination) and have utilized them to depict and startup discretionary and complex hierarchical relations. We have given a calculation to discovering answers for cooperating requirements utilized in descriptions of relations. At long last, we have tried these strategies by fusing them in an authoritative launch language, called EFIGE, and its mediator. Some limitations were conveyed after indagating. It included the factors which would impact its capability.

REFERENCES

1. Pattison, H. Edward, Daniel D. Corkill, and Victor R. Lesser. "Instantiating descriptions of organizational structures." *Distributed Artificial Intelligence* 1 (1987): 59–96.
2. Horling, Bryan, and Victor Lesser. "A survey of multiagent organizational paradigms." *Knowledge Engineering Review* 19.4 (2004): 281–316.
3. Sims, Mark, Daniel Corkill, and Victor Lesser. "Separating domain and coordination knowledge in multiagent organizational design and instantiation." *Proceedings of the AAAI-04 Workshop on Agent Organizations: Theory and Practice.* AAAI Press, California, 2004: 1–7.
4. de Kinderen, Sybren, and Monika Kaczmarek-Heß. "On model-based analysis of organizational structures: an assessment of current modeling approaches and application of multi-level modeling in support of design and analysis of organizational structures." *Software and Systems Modeling* 19.2 (2020): 313–343.
5. Les Gasser, et al. "Representing and using organizational knowledge in distributed AI systems." *Distributed Artificial Intelligence.* Morgan Kaufmann, 1989: 55–78.
6. Hannoun, Mahdi, et al. "MOISE: an organizational model for multiagent systems.". In: Monard M.C., Sichman J.S. (eds) *Advances in Artificial Intelligence.* IBERAMIA 2000, SBIA 2000. Lecture Notes in Computer Science, vol 1952. Springer, Berlin, 2000: 156–165.
7. Malone, Thomas W., and Stephen A. Smith. "Modeling the performance of organizational structures" *Operations Research* 36.3 (1988): 421–436.
8. Kaufmann, Wesley, Erin L. Borry, and Leisha DeHart-Davis. "More than pathological formalization: understanding organizational structure and red tape." *Public Administration Review* 79.2 (2019): 236–245.

10 Agora Architecture

Nidhi Gupta, Shailesh Singh, and Sonia Gupta

CONTENTS

10.1 INTRODUCTION

To implement a custom programming environment, R.Bisiani [1] has designed a set of methodologies, tools, and architecture, known as Agora. The need to develop such an environment was to build and evaluate heterogeneous artificial intelligence applications effectively and quickly.

10.1.1 CHARACTERISTICS OF SYSTEM FOR WHICH AGORA IS USEFUL

1. System should be heterogeneous: this means no single computational model, machine or languages can be used.
2. System should be in rapid evolution: this means the algorithm could often change but a fraction of the system remains invariable, as in the research system.
3. System should be computationally extravagant: this means not a single processor is sufficient to obtained required work.

Agora authenticates heterogeneous systems through virtual machines, which does not depend upon language, allowing a variation of the computational model. This could be accurately mapped into large numbers of different computer architectures. The feature of rapid evolution is assisted by Agora in the same way as in the Lisp environment, i.e., by incremental programming.

Programs running on a concurrent virtual machine can be combined with the environment and shares similar data with the programs which are created independently. This technique provides an endless set of custom environments that were designed according to the needs of the user.

By all means, parallelism is supported as the systems will always be described as parallel computation no matter how they are running on a single processor. We can say that Agora is by no means an environment in the inspection of application; basically, it was given by exigency, which comes from designing and implementing CMU Speech Recognition System [4].

10.2 ARCHITECTURE OF AGORA

Agora has layered architecture and a layered model is used to explain its structure shown in Figure 10.1 from bottom to top flow.

The first layer or bottom layer is a mesh of wide-ranging (heterogeneous) processors. These heterogeneous processors may be single processors, loosely connected multiprocessors, and shared memory multiprocessors or custom hardware accelerators. The basic software required to execute computations all above heterogeneous processors is provided by Mach Operating System. Tools required for mapping of Mash hypothesis into a real machine is provided by Agora. Mach is a UNIX compatible OS which runs on multiprocessors.

The second layer is the Mach layer. The three major functions of this layer are(1) message passing, (2) shared memory, and (3) threads. The fundamental delivery

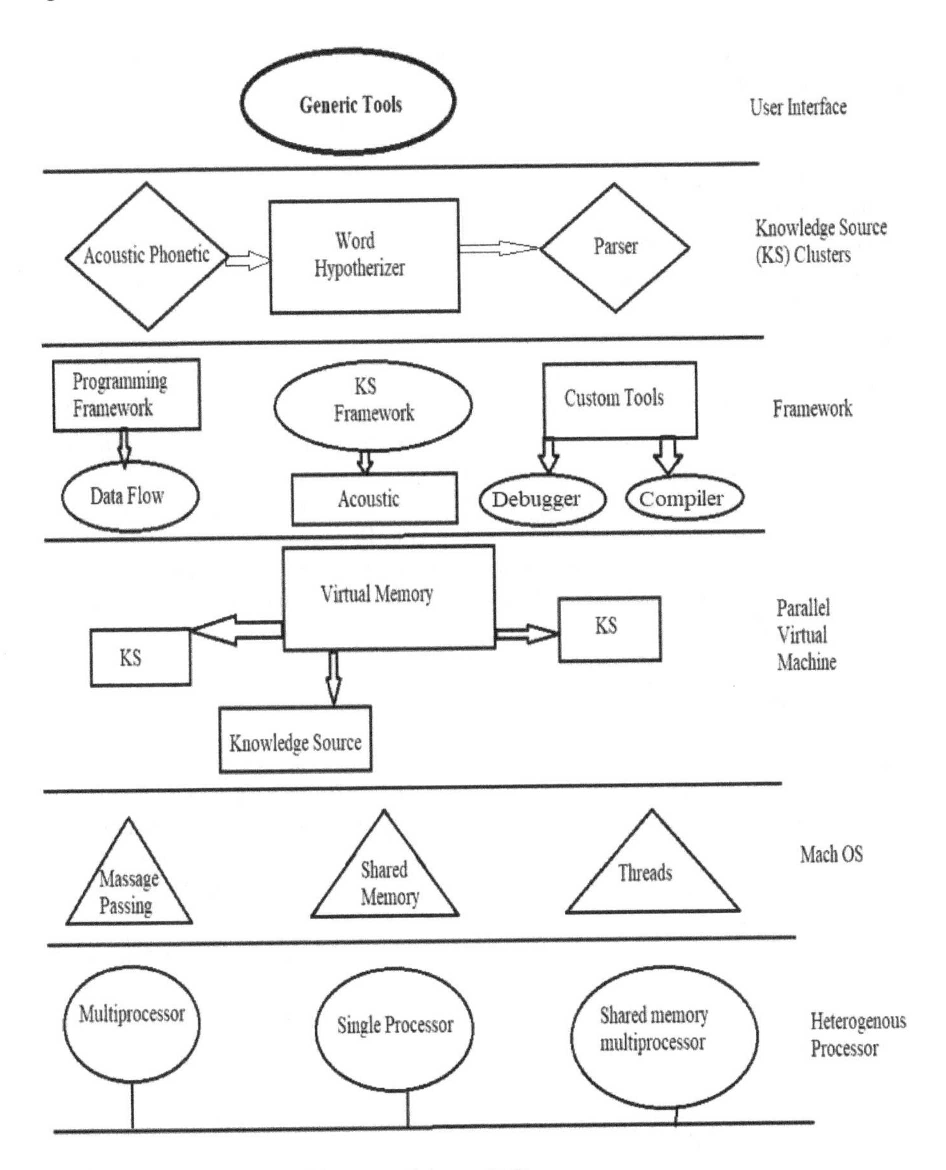

FIGURE 10.1 Layered architecture of Agora [1,3].

process is passing the message. The unique transmission mechanism of message passing is obtained by running all Agora applications on machines. To enhance performance shared memory is used. Threads are also called lightweight processes. It consists of its own program counter, stack, and set of registers. These are the processes that share address space with another process. Threads help in the fast creation of new computation.

The third layer is the concurrent or parallel virtual machine layer. This layer deals with the assembly language level of Agora. This layer's function is to choose the most suitable Mach primitives, the codes are assembled or compose and then linked, and various tasks will be assigned to machines. All the mathematical computations performed will be machine-independent. Systems are completely described in this layer. A parallel virtual machine layer describes the framework in spite of programming user computation.

The next layer, i.e., the fourth layer, is the framework layer. This is the level where there is most of the research program application resides. A special environment is created to communicate among users through familiar terms. Associated frameworks perform the function of representation or explanation, assembling, debugging and production run of an application system. ABE's frameworks [6] are similar to the framework used in Agora. Tools for generation and maintenance for any kind of program is provided by the framework. For example, the framework has a virtual machine code for implementing the data flow or remote procedure call transmission mechanism. The framework uses a tool for merging existing virtual machine code with user-supplied code.

Frameworks use a structured graphical editor that allows users to understand programming with regard to the data flow graph. Researchers make use of frameworks for generating framework instantiations (which includes user-provided data and code). One framework instantiations will be linked with other frameworks to create a complex framework.

10.3 AGORA'S VIRTUAL MACHINE

Agora's virtual machine will be designed keeping two things in mind. First, the machine should be capable of accurately executing a variety of computational models; second, the machine should avoid restricting the implementations for computer architecture. Agora represents data as a set of elements of a similar type. Global structures called cliques store elements. Every clique has an individual name that defines its identity and type.

Agora enforces users to break computation into independent components known as a knowledge source (KS), which operates parallelly. KS transfers data using cliques and becomes active whenever specific patterns of elements are generated. The name of each clique is registered in Agora at the time of its creation. A particular KS that knows a clique's name will execute operations. As the name is global sharing of click between cases only requires that clique be created by means of KS and another case declared that it is shared.

10.3.1 Element Cliques (EC)

Sets of elements each of the same type are known as element cliques. The element types are defined within the KS code by using language which helps in programming cares along with some extra information. Agora removes this supplementary information from source code prior to code being delivered to the compiler or interpreter. The additional information added in type declaration is required for scheduling and debugging purposes, for example, the calculated number of accesses per second, displaying procedure etc. By making use of capabilities, KS accesses elements in EC.

Agora's capabilities will be modified and used to copy from cliques into the address space of KS and vice versa. Read only and add elements to two accessing modes. In spite of the fact that anyone can read from an EC, KS adds element only if they have capabilities to EC with add element access excess. In any event, once added element cannot be deleted or modified.

10.3.2 Knowledge Source (KS)

Every KS consists of single or more functions that were completely independent of the system. Calling Agora primitive create KSs on every call to this function, it generates a number of instances to similar KS. On the creation of KS instance, some patterns will be determined. Once the pattern is fulfilled, then the KS function will be activated. The pattern is determined based on arrival events and the value of data stored in the elements.

10.3.3 Mapping of KS into Mach layer

After the creation of KSs, they were mapped into Mach primitives [5]. With this mapping, a KS can be clustered with another KS that can profit from sharing computer resources. Clusters might be implemented as a multiple process that shares memory or multiple processors connected through messages. Multiple citations of similar KS are very effortless or effective implementation on Mach OS like a single process where multiple Threads of data processing implement KS. Any KS can be cluster by making use of Agora primitives controls the computation cover with KS.

Finally, Agora Virtual Machine gives a mechanism for controlling multiprocessing dynamically and statistically: KS will be grouped in a variety of ways and will be implemented on different configuration processors dynamically controls the allotment of processes.

10.3.4 Frameworks

The primary interface for describing, building, and executing of any application is a Framework level. A framework gives an environment which adapts according to the requirement of the system components for which it is used. A framework consists of a set of customized tools, confined into a single environment. This permits the tool to work collectively and benefits by sharing information of tools. Agora provides a simple frame language CFrame, to facilitate sharing.

10.3.4.1 Typical Framework Tools

A tool in the Framework is a kind of program which the user can call from within the framework. A tool is either an external program like C compiler or screen editor or maybe any other application created using Agora. Every Framework contains a minimum set of tools. Tools should provide [2]:

1. Task description: mechanism user defines system. The description includes defining the system's structure, if it is not predefined by the framework, and providing user-provided code.

2. Preprocessors: preprocessor converts the user-supplied code into parallel virtual machine primitives. This includes merging of user-supplied code with the code described within Framework so as to implement the required programming model.
3. Multiprocess running and debugging: the user interface allows users to start and run the system on different real machines. Highly advanced Framework provides multiprocessors debugging which continuously monitors memory activities and carry out controlled reply of start in the number of EC, language debuggers, and particular sequence of KS, etc.
4. EC display: a mechanism that permits the user to view the EC of running the system's data.
5. EC management: at least there must be some procedure for adding elements from a file into EC and vice versa. For this, an element editor is provided to view elements in EC and add new elements to EC.

10.3.4.2 Knowledge Base: CFrame

Gora provides a simple frame language known as CFrame which is designed after SRL language [10]. C frame allows program and user stored and access information in the form of a semantic net. EC stores semantic net and can be shared by different tools, even if tools are running on different machines. C frame provides effective Set of primitives allows the tools in framework to modify a network of a frame that holds slot with value. These shots represent either attribute of frame or relation.

10.4 EXAMPLES OF SYSTEMS BUILT USING AGORA

10.4.1 Intelligent Transport System (ITS)

Mohammad A. Salahuddin and Ala Al-Fuqaha [3] present Agora for ITS application, i.e., intelligent transport system application. In this system, vehicles, and pedestrians, vehicles and pedestrians are equipped with cellular or wireless devices that are interrelated with each other through Agora infrastructure and its cloud services. Here, Agora provides a series of safety, efficiency, and commercial applications. The safety applications are highest priority application which includes how to adjust to reduce traffic and accidents. The efficiency applications include intelligent traffic flow management and navigation, gas consumption, vehicle carbon footprint etc. while the commercial application includes entertainment and Information applications like gas price, VOIP, and ant messaging [8].

Agora Framework constitutes regional manager (RM), super nodes (SN), traffic management centers (TMC) cloud services, mobile integrated system (MIS), and vehicle integrated services (VIS) [7].

VIS and MIS communicat with Agora super nodes and RM to obtain and upgrade their neighborhood information and create V2V vehicle to vehicle and V2M vehicle to mobile communication. V2M communication combines wireless and cellular networks to promote ITS application. It has two advantages:

1. Vehicles equipped with smart devices used Agora application without any additional onboard unit
2. Pedestrians equipped with smartphones use Agora application.

TMC (traffic management centers) provide extra safety and efficiency applications like advisory alert, weather summary, variable messages etc.

10.4.1.1 Architecture of Agora ITS Framework

Architecture of Agora consists of hardware and software components (Figure 10.2).

10.4.1.1.1 Hardware Components

The hardware part consists of onboard units (OBU) and roadside units (RSU). RSU has intelligent access points (IAP), mobile internet switching controllers (MiSC), mesh wireless router (MWR), and AP canopy advantage access point. Figure 10.3 shows RSU components setup and configured into 2.4 gigahertz mesh network. The MWR mobile wireless router enables OBU onboard unit components to Interface to the internet through IAP and or the switching controllers. The MWR is positioned to increase the range between IAP and vehicle-mounted modem VMM in the OBU and simultaneously increases network spectral efficiency77]. The MiSC makes possible network

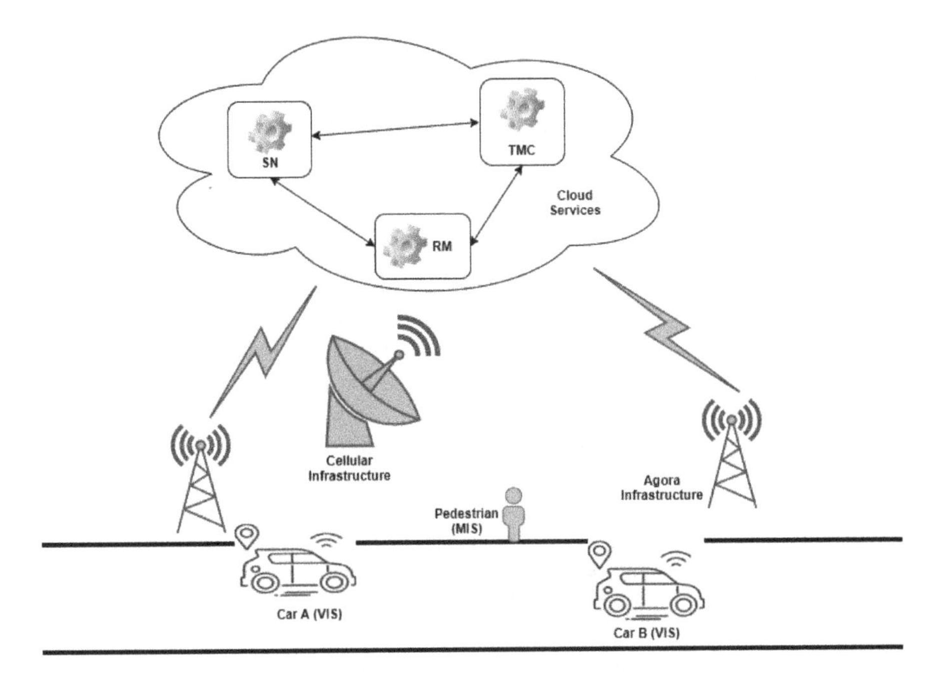

FIGURE 10.2 Agora architecture for ITS framework [7].

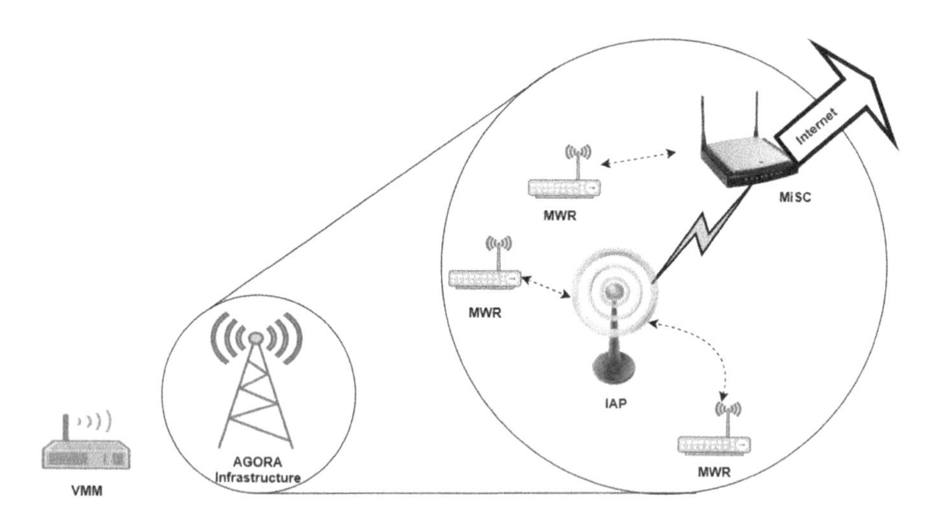

FIGURE 10.3 VMM in OBU connected to RSU in Agora structure [7].

management functions like provisioning, allocation, management and fault management monitoring. OBU components consist of GPS (global positioning system), OBD (onboard diagnostic) system scanner, Windows-based nettop with touchscreen display, and VMM (vehicle-mounted modem), as shown in Figures 10.4 and 10.5.

10.4.1.1.2 Software Components

The software component for ITS application has server-side and client-side applications. The server side consists of super node (SN), regional manager (RM), and Web Server manages TMC. Client side applications are divided into VIS and MIS for vehicular and cellular users in Agora (Figure 10.6).

10.4.1.1.2.1 Server-Side Components

The region of interest in Agora is split into the centroid shown in Figure 10.6. Centroids are location coordinates. Supernode SN controls Inventory of centroid sets to manage Global view of available centroids sets. These sets of centroids are allotted to RM through SM. Whenever SM receives a ping from any RM, it allots a new centroid and updates the global view. But if SN peer has already allotted the centroid, then rollback execute.

The sequence of action needed to create V2V and V2M communication is shown in Figure 10.7.

TMC is a storehouse of transportation network data received from vehicles and pedestrians. TMC architecture is shown in Figure 10.8.

Various steps have been taken to make a to be expandable, load balanced, fault-tolerant Framework with low potential, these are:

1. Firstly, RM, backup of RM, and SN should be arranged in a hierarchical manner to eliminate the inherent disadvantage of a tired architecture.

1. Intelligent Access Point(IAP)

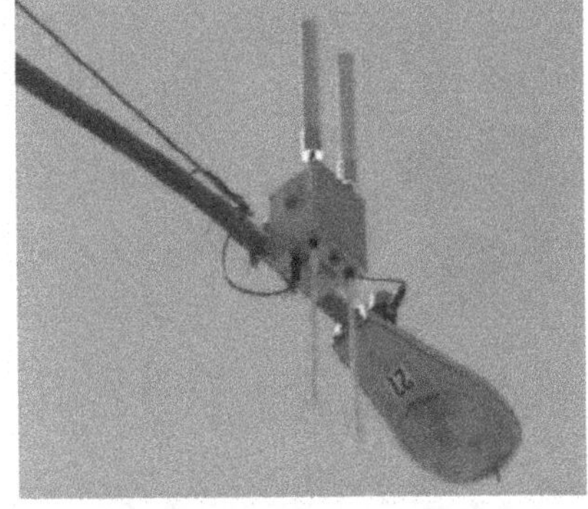

2 Mesh Wireless Router(MWR)

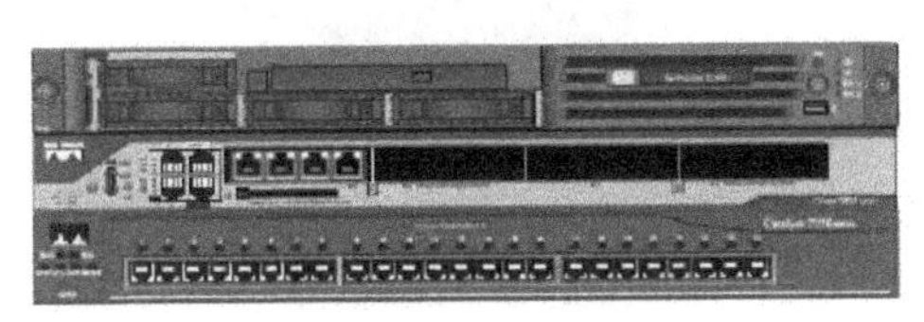

3 Mobile internet Switching Controller (MiSC)

4 Canopy Advantage Access Point

FIGURE 10.4 Agora RSU components [7].

2. Secondly, multiplicity SN, RM, and backup of RM (RM BAK) numbers may introduce redundancy, divide workload, and give low latency for application request.
3. Thirdly, centroid sets present in a high traffic density area will be blocked from having more than two centroids.
4. Fourthly, V2V and V2M Agora packets are first encrypted and then communicated over UDP to eliminate the disadvantage of inherent low packet delay.

10.4.1.1.2.2 Client-Side Components

Client side components are classified into VIS and MIS for vehicular and mobile Agora users respectively. Both VIS and MIS are XML configured interphase, and perform numerous safety efficiency and infotainment applications like road sign notification, VOIP calls for emerging response, contextual information etc.

The real time situation application is an integral part of VIS and MIS which provides detail of the closeness of neighboring vehicles. The neighboring vehicles which are in safe distance are shown as green blinking square while there is an alert for possible collision and red blinking square for close distance vehicles (Figure 10.9).

1 Global Positioning System

2. On-Board Diagnostic System Scanner

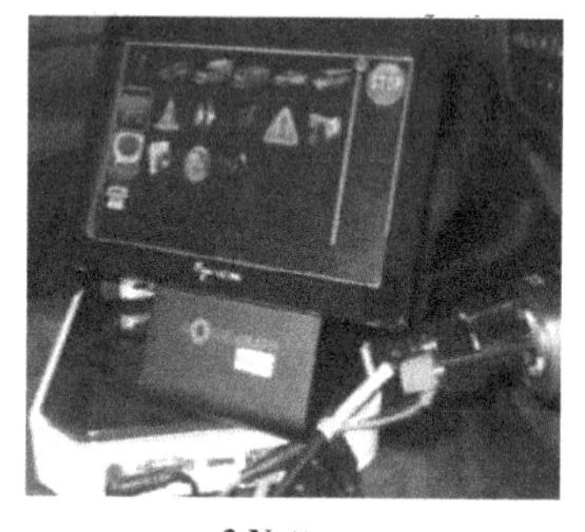

3 Nettop

4 Vehicle Mounted Modem

FIGURE 10.5 Agora OBU components [7].

10.4.1.2 Agora ITS Applications

Agora provides various items application for VIS and MIS, these applications are classified under safety, efficiency, and infotainment.

10.4.1.2.1 Safety Applications

The main purpose of designing safety applications is to prevent damages from occurring on the road in terms of life and property. It includes pedestrian alert, advisory alert, and government 2.0 which help to report dangerous road conditions based on current geographic location to the TMC. Agora provides security application by using RFID tag base user restriction application (Figure 10.10).

10.4.1.2.2 Efficiency Application

Efficiency applications, as the name suggests, increase the efficiency of transportation and vehicles network. This application includes traffic alert, fire truck alert, ambulance alert etc. This application uses TMC Web Services to fetch data based on the geographical location of the vehicle. this helps the driver not to take these areas and opt for an alternate route, reducing conditions and traffic in an already backlogged area (Figure 10.11).

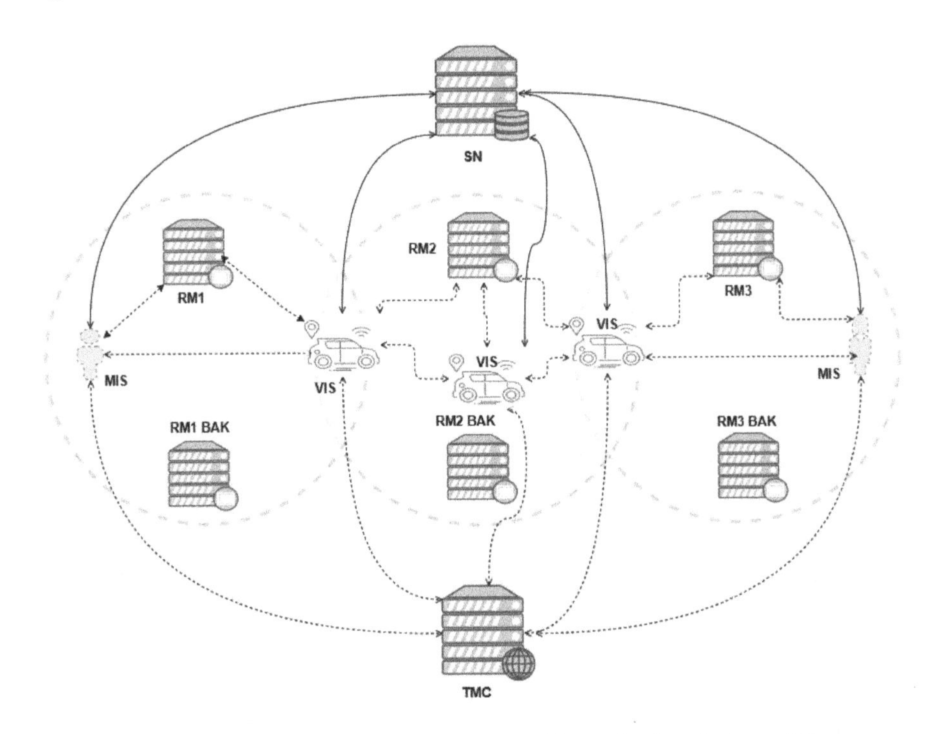

FIGURE 10.6 Integration of client-side and server-side applications in Agora [7].

10.4.1.2.3 Infotainment Application

This application includes finding dining restaurants, weather, checking gas prices, and carbon emissions etc. This application aims in providing information, entertainment, and convenience for Agora users (Figure 10.12).

10.4.2 CMU Speech Recognition System

Figure 10.13 shows the use of Agora in designing of CMU speech recognition system, ANGEL. This system uses a high number of computations as compared to a single processor, it is programmed in two languages, C and common Lisp code, and it also makes use of different styles of computations. It is an evolutionary system. Transfer of both data and control is indicated through arrow as shown in the figure. The majority of components at the top level communicate using a data flow paradigm while few modules make use of remote procedure calls.

The acoustic and phonetic components use both data flow and blackboard paradigm. A framework at the top level provides a graphic editor for programming data flow and remote procedure call computation. Each subcomponent of top level framework is designed through different frameworks. The work hypothizer is used as a subcomponent [9]; it applies a beam search algorithm during the selected time in reply and generates word hypothesis. Phonetic hypothesis and vocabulary are the inputs of search. Markers called anchors are used to indicate time.

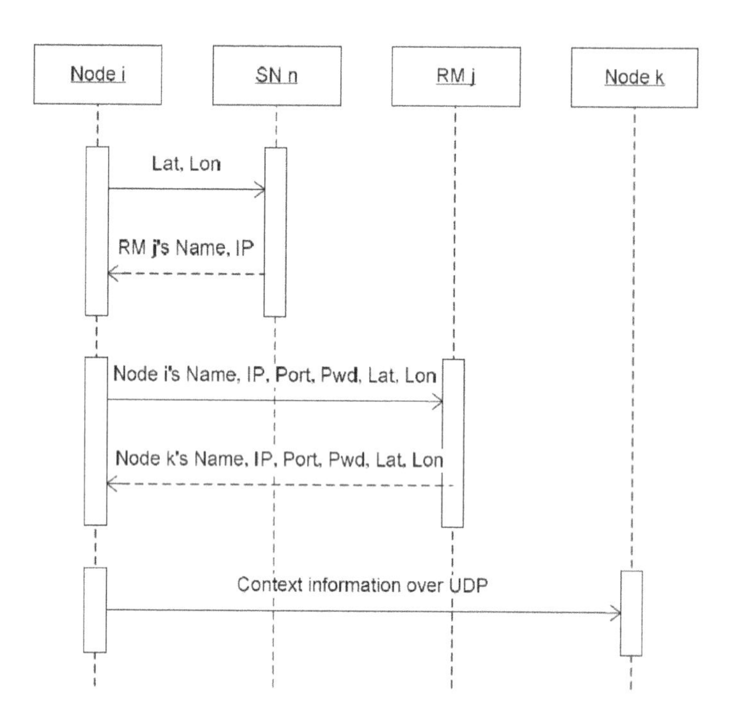

FIGURE 10.7 Sequence diagram for pedestrian/vehicle connection initiation with Agora [7].

A word hypothizer should accept phonemes and anchors in random order. It performs a search around each anchor after checking whether the entire acoustic-phonetic hypothesis is within their Delta time's unit from anchor are available. Phonetic hypothesis acts on uncertain time and in any order.

The word hypothizer needs 2 functions:

1. Matching function (match ()) which hypothesis word from phonemes
2. Condition function (enough phonemes()) will check whether there will be ample phoneme within a specified time interval from anchor.

The word hypothizer editor allows researchers to specify the above said two functions and bind them through virtual machine level description used within a word hypothizer frames is described by stylized code, as shown in Figure 10.13. This description may be within any language which is supported by Agora C and common Lisp. This system uses three KS [2].

KS setup generates instantiation of a KS word hypothizer that becomes active whenever anchors arrive. With any of three KSs accepts anchor, that particular KS checks whether there is enough phonetic hypothesis available or not. If enough phonetic hypothesis is available then the KS executes the match function; if there is not enough phonetic hypothesis available then it generates instantiation of KS, which waits until all the necessary phonemes are available before executing match function. When the target machine does not have shared memory, then the KS word hypothizer will be compiled

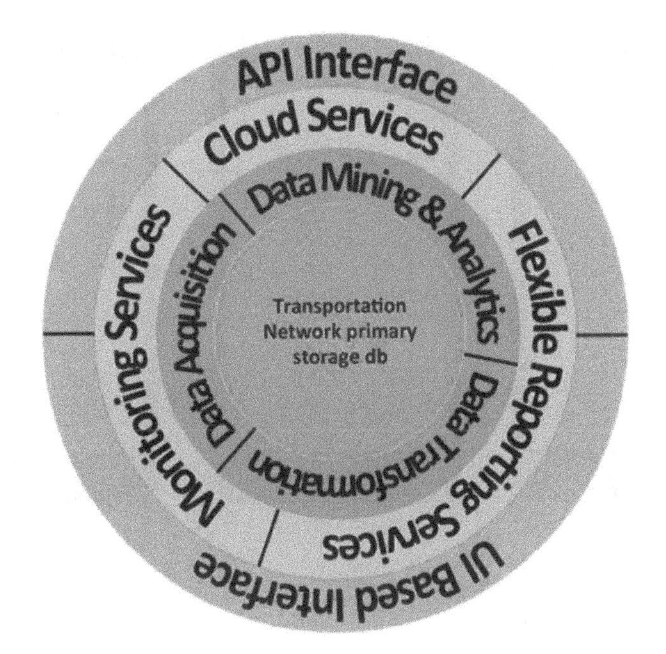

FIGURE 10.8 Architecture of traffic management center (TMC) [7].

into tasks. The number of cases that the framework designers need to efficiently use is indicated by the parameter to Agora during the KS creation procedure.

10.5 APPLICATION OF AGORA AS A MINIMAL DISTRIBUTED PROTOCOL FOR E-COMMERCE

To support bulk transaction with low accepted cost in Electronic Commerce, a predetermined protocol named Agora is used as a web protocol because of its four noble properties:

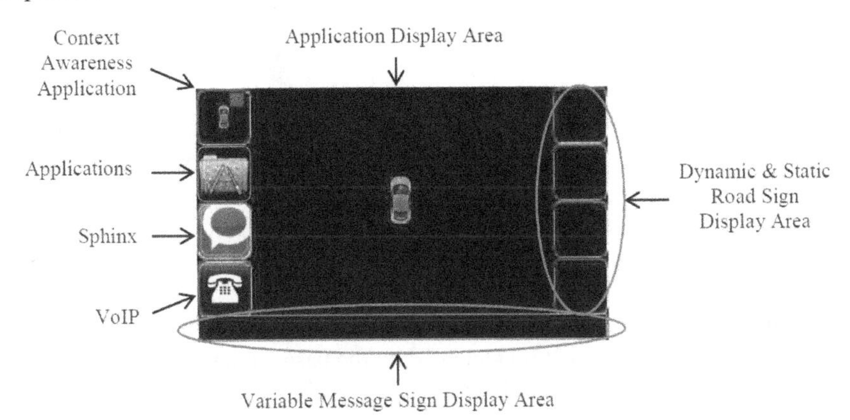

FIGURE 10.9 Vehicle Integrated System (VIS) interface [7].

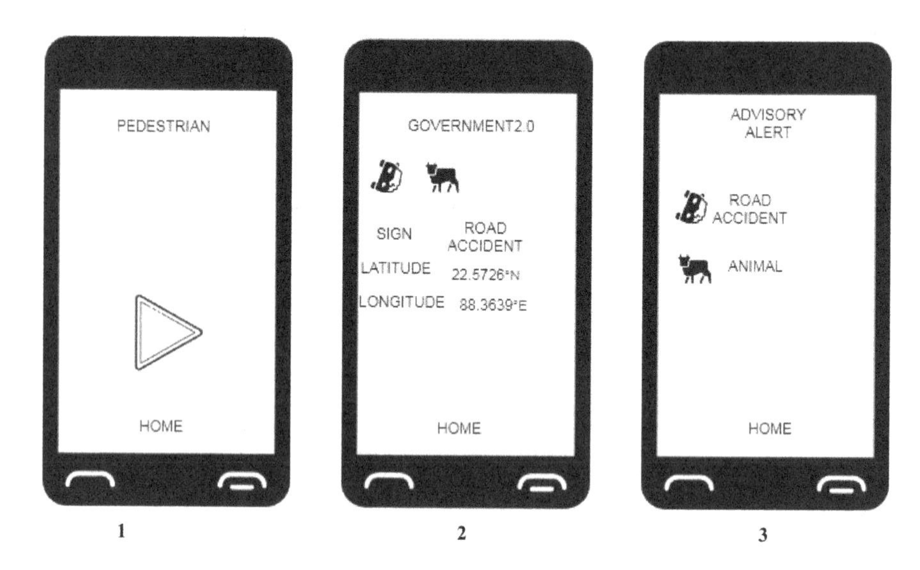

FIGURE 10.10 1. Pedestrian alert; 2. Government 2.0; 3. Advisory alert application on MIS [7].

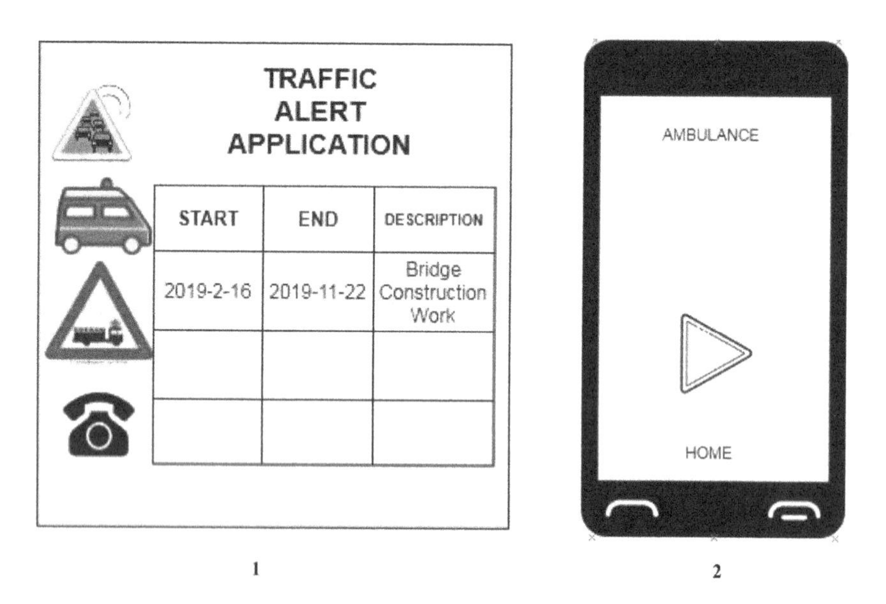

FIGURE 10.11 1. Traffic alerts; 2. Ambulance alerts as efficiency application in Agora [7].

1. Minimal: The accepted cost of transaction in Agora is nearly free web browsing, where the cost is calculated on the basis of number of messages
2. Distributed: Agora is fully classified
3. Online arbitration: Certain arguments between customers and merchants can be resolved by online arbiter

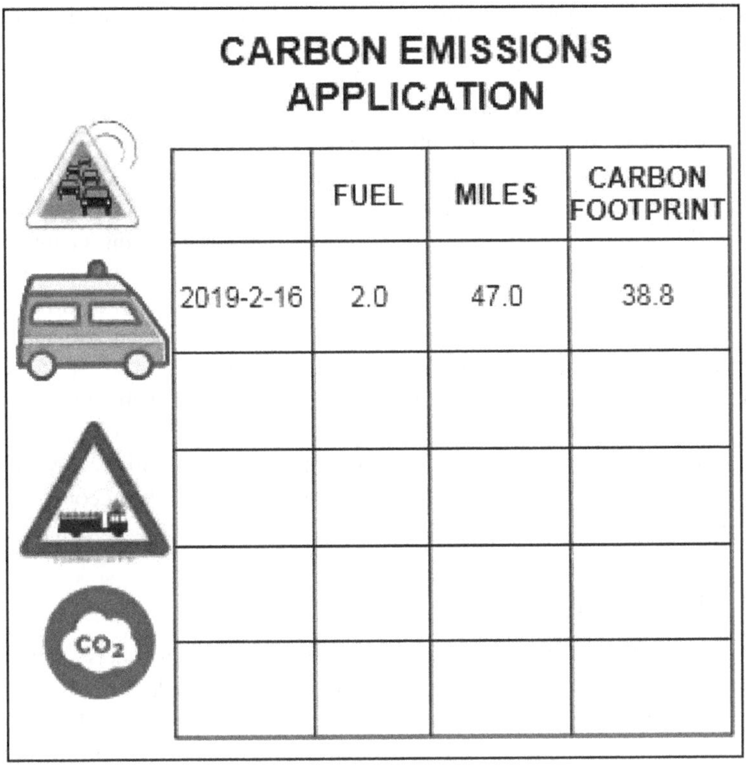

CARBON EMISSIONS APPLICATION

	FUEL	MILES	CARBON FOOTPRINT
2019-2-16	2.0	47.0	38.8

FIGURE 10.12 Carbon emission infotainment application in VIS [7].

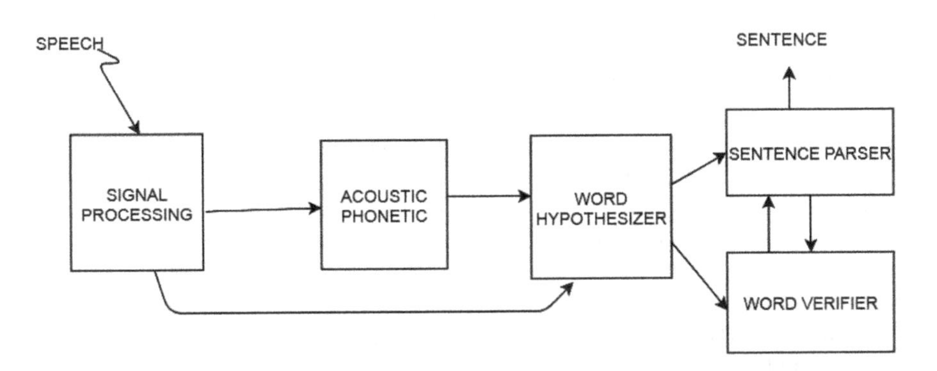

FIGURE 10.13 ANGEL speech recognition system structure [1,3].

4. Fraud control: Agora can restrict the level of fraud to some pre-decided level

Hence, we can say that an Agora is a secured authenticated and non-repudiated protocol and can be used in communications which are insecure.

Gabber and Silberschatz [11] in their study considered banks as financial institutes that control accounts of customers and merchants. Merchants are basically the vendors electronically sell their goods. Banks furnish transaction processing facilities, as do credit card companies. Customers are the end-users who will buy goods electronically.

Gabber and Silberschatz [11] assume that banks are trusted body whereas customers and merchants are not trusted. Arbitrators are a fourth type of entity which resolves the issues between the customers and merchants.

Agora is a web protocol for E-Commerce that is fully distributed and needs 4 messages to complete a transaction (includes good transfer) without central authority access. Agora messages will be cart on existing http messages without any additional message. In this protocol the transaction demands transfer of the account identifier, not the actual funds, which is why Agora is called credit-based protocol.

10.5.1 Basic Protocol

There are 4 entities in Agora protocol banks, merchants, customers, and arbiters. Here concepts of multiple banks are used. There is a unique identifier Bid, a private key K_b^s public key K_b^p for each bank. Before the initiation of any sale or purchase by the customer and merchants, the important requirement is that the customer and merchants must have bank accounts.

Cacct is used to denote customer account while Macct is used to designate a merchant account. Mname is a unique public name to identify Merchant and this unique public name can be used as IP address or domain name of the merchant. K_c^s and K_m^s all the private keys of the customer and merchants respectively and K_c^p and K_m^p are their public keys respectively.

Bank provides ID to customers and merchants which customers and merchants can preset so as to identify themselves. Customer sign ID is SCid while SMid is merchants sign ID. Arbiters are responsible for resolving any disputes between customers and merchants and not necessarily the bank, but they are the trusted entities. Arbiters have the power to repudiate the committed transaction. Aid, K_a^s, K_a^p are the unique identifiers, private key, and public key respectively for each arbiter. A digital signature is used to sign messages in Agora protocol.

$S_x(msg) = E_k(H(msg))$, E is an encryption function, H is a secure one-way Hash Function and k is the signers private key K_x^s

When $H(msg) = D_k\ (S_x\ (msg))$, then only the signature S_x (msg) is verified, D is a description function, k is the signers public key K_x^p.

For message signing, following notations will be used

$S_a(msg)$ signing a message with arbiters private key K_a^s
$S_b(msg)$ signing a message with bank private key K_b^s

S_c (msg) signing a message with customer private key K_c^s
S_m (msg) signing a message with merchants private key K_m^s

10.5.2 Accounts

A bank provides IDs to customer and merchants, which means customers and merchants must have accounts in some bank. Customer and merchant IDs validate only during the billing period and the bank generates new IDs after every billing period. By the end of the current billing period or by the end of next period, which is also known as the grace period, these IDs expire. Through this procedure the bank forces the customer to pay their bills on time.

This protocol imagines that in order to prevent brute force attacks, all participants modify their public and private keys in every billing period. Table 10.1 describes the generation of merchant and customer IDs where x||y||z is used to explain the concatenation of x, y, and z.

10.5.3 Transactions

Agora transactions include for messages whose control flow is shown in Figure 10.14.

M0—This is the beginning of any transaction. This message indicates that the customer demands a price quotation from the merchant.
M1—This is the reply message from the merchant.

$$M1 = SMid \| Seq \| \overline{price} \| S_m\left(H\left(Mid \| Seq \| \overline{price}\right)\right)$$

TABLE 10.1
Customer and Merchant Identification [11]

Name	Description	Format
Cid	Customer ID	Bid ‖ expiration_date ‖ Cacct ‖ K_c^p
SCid	Signed Customer ID	Cid ‖ S_b (H(Cid))
Mid	Merchant ID	Bid ‖ expiration_date ‖ Mname ‖ Macct ‖ K_m^p
SMid	Signed Merchant ID	Mid ‖ S_b (H(Cid))

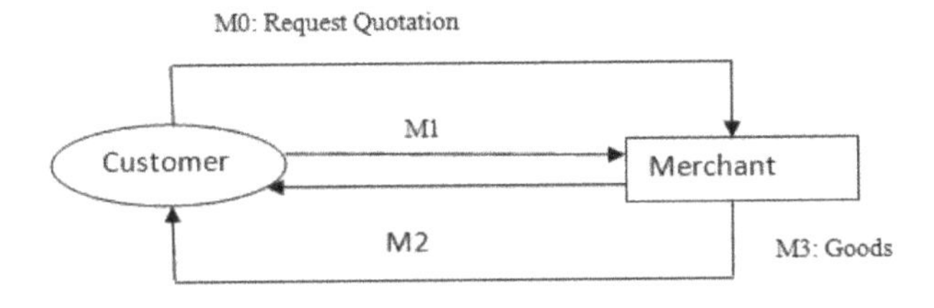

FIGURE 10.14 An Agora transaction.

Seq is the unique transaction ID; price is requested price of goods.

M2—In this message, the customer first verifies SMid., checks the expiry date of Mid, and compares Mname with the presumed merchant name, extracts K_m^p from Mid, and lastly verifies Seq and price.

Reply from customer.

$$M2 = SCid \| Seq \| price \| S_c\left(H\left(Cid \| Mid \| Seq \| price\right)\right)$$

M3—Here the merchant verifies SCid, checks expiry date of Cid, extracts K_c^p from Cid, verifies Seq and price, and compares Seq and price with M1. Merchant commits the transaction and supplies the goods to the customer once M2 passes all the tests. Finally the merchant raises the transaction to the bank for billing by the end of the billing period; the M1 and M2 pair constitutes proof of transaction.

10.5.4 Properties of Agora Protocol

10.5.4.1 Minimal

Any web commerce protocol is minimal when it generates the same number of messages as free web browsing. HTTP messages need to retrieve a page in free browsing as shown in Figure 10.15. The pages which refer to pay per view pages when payment is enforced by protocol are called menu pages. The regular HTTP messages generated during free browsing can be piggyback by Agora protocol, which is shown in Figure 10.15.

Here, the message M0 is a normal GET request for the menu page.

M1 message is generated by the merchant for all pay per view pages as quoted by menu. Menu encloses M1 in such a way that they are not visible in browser.

M2 message is generated from the correct M1 message as soon as the customer decides to purchase a page. The following GET request for the pay per view page contains M2 as a parameter. According to the content of the pay per view page, the server responds.

We can say that Agora is a minimal protocol because it is completely embedded in HTTP messages. No additional message will be generated in Agora as compared to free web browsing means, while some of the messages are longer mainly menu pages containing various M1 messages.

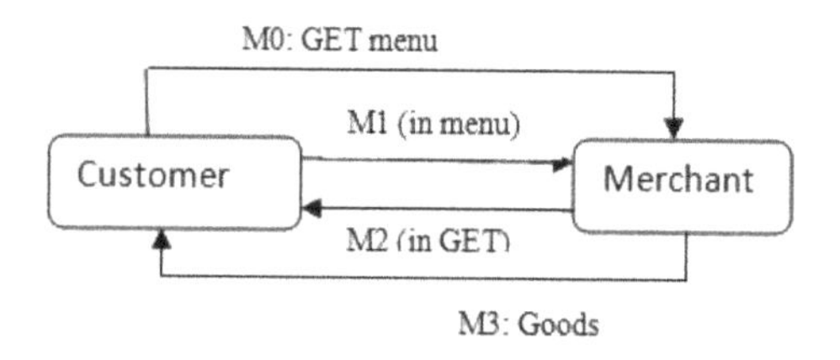

FIGURE 10.15 Piggybacking on HTTP messages [10].

10.5.4.2 Distribution

As both customers and merchants will verify the identity of locally using $K_b^{p,}$ which is the back public key; therefore, Agora is fully distributed. No prior preparation is required for the transaction and there is no need to access the bank. Merchants contact the bank only once during the billing period to transfer the accumulated transactions.

10.5.4.3 Authentication

Merchants and customers both sign M1 and M2 respectively. K_m^p and K_c^p are used by merchants and customers to verify signatures. Customers cannot repudiate M2 once it is verified. Fraud accounts are not created by customers and merchants as the private key of the bank signs all the ID.

10.5.4.4 Security

Agora is secure against any intruder who can intercept, replay, or modify the message.

10.5.4.4.1 Replay

Since M2 contains unique transaction ID and merchant ID, therefore an old M2 message cannot be used for any other transaction. Replay M0 will generate new M1 while old M2 message will never match the new M1.

10.5.4.4.2 Double Charging

Since each transaction has a unique sequence number Seq, the merchant cannot double charge the customer. This double charging would be immediately detected by the bank.

10.5.4.4.3 Alteration

Once the customer has signed M2 that contains the original price, an intruder cannot alter the amount charged for goods.

10.5.4.4.4 Man in Middle

Man in middle M can intercept and change all messages in between the merchant and the customer. M replace M1 with new M1 which contains high price quotation, Smid, and sign with his own key. M submits M2 to the bank pass merchant according to original price and delivers goods to the customer, as soon as the customer agrees to purchase. M takes the benefit of his own pocket with the difference between the original price and raise price. But this man in middle attack is impossible as the customer can verify M1 which contains the expected Mname in the SMid part.

10.5.5 Enhanced Protocol to Regulate Fraud

The basic Agora protocol is susceptible to two major types of fraud: first, fraud by customer; and secondly, fraudulent use of stolen ID. Because of distributed verification of customer IDs without access to bank, the above two frauds can happen.

The E-Agora is used to designate the enhanced agora protocol and this protocol allows low-level fraud. Following parameters are used in this enhanced protocol, L indicates the customer liability limit, T indicates the notification period, R indicates maximal purchase rate per period T, P is the probability $0 \le p \le 1$, and M indicates the price threshold.

10.5.5.1 New Message

E-Agora protocol is the enhanced Agora protocol and addresses fraud problem by following addition to basic protocol. Control flow and messages of protocol are described in Figure 10.16. Bank broadcast list of revoked CIDs to merchants, merchants communicate with banks occasionally. Merchants need to maintain a current list of revoked CIDs while denying any transaction that belongs to revoked CID.

The merchant should contact the bank after verification of M2, if any one of the following two conditions is true:

1. A uniformly distributed random variable r in the range of $[0, 1] \le p$
2. The addition of previous and current purchase by the same customer exceeds the threshold value M

The merchant sends a message M4 to the bank if any of the above two conditions is true.

$$M4 = Mid \| Cid \| \overline{price} \| S_m \left(Mid \| Cid \| \overline{price} \right)$$

$\overline{price}$ is the addition of previous and current purchase by the customer.

The bank updates customer balance after receiving M4. An increase in customer purchase rate exceeds the rate R per period T or if the customer exceeds credit limit, the bank will not approve the transaction and revoke CID, otherwise the bank accepts the transaction. After completion of M4 message processing, the bank replies with M5 message to the merchant.

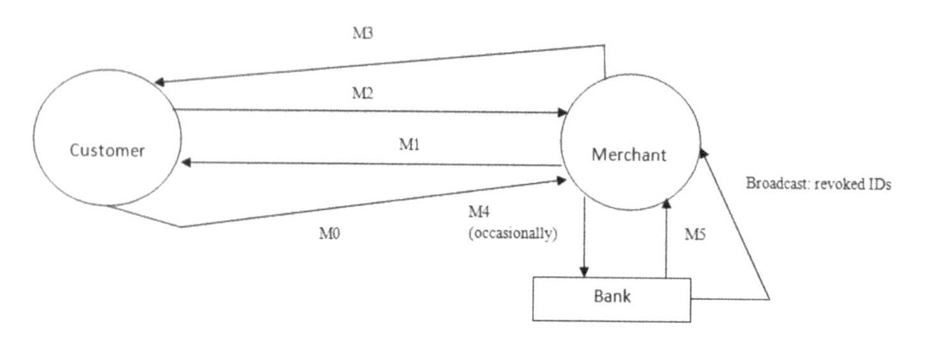

FIGURE 10.16 E-Agora [10].

M5 = Cid || code || S_b(Cid || code), code indicates whether the bank rejects or accepts the transaction.

10.5.5.2 Batch Processing

Once in every period T, the merchant needs to communicate with the bank and transfer the batches of transactions and this communication will be done at offpeak hours. Due to this, customer balance will never be out-of-date by more than T. The bank may add a list of removed CIDs on any message to the merchant so as to reduce overheads.

10.5.5.3 Selection of Parameter

$$M = p * L$$

$$R = M / T$$

The value of p is selected in such a way so that it reduces the overhead of sending M4 and M5. To keep M reasonably large p should not be too small. R is an acceptable purchase rate; M is the transaction value which is the cost of effective communication to the bank.

10.5.5.4 Online Arbitration

Arbitration protocol is introduced to remove certain disputes occuring in the E-Agora protocol. The following modification occurs in the E-Agora protocol after the introduction of an arbitration protocol.

1. The merchant introduces a secure hash value of goods in M1.
2. After sending M2 to the merchant, the customer expects the good; when the customer does not receive the good or if the sum of the received good is different than expected in M1, the customer has to contact an arbiter by sending message A1.

$$\text{A1} = \text{M1} \| \text{M2}$$

3. By sending message A2 to the merchant, the arbiter asks for the goods. Arbiter signature is required so the merchant can validate it. By doing this, eavesdropping can be prevented from asking for goods which somebody else paid for.
4. If the merchant replies in message A3 with the correct goods, the arbiter forward the same to the customer through message A4.
5. If the merchant does not reply or replies with incorrect good, the arbiter repudiates the transaction and sends a repudiation message A5 to the bank with a copy of A4’ to the customer.

REFERENCES

1. R. Bisiani, *A Software and Hardware Environment for Developing AI Applications on Parallel Processors.* Computer Science Department, Carnegie Mellon University Pittsburgh, PA, 268–3072 1986.
2. R. Bisiani, et al., "The Architecture of Agora Environment" Chapter 4, Distributed Artificial Intelligence, 1, 99–117, 1987. www.microsoft.com/en-us/research/publication/architecture-agora-environment/
3. M.A. Salahuddin, A. Al-Fuqaha, "Agora: A Versatile Framework for the Development of Intelligent Transportation System Application", in D. Benhaddou, A. Al-Fuqaha (ed.), *Wireless Sensor and Mobile Ad-Hoc Networks: Vehicular and Space Applications*, Springer, (pp. 163–184), January 2015.
4. R. Bisiani, A.D Adams, "The CMU Distributed Speech Recognition System", *11th DARPA Strategic System Symposium*, Naval Postgraduate School, Monterery, CA, October 1985.
5. R. Baron, et al., "Mach-1: An Operating System Environment for Large Scale Multiprocessors Application", *IEEE Software*, 2(4), 65–67, July 1985.
6. L. Erman, et al., "ABE: Architecture Overview", *Proceeding of Distributed AI Workshop*, Sea Ranch, CA, 1985.
7. Motorola. Motomesh 1.2 Network Setup and Installation Guide. Motorola, 2006. https://manualzz.com/doc/25626506/motomesh-1.2-network-setup-and-installation-guide
8. Y. Khaled, et al., "On the Design of the Efficient Vehicular Application", IEEE Vehicular Technology Conference (VTC), Barcelona, 2009.
9. P.E. Green, "AF: A Framework for Real Time Distributed Cooperative Problem Solving", *Proceeding of 1985 DAI, Workshop*, Sea Ranch, CA, December 1985.
10. M.S. Fox, J.M. Dermoh, The Role of Databases in Knowledge Based Systems. Robotics Institute TR, Carnegie Mellon University, 1986. www.researchgate.net/publication/2557994_The_Role_of_Databases_in_Knowledge-Based_Systems
11. E. Gabber, A. Silberschatz, "Agora: A Minimal Distributed Protocol for E-Commerce", Proc. of the 2nd USENIX Workshop on Electronic Commerce, Murray Hill, NJ, 1996.

11 Test Beds for Distributed AI Research

Shubhangi Sankhyadhar and Mohit Pandey

CONTENTS

11.1 INTRODUCTION

Distributed artificial intelligence (DAI) is a subfield of artificial intelligence, which is now more than two decades old. This field is related to the concept of the multiagent system. DAI is an approach that considers agents as computational units and how they communicate with each other by coordinating knowledge towards achieving a specified goal. More generally, DAI is concerned with situations in which several systems interact in order to solve a common problem: computers and persons, sensors, aircraft, intelligent vehicles, etc. For developing such kind of intelligent model, very extensive research is required. DAI research is multidisciplinary and cumulates the philosophy from several other fields such as artificial intelligence, distributed computing, social and organizational sciences, NLP, multi-agent system, cognitive sciences, and related fields.

Without experiments and testing any research is meaningless. In all areas of the sciences, an early stage of evolution as well as experimentation plays an important event in developing the new concepts and transformation into novel theories. The considered playground for experiment and testing is known as testbeds.

As we know, experimentation is a very crucial and important part of DAI research. By considering the importance of experimentation, many software testbeds, programming languages, and some other individual platforms for DAI research work have been developed.

Some testbeds are discussed in this chapter, including actor-style language and software based on actor MAGES and ABCL; LISP and LISP-based software MICE and MACE; Prolog and Prolog-based software MCS; agent-oriented programming (AOP) and MY World; procedural reasoning system (PRS), and others, including DAISY, METATEM, DVMT, and ARCHON.

11.2 BACKGROUND

DAI testbed plays a crucial role in DAI applications by providing application designers of DAI a platform to create agents, facilitating communication between agents, providing an environment where the agent can perform certain actions as well as providing computer simulations of the application.

Investigators at DAI are concerned with understanding and generating models and information in affiliated businesses. People tend to distinguish two main research areas in DAI [1]: problem-solving distribution and multiagent systems.

Distributed Problem Solving (DPS) looks at how the problem-solving function can be differentiated between multiple modules (or "nodes") that collaborate in segmenting and sharing problem information and its emerging solutions. In a pure DPS program, all interactions (collaboration, communication if any) are included as part of the plan.

Multidisciplinary (MAS) research is concerned with the behavior of a set of independent (potentially pre-existing) agencies aimed at solving a given problem. MAS can be defined as "an integrated network of problem-solving networks that work together to solve problems beyond their control" [2]. These problem-solvers are often called agencies; they are automated and can be complicated by nature (illustrated by various levels of problem-solving).

DAI does not care about communication issues between similar processes at problem-solving and presentation levels. That is, we do not care for the same processing of AI for reasons of good AI computations per cell" [3].

DAI brings a new perspective on knowledge representation and problem-solving, by providing some scientific constructions and practical presentations in practice. We are shining a new light on the science of understanding and artificial intelligence. Konolige and Nilsson believed that DAI could be crucial to our understanding of artificial intelligence [4]. There are many arguments to support this claim:

1. A system can be so complex and contain so much information that it is best to break it down into different parts of a system to work properly (e.g., flexibility, modularity, and efficiency).

2. DAI provides a practical framework for testing our clairvoyance about reasoning processes based on knowledge, actions, and planning.
3. The methods used by a rational system to communicate with other system components can also be used to communicate with other types of subtle dynamic processes.
4. Research on DAI can contribute to our understanding of natural language-based communication processes [5].

   Conventional distributed computing and simulation algorithms allow the programmer to solve a communication problem and provide language structures and contract control processes through which the system can use its solution. In distributed AI, we try to improve flexibility by designing problem-solvers through both solving the coordinate problem and then solving the solution [6].

A testbed task for DAI application is to provide the designer of such an application with the tools necessary to construct agents, describe their interaction, construct the environment in which they are to act, and support the simulation of the application on a computer. Different testbeds for AI systems have been proposed and implemented such as DVMT, MACE, MCS/IPEM, Actor, Concurrent MetaTem, Archon, Procedural Reasoning System, PHOENIX, TRUCKWORLD, TILEWORLD, ABCL/1, ABE MICE, blackboard architecture. Examples of independent domains in which DAI is tested include traffic control, air traffic control, process and production control, telecommunications, and personal computer communication. Several applied linkages and problem-solving frameworks have been developed, subject to different hierarchy, responsibility and knowledge, and the sharing of specific outcomes. These include [6]:

- Blackboard frameworks
- Contracting framework
- Integrative framework
- Open system frameworks

11.3 TOOLS AND METHODOLOGY

In all sciences at an early stage of evolution, experimentation plays an important part in developing new concepts and transforming them into novel theories. The considered playground for experiment and testing is known as testbeds. Different testbeds for AI systems have been proposed and implemented such as DVMT, MACE, MCS/IPEM, Actor, Concurrent MetaTem, Archon, Procedural Reasoning System, PHOENIX, TRUCKWORLD, TILEWORLD, ABCL/1, ABE MICE, Blackboard architecture.

11.3.1 MACE

A 1985 workshop on DISTRIBUTED ARTIFICIAL INTELLIGENCE (DAI) was held between December 3 and 6 at Sea Ranch, California. Twenty-eight participants gathered to debate on the theory and practice of DAI. Les and [7] discussed the theoretical nature of Distributed AI and the implementation of a multiprocessor DA1 testbed and language called MACE (for Multi-Agent Computing Environment). MACE provides high-level support for DAI testing with interdisciplinary diversity elements and in a range of problem domains. The main objective of MACE is to provide a programming environment. A description language, simulator, and run time environments (such as processors, processor speed, network interconnections, and communications) are required for DAI [8]. MACE supports examining different styles of DAI at different complexity levels; moreover, it helps in building an extensive range of DAI systems. The MACE environment maps agents with processors and provides communication between multiple agents via messages.

The MACE system consists of agents and their communication as shown in Figure 11.1. The MACE system has the following components [9]:

Collection of agents: The basic computational units of the MACE system are agents. The MACE agent description language describes organized clusters of agents. MACE agents are coinciding objects, that consists of user-defined procedural part called and engine, along with a collection of databases.

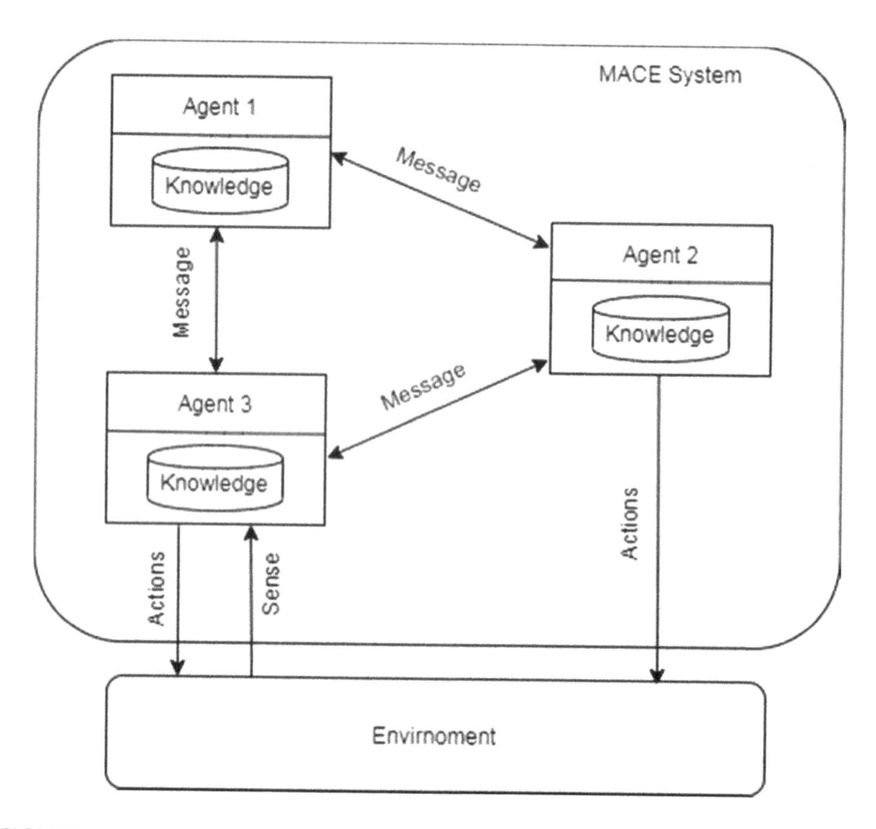

FIGURE 11.1 The MACE system.

Community of system agent: MACE command interpretation, user interface, agent-builder cluster, execution monitoring, tracing, error handling are provided by a number of predefined system agents.

Collection of facilities: All agents use the facilities that include a pattern matcher, a simulator, several standard agent engines, handlers for standard errors, and standard messages that are understood by all agents.

Instrumentation: MACE is an instrumented testbed that allows characteristic measurement for problem-solving during a wide range of experiments. The most common measurements are database size, message congestion, work done by agent and processor node's load.

Description database: It contains the descriptions of the agent that is maintained by a system agent cluster. A system agent cluster performs the following tasks:

- Constructs new description
- Verifies description
- Executable agents are constructed from description

Collection of kernels: MACE kernels perform functions like scheduling agents for execution, handling communication among agents, mapping of agents with processor, routing of a message between agents.

Reflection capabilities: The knowledge database of MACE agent consist of plans, roles, agent implementation, course, high level, and behavioral reflection capabilities.

Evolutionary philosophy: MACE agent changes and expertise over time. MACE agents are self-contained objects which have acquaintance database, attributes and takes action on messages they receive.

11.3.1.1 MACE System

11.3.1.1.1 MACE Agents

MACE agents are self-contained objects that do not inherit anything from other objects, active objects that are "social" in nature and communicate with other agents and the environment via messages. Every agent is uniquely classified and is assigned a unique address to each of these classifications. This address contains:

- Agent's name
- Agent's class
- Residential node
- Machine on which it runs

MACE has some predefined system agents such as garbage collectors, staffing ("personal") specialists, fault control specialists, a "postmaster" who knows routes to other agents, collection of "allocators" which perform dynamic load balancing, etc.

Each agent contains attributes that are not directly visible to other agents, but their contents can be sent in messages to other agents.

Attributes of agent = Engine + Database.

Some attributes include:

- Type A: It defines the class of an agent.
- Status: It describes the state of an event, i.e., Running, waiting, new, destroyed, or stopped
- Engine: The engine is a function that tells how messages are interpreted and changes its local state.
- World Model: Each agent has a unique address, which is extracted from the world model.
- Acquaintance: It encapsulates agents and knowledge in MACE.
- Incoming Messages: It is a database that contains messages to be received.

Note: Messages in MACE are LISP lists that include sender and receiver addresses.

11.3.1.1.2 Sensing in MACE

Agents perceive the world by receiving messages and are explicitly notified of internal events and system-level events. A MACE agent can monitor two types of events:

- Events that are the result of some visible actions. These include either sending a message or receiving a message, any change in status, or if a new event is created or destroyed. Such events are monitored by *Demons* (mappings from event description to messages).
- Events that are due to kernel actions or monitoring an agent notices change in state. These events are monitored by *imps* (predicate-action pairs).

11.3.1.1.3 Actions Taken in MACE

In an agent, there are two executable parts: its initialization code and the agent's engine. MACE agent takes an action when the kernel evaluates the agent's engine. Message sending is also an action; the message can either be unicasted, multicast, or broadcasted to agents of a particular class.

Agents can take three kinds of actions:

- Changing internal state by modifying their attributes
- Sending messages to agents or outside environment
- Sending *monitoring* request to the MACE kernel

11.3.1.1.4 MACE ADL

- To describe agents MACE provides a language known as ADL (Agent Description Language), while to describe aspects of simulation and executable environment, it provides a language known as EDL (Environment Description Language).
- It is an extension of LISP. It provides a description of generic agent types, which are nonexecutable, and executable instances known as actual agents.

- In MACE, agents are grouped into organizations, which in turn provides a single message entry point.
- Organization provides encapsulation and conceptual modularity.

11.3.2 Actor Model

The actor model is a mathematical theory for concurrent computations that considers "actors" as universal primitives of concurrent computation. An actor only communicates with permissible actors for communication as shown in Figure 11.2.

An actor performs the following functions when it receives a message:

- An actor communicates with other actors by passing messages using a mailbox
- This creates more actors
- This also decides what behavior should be adopted for the next message to be received

An actor does not share memory, i.e., a private state can be maintained by an actor that cannot be altered by another actor.

An actor model employs various languages:

- Act 1
- Act 2
- Act 3

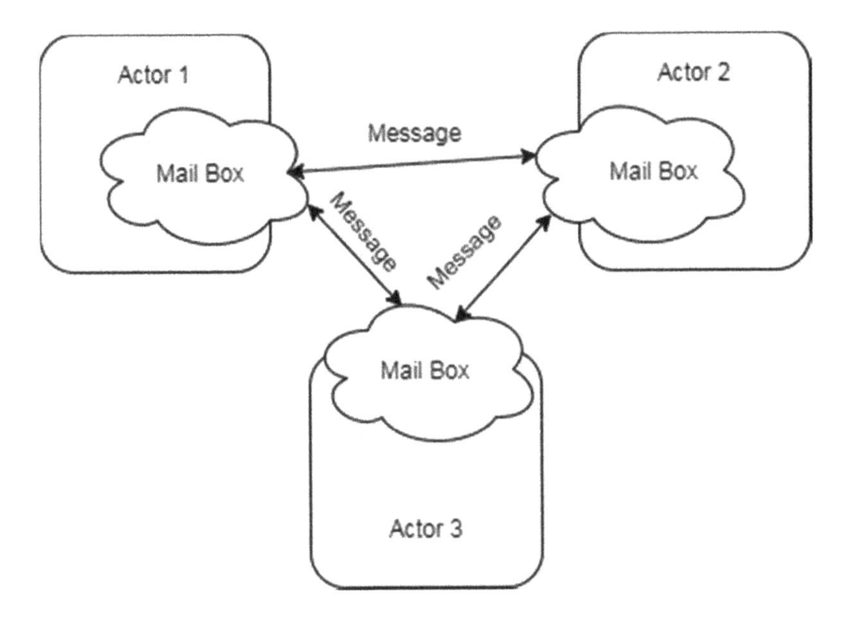

FIGURE 11.2 Actor communication.

- ActTalk
- ABCL
- Erlang
- Io
- Scala
- And many more

In this section we will describe certain languages used as testbeds for DAI.

A. ABCL/1: This is an actor-based object-oriented programming language developed in 1986 by Akinori Yonezawa
   - **Features of ABCL/1:**
     I. **Objects:** An object it is slightly different from that described in the actor model. Each object has its own processing power as well as local memory.
     An object is among the following three modes:
       i. **Dormant:** This is the initial state of an object.
       ii. **Active:** Whenever an object receives a message, it comes in active mode.
       iii. **Waiting:** When an object in an active state requires some specific message, it enters in its waiting mode.
     The working of objects can be easily understood with the help of Figure 11.3.
     An object in ABCL/1 is defined as:
     [object*object-name*
     (state*representation-of-local memory*)
     (script (=> message-pattern where constraint…..action….))]

Note: "where constraint" and object name are optional.

     II. Message Passing:
     Messages between actors are classed as point-to-point communication, and these messages cannot be broadcasted or multicast. To

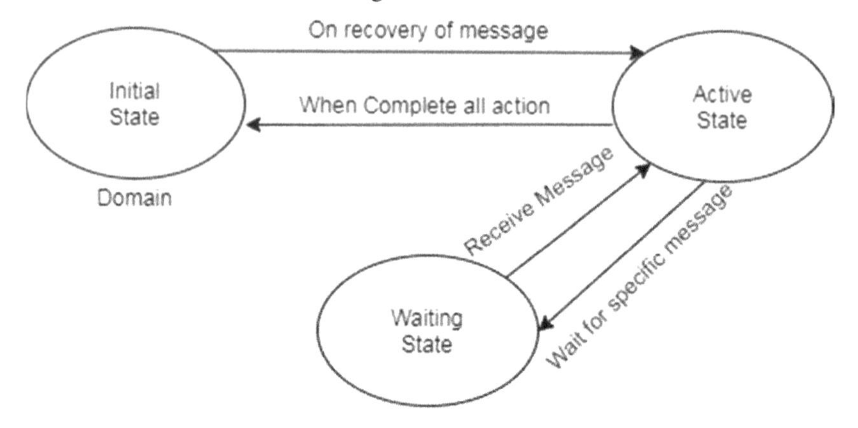

FIGURE 11.3 Objects working protocol.

send the message to the object, the sending object must know the receiving object's name as on forgetting the name by the sender object results in following an indirect message sending strategy. ABCL/1 has two modes of message passing:

i. Ordinary Mode: When a message(say M) is sent to an object(O) in the ordinary mode following conditions may arise:
    - If O is in dormant mode, then if message M is acceptable, O becomes active and starts performing else M is discarded.
    - If O is in active mode, then M is in the queue.
    - If O is in a waiting state, then M is checked for acceptability, if M is found acceptable, O is reactivated; otherwise, M is put in the queue.
ii. Express Mode: It works just like an interruption, i.e., it controls and monitors objects, whether dormant, active, or waiting.

- **Design Principles of ABCL/1 [10,11]:**
    I. Clear Semantics of Message Passing: Semantics should be clear and transparent among objects between whom message to be passed.
    II. Interobject message passing is based on an underlying object-oriented computation model.

B. ActTalk:

ActTalk is not a new language it is only actors embedded in Smaltalk-80. ActTalk is an actor model implementation so it must support concurrency. For this, it takes processes and semaphores provided by SmallTalk-80 and additionally actors taken are objects, classes, and messages [12]. ActTalk kernel is the basis that provides Object oriented Concurrency Programming semantics. Set of Kernel component classes makes a kernel. The basic classes are:

- **Actor class:** It defines an object's semantics.
- **Actor behavior class:** It defines an actor's behavior semantics.
- **Activity class**: It defines the internal activity of an active object

ActTalk design Principles [13]:

i. ActTalk is integrated within the programming environment of SmallTalk-80.
ii. ActTalk has a rich library of certain component classes which provides certain activity models, language constructs, communication between objects, synchronization/
iii. ActTalk provides a modular approach as the kernel has various sets of kernel components.
iv. ActTalk provides a very basic and standard semantics of object-oriented concurrency programming.

11.3.3 MICE Testbed

MICE (Michigan Intelligent Coordination Experiment) overcome the problem of coordination techniques in heterogeneous environments. The characteristics of various different environments can be simulated using MICE [14,15].

When multiple agents coordinate in a homogeneous environment, our previous testbeds provide a satisfactory result; however, when it comes to coordinate properly in a heterogeneous environment with certain characteristic features, these agents result in failure.

Our earlier languages like MACE, ACD, focus on agent communication with the environment or other agents even if it focuses on other tasks such as error handling, monitoring etc.

On the contrary, MICE is the future that focuses only on the environment in which the agent acts. As we know, real-time environments are dynamic, so a flexible approach needs to be developed to act realistic in dynamic environments.

11.3.4 ARCHON

ARCHON is a software framework that gives a platform to DAI application components for interaction. It also provides design methodology to structure these interactions. The Archon project applies to several real-world applications like electricity-transportation management systems, particle-accelerator control, electricity distribution, supply control of cement kiln complex, and control of a flexible robotic cell.

11.3.4.1 Multiagent Environment

In the ARCHON multiagent environment, an agent is the smallest entity for problem-solving purposes. Agents can handle their own problems as well as able to communicate with other agents in the community. Agent cooperation and communication with other community members is useful to improve the problem solution of individual and as well as overall application. The agent is a composition of Archon layer and application program also known as an intelligent system The process design involves multiagent environments and has two main phases: problem analysis and actual design. Documentation and identification of interactions between existing systems, existing systems, and their users and between different users. In the analysis phase for maintainability and attainable parallelism, an existing system might be split into several agents. The actual design phase considers both the existing systems and a new system. The existing systems and their users create a partially automated multi-agent system. By discovering all additional cooperative actions in the analysis phase and applying ARCHON architecture tools, designers extend and enhance the whole system.

Depending on its computer and software requirements, Originally ARCHON layer was Implemented using LISP; later extended to C++ and now running under UNIX on Sun4 architectures and under Linux on i386 machines

11.3.4.2 The ARCHON Architecture

ARCHON software has been used to integrate distinct applications of types in which agents will be loosely coupled and semi-autonomous. Because of loosely coupled agents, the count of interdependencies kept to a minimum between their respective ISs; control is decentralized as if agents are semi-autonomous. In an ARCHON community each agent has its own local authority to control its own IS and interactions with other agents, i.e., global authority does not exist. To achieve a global interrelated goal, agents interactions are required and those interactions are governed by the agent's AL. For example asking for information from people you know, asking for services from people you know, and volunteering information that you believe is relevant to others.

The agent needs to: control the operations within its local IS, determine when to interact with other agencies (where it needs to demonstrate the capabilities of it IS and IS's other agencies), and communicate with others. These basic requirements are incorporated into the architecture and execution of ARCHON (Figure 11.4). This construction task consists of four modules:

- Planning and Networking Module (PCM)
- Advanced Communication Level
- Agent Information Management (AIM)
- High- Level Communication(HLCM)

Every information system function is represented in the monitor unit (MU). The MU controls the local Information System. Many present a standard interface in Monitor or any other programmable management language and IS hardware. Planning and Networking Module (PCM) is an integral part of the ArCHON Layer, which discusses the agency's roles in a broader community of partnerships [16,17]. This

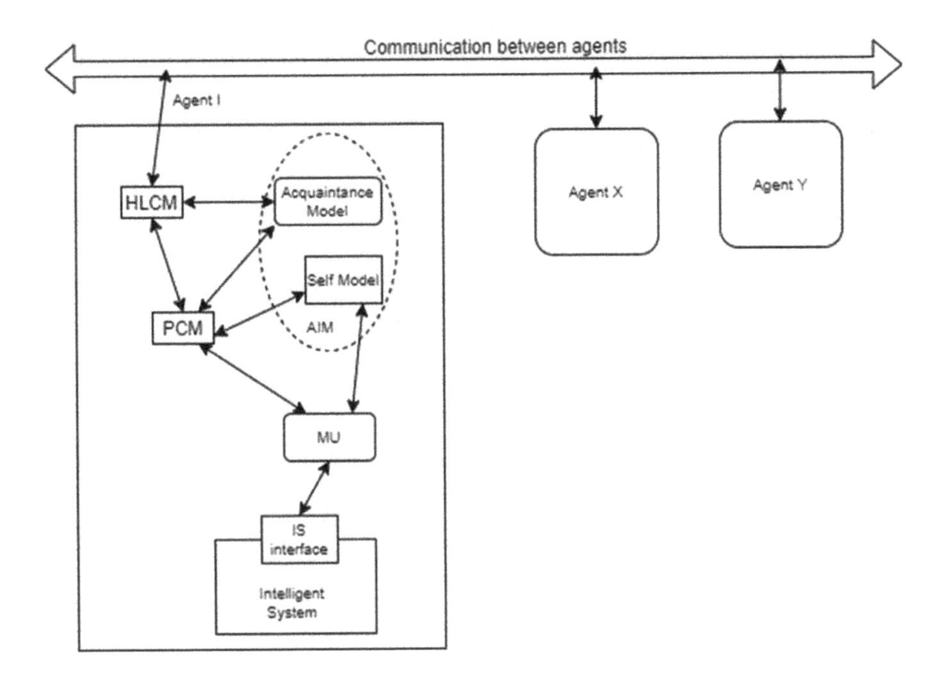

FIGURE 11.4 ARCHON architecture.

module should evaluate the agency's current status and determine what steps should be taken to facilitate collaboration with others while ensuring that the agency contributes to the community's welfare. The Agent Information Management (AIM) module is a distributed management system designed to provide information management services to affiliated organizations [18]. Within ARCHON, it is used to store agent models and domain level data. High-Level Communication (HLCM) allows agencies to communicate using resources according to the TCP / IP protocol. HLCM incorporates ISO / OSI Session Layer functionality that continuously monitors the communication environment and automatically adjusts connection breaks. Information can be sent to named agencies or relevant organizations.

Software and ARCHON have been used to develop real-time DAI strategies in many different industrial areas, including electric distribution and supply, distribution and distribution, cement control, particle acceleration control, dynamic cell control, and more.

11.3.5 Distributed Vehicle Monitoring Testbed (DVMST)

The distributed vehicle monitoring testbed produces a computation model of a distributed knowledge-based problem-solving system operating on a conceptual version of vehicle monitoring task. The working of basic DVMST is as follows:

1. The system contains several processing nodes.
2. Each node has a sound detecting sensor.
3. Communication between nodes is done via a radio communication network.
4. The vehicle's movement results in generating acoustic signals that are perceived by the noise detecting sensors.
5. To detect the vehicle's location, we can information from various sensors is correlated so that it results in accuracy.

DVMST is an appropriate tool to solve distributed problem-solving tasks due to the following reasons:

1. It collects a massive amount of real-time data via noise sensors and segregates it into an abstract map.
2. Information is incrementally segregated to a dynamic answer map.
3. Complexity is easy to calculate as it depends on vehicle numbers, movement patterns, and errors in collecting sensory data.
4. It results in a wide range of temporal (related to time), spatial (related to space), and functional networks (related to functionality of network) decomposition (Figure 11.5).

11.3.6 AGenDA Testbed

AGenDA testbed was developed by Deutsches Forschungszentrum für Künstliche Intelligenz (DFKI), which is a German Research Centre for Artificial Intelligence

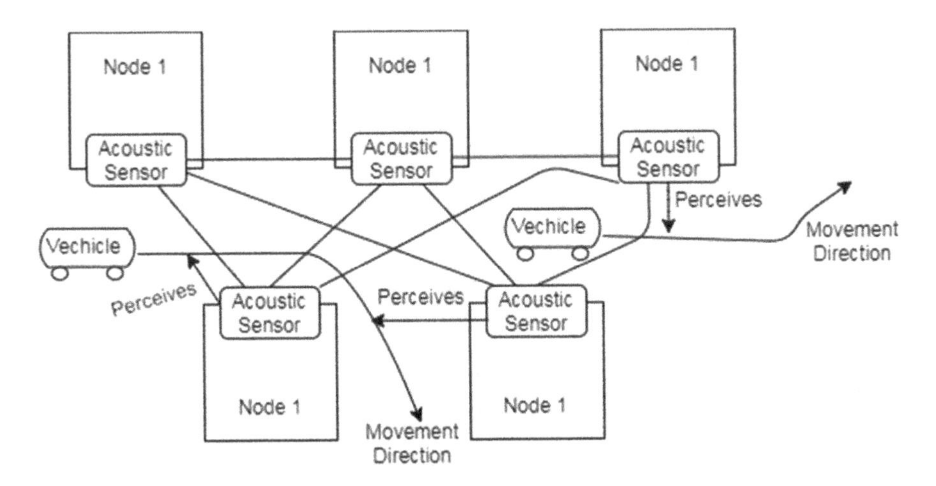

FIGURE 11.5 DVMT.

[19]. This agent focuses on how a group of agents in a system interacts with each other to achieve a common goal in a real-time environment. AGenDA testbed is designed in two levels:

1. Architectural Level
2. System Development Level

We will describe each of this level in detail:

11.3.6.1 Architectural Level

The architectural level in AGenDA testbed is designed by using InteRRap agent architecture. InteRRAP architecture aims to organize and structure the agents and provide the functionalities that help the agents interact in an unstable environment. InteRRAP architecture provides the following features:

- Agents should know how to react in an unexpected situation.
- Agents should work in a goal-oriented direction.
- Agents must solve any task in such a manner that they should be able to meet real-time constraints
- Agents should be able to manage with other existing agents

InteRRap defines a clear boundary between the knowledge base and functional parts, whereas it holds the hierarchical structure of the model. The major components of InteRRap are as shown in the diagram. The information flows between the agent control unit and the hierarchical knowledge base. The cooperation component of the control unit communicates the information from the cooperation component of the knowledge base. The plan-based component draws knowledge from local plans components of the knowledge base, while behavior-based component draws

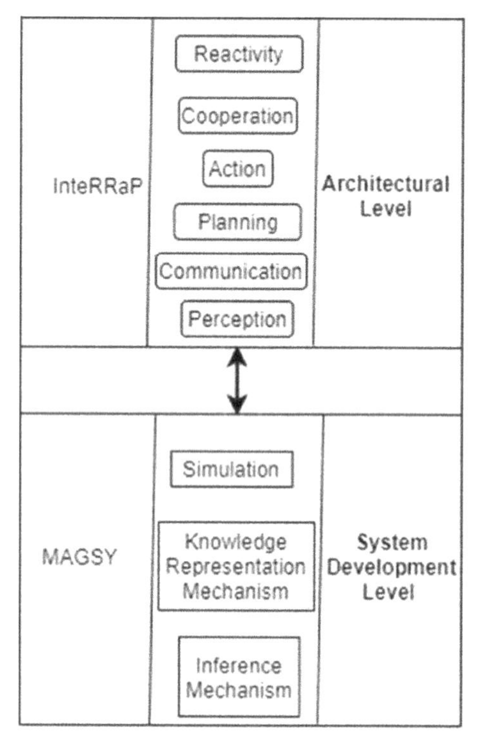

FIGURE 11.6 AGenDA.

knowledge from the patterns of behavior component of the knowledge base, and the fourth component, i.e. world interface component, draws knowledge from the work model component of the knowledge base (Figure 11.6).

11.3.6.2 System Development Level

This level is implemented using MAGSY System, which is a rule-based multiagent system.

It provides tools for construction, visualization as well as evaluation of DAI systems. In MAGSY an agent is described as (F,R,S), where

F = Facts that represent the agent's local knowledge
R = Set of rules
S = Set of services provided by the agent

11.3.6.3 Other Testbeds for DAI

1. MCS Testbed

   Multiple Agent Core System (MCS) is a research tool in DAI developed at the University of Essex.

MCS is PROLOG based and runs on UNIX. MCS is experimental-based rather conceptual. The experiments notably concern the connection between a "situated" action and action ensuing from a "predictive" or "strategic" designing in multiagent communities and the impact of adjusting the content of agents' social models on their behavior and collective effectiveness.

2. Phoenix Testbed

   Phoenix is a LISP-based testbed. It provides a flexible architecture for agents and provides communication facilities. Phoenix testbeds provide more than a translation facility.

3. CADDIE Testbed

   Control and Direction of Distributed Intelligent Agents (CADDIE). It is a C++ and PROLOLOG based testbed. It provides a system where agents, as well as communication between them, is described. It provides tools that define the scenario, and collects as well as analyzes data.

4. Concurrent METATEM

   The formal language of the DAI in which each agent is defined by expressing its desired functionality is a transient mental belie created directly to produce the agent's actual functioning.

5. Procedural Reasoning System (PRS)

   It is a common way of building an agency to try to marry goals from practical and meaningful plans; it has been used in many DAI experiments.

11.4 CONCLUSION

There are several testbeds available that solve a distributed approach to problem-solving. The main aim of the testbed is to provide multiagent communication in an inconsistent environment. Testbeds like MACE, ABCL, ActTAlk focuses on agent communication, whereas MICE is a testbed that focuses not only on communication but the environment in which agents communicate. Some testbeds focus on one application application-solving approach, such as DVMST which has its own advantages of effectively solving a particular problem, but at the same time, following such an approach would generate testbeds for N problems. In a real-life environment, things are unstable, so it is necessary to have such experimental implementations of DAI that can deal with the ever-changing nature of the environment and provide a controlled multiagent communication among agents and with the environment.

REFERENCES

1. Bond, H., and L. Gasser. "What is DAI." In Alan H. Bond, Leslie George Gasser (eds.), *Reading in Distributed Artificial Intellingence.* Morgan Kaufman, San Francisco, CA, 1988.
2. Durfee, Edmund H., Victor R. Lesser, and Daniel D. Corkill. "Trends in cooperative distributed problem solving." *IEEE Transactions on Knowledge & Data Engineering* 1 (1989): 63–83.

3. Gasser, Les. "DAI approaches to coordination." *Distributed Artificial Intelligence: Theory and Praxis* (1992): 31–51.
4. Konolige, Kurt, and Nils J. Nilsson. "Multiple-agent planning systems." *AAAI* 80: 138–142 (1980).
5. O'Hare, Greg M.P., and Nicholas R. Jennings, eds. *Foundations of Distributed Artificial Intelligence*, Vol. 9. John Wiley & Sons, London, UK, 1996.
6. Gasser, Les. "Large-scale concurrent computing in artificial intelligence research." *Proceedings of the 3rd Conference on Hypercube Concurrent Computers and Applications, Volume 2*, California, 1989.
7. Gasser, Les. "The 1985 workshop on distributed intelligence." *AI Magazine* 8(2): 91–97 (1987).
8. Heath, Michael T., ed. *Hypercube Multiprocessors, 1987: Proceedings of the Second Conference on Hypercube Multiprocessors, Knoxville, Tennessee, September 29–October 1, 1986*, Vol. 29. Siam, 1987.
9. Gasser, Les, Carl Braganza, and Nava Herman. "Implementing distributed AI systems using MACE." In Alan H. Bond, Leslie George Gasser (eds.), *Readings in Distributed Artificial Intelligence*. Morgan Kaufmann, San Francisco, CA, 1988, 445–450.
10. Briot, J.-P., and Jean de Ratuld. "Design of a distributed implementation of ABCL/I." *Proceedings of the 1988 ACM SIGPLAN Workshop on Object-Based Concurrent Programming*, New York, 1988.
11. Yonezawa, Akinori, Jean-Pierre Briot, and Etsuya Shibayama. "Object-oriented concurrent programming in ABCL/1." *ACM SIGPLAN Notices* 21.11 (1986): 258–268.
12. Briot, Jean-Pierre. "Actalk: a testbed for classifying and designing actor languages in the smalltalk-80 environment." *ECOOP* 89: 109–129 (1989).
13. Briot, Jean-Pierre. "Actalk: a framework for object-oriented concurrent programming-design and experience." *Object-Oriented Parallel and Distributed Computing II*—Proceedings of the 2nd France-Japan workshop. (1999): 209–231.
14. Durfee, Edmund H., and Thomas A. Montgomery. "MICE: a flexible testbed for intelligent coordination experiments." *Proceedings of the 9th Workshop on Distributed Artificial Intelligence*, Bellevue, Washington. 1989.
15. Montgomery, Thomas A., and Edmund H. Durfee. "Using MICE to study intelligent dynamic coordination." *Proceedings of the 2nd International IEEE Conference on Tools for Artificial Intelligence*. IEEE, Herndon, VA, 1990.
16. Jennings, Nicholas R."The ARCHON system and its applications." *2nd International Conference on Cooperating Knowledge Based Systems (CKBS-94)*. Keele, United Kingdom, 1994, 13–29.
17. Jennings, Nicholas R., and Jeff A. Pople."Design and implementation of ARCHON's coordination module." *Proceedings Workshop on Cooperating Knowledge Based Systems*, Keele, UK, (1993): 61–82.
18. Visser, A., et al. *Application Study: Robot Arm Control, the Outcome of Using Distributed AIM*. Technical Report Archon UvA/TN014/11–93, Department of Computer Systems, University of Amsterdam, 1993.
19. Bajaj, Sandeep, et al. *Virtual Internetwork Testbed: Status and Research Agenda*. Technical Report 98-678, University of Southern California, 1998.

12 Real-Time Framework Competitive Distributed Dilemma

Vijay Yadav, Raghuraj Singh, and Vibhash Yadav

CONTENTS

12.1 INTRODUCTION

Distributed issue elimination is the beneficial arrangement of decentralized problems and approximately combined assortment of information agencies (methods, sets of rules, and so on) situated during various particular processor hubs. The Kallman syndrome (KS)s participate with the assumption that nobody among them has adequate data to disentangle the entire problem; common knowledge exchange is essential to permit the gathering as a whole to flexibly an answer. By democratization, we imply that both checks and controls information is legitimately and once in a while geologically distributed; neither worldwide control nor worldwide information is stockpiling. Inexactly coupled methods, singular KSs spend a decent level of their time in calculation rather than correspondence. A Distributed Problem Solver offers better speed, unwavering quality, extensibility, the ability to deal with applications with a characteristic spatial dispersion, and in this way, the capacity to endure unknown data and information. Since such frameworks are exceptionally particular, they likewise offer applied clearness and straightforwardness of plan.

Although much work has been cleared out distributed handling, the greater part of the applications hasn't tended to issues that are significant for arranging artificially intelligent (AI) problem solvers. For example, most of the preparation is commonly done at a focal site with remote processors restricted to fundamental information assortment (e.g., Mastercard check). While it's not unexpected to appropriate information and preparing, rarely disperse control, and consequently, the processors don't collaborate considerably. Scientists inside the territory of distributed handling haven't taken problem-solving as their essential core interest. It has commonly been expected, for example, that an all-around characterized problem exists in which the primary concerns abide by an ideal static circulation of assignments, techniques for interconnecting processor hubs, asset designations, and counteraction of gridlock. Complete information on timing and priority relations between assignments has commonly been expected, and consequently, the significant purpose behind conveyance has been taken to be load adjusting (see, for example [1, 2]). Distributed problem solving, on the other hand, included as a piece of its essential purpose the parceling of the problem. The principal significant qualification between distributed problem solving and distributed preparing frameworks is frequently found by inspecting the starting point of the frameworks and, along these lines, the inspirations for interconnecting machines. Dis-tributed handling frameworks regularly have their root with an end goal to incorporate a system of machines equipped for finishing an assortment of broadly different assignments. A few specific applications are regularly imagined. Every application assembled at one hub (concerning occasion during a three-hub framework proposed to attempt to finance, request passage, and procedure control). The point is to search out how to accommodate any contentions and downsides emerging from the will to hold out different undertakings to understand the upsides of utilizing various machines (sharing of information bases, agile corruption, and so forth). Lamentably, the contentions that emerge are regularly specialized (e.g., word sizes and database groups) and incorporate sociological and political problems [33]. The arrangement to orchestrate an assortment of different errands brings about a need with issues like access control and security and prompts seeing participation as a kind of bargain between conceivably clashing viewpoints and wants at the degree of framework plan and design. In distributed problem solving, on the other hand, one assignment is imagined for the framework, and accordingly, the assets to be applied don't have any other predefined jobs to perform. A framework is made once more. Subsequently, the equipment and programming are frequently picked in light of one point: the decision that outcomes in the most powerful condition for helpful conduct. This additionally implies collaboration is seen as far as good problem-solving conduct; that is, by what means can frameworks that are ready to oblige each other be a proficient group? Our interests are in this way with creating systems for agreeable conduct between willing substances, rather than structures for implementing participation as a kind of bargain between conceivably contradictory elements.

This prompts us to inquire about the form of communication between collaborating hubs. We are principally worried about the substance of the information to be imparted among hubs and, accordingly the utilization of the data by a hub for helpful problem-solving. We are less worried about the exact structure during which the correspondence is affected.

In this paper, two sorts of collaboration in distributed problem solving are task-sharing and result-sharing. Inside the previous hubs help share the numerical data burden for the execution of subtasks of the general problem. Hubs help each other by sharing incomplete outcomes upheld to alternate points of view on the general problem. An alternate point of view emerges because the hubs utilize distinctive KS's (e.g., linguistic structure versus acoustics inside the instance of a discourse getting framework) or various information (e.g., information that is detected at various areas inside the instance of a distributed detecting framework). For each structure, the fundamental strategy is introduced, and frameworks in which it has been utilized are depicted. The utility of the two structures is analyzed, and their correlative nature is talked about.

The physical design of the issue solver isn't of essential enthusiasm here. It is thought to be a system of inexactly coupled, offbeat hubs. Every hub contains an assortment of unmistakable KS's. The hubs are interconnected, meaning that by sending messages, each hub can communicate with each other. Hubs don't exchange memories.

12.2 REAL-TIME ROUTE GUIDANCE DISTRIBUTED SYSTEM FRAMEWORK

A typical methodology for course direction imagines a focal controller with the ability to anticipate driver starting point goal (O-D) trip wants, to ideally appoint a way to each driver from start to finish, additionally on re-course as justified [4, 5]. The announced impediments of brought together frameworks incorporate the tremendous preparation and correspondence needs between the TMC and many clients one after another. Processing, stockpiling, and correspondence limits are required at the TMC. Therefore, the TMC should as much as possible be over-burdened [3]. Moreover, such frameworks were accounted for to have high framework working expenses [6].

Conversely, various leveled distributed models accommodate privately situated ongoing receptive methodologies for vehicle steering that accept restricted accessible data [7, 8]. In huge scope organizations, the need for quick control activity in light of nearby data sources and bothers unequivocally proposes the utilization of distributed data and control structures. While distributed frameworks are widely abused in zones like broadcast communications and figuring system control, recently distributed frameworks have been considered as a good reason for course direction in vehicle traffic systems. Hawas and Mahmassani [9] built up a non-helpful decentralized structure and a group of heuristic-based principles for responsive continuous course direction. This decentralized structure is the capacity to influence fluctuating degrees of information, spatially, and transiently.

In contrast to this methodology, it doesn't require from the earlier information (or expectation) of the time-subordinate OD request wants. This structure accepts a gathering of neighborhood controllers distributed over the system. Every nearby controller is at risk for giving receptive course direction to vehicles in its domain. The controllers are non-agreeable because they are not trading information on the traffic states in their particular domains. Nearby choice guidelines that join heuristic assessment capacities are indicated, reflecting shifting degrees of insight. The non-helpful decentralized design has been computationally proficient and genuinely powerful and successful under intermittent, likewise as occurrence circumstances [9]. The utilization of distributed multi-operator frameworks to improve dynamic course direction and traffic the executives is accounted for in Adler et al. [10]. Bury vehicular correspondence (IVC) systems give decentralized answers for traffic the board problems [11–14]. IVC systems are launches of versatile, spontaneous systems with no fixed foundation and rather accept common hubs to organize board capacities.

There are a few ITS activities upheld IVC systems. FleetNet [12] utilizes an IVC system to upgrade drivers' and travelers' security and extravagance. VGrid [11] proposes tackling vehicle traffic stream control issues self-governing. TrafficView [13] characterizes a structure to disperse and accumulate data about the vehicles bolstered by IVC.

During a traffic arrangement Hawas, Napeñas, and Hamdouch [7] created two calculations for inter-vehicular correspondence (IVC)-based course dires. Although the exhibition of such IVC-based calculations is sensible when contrasted with the incorporated frameworks, there are as yet numerous difficulties like the quick topology changes, the incessant discontinuities, and the little successful system distance across. Because of the high relative speed of vehicles, the IVC organize encounters exceptionally quick changes in topology. Likewise, because of the low organization of vehicles having IVC, the IVC arrangement is liable to visit fracture. At last, because of the poor availability, the successful system measurement is commonly small. These viewpoints force limitations whenever sent using IVC advancements. For instance, one should bargain the extra viability of getting more extensive scopes of correspondence against the conceivable debasement in execution on account of poor correspondence.

Remembering the gigantic preparation and high operational cost identified with the robust frameworks, the insecurity, and correspondence requirements identified with the IVC-based frameworks, this chapter looks to gracefully improve on the earlier work of Hawas and Mahmassni [8]. The improvement is intended to determine the announced cycling issues ordinarily experienced inside the average unadulterated circulated frameworks. The advancement is looked for through permitting data trade (or collaboration) among the changed decentralized controllers. In one case, we study the possibility of using controller correspondence to exchange data about the traffic conditions of their domains. Such improvement is expected to defeat the limitations of the quick topology changes, the successive discontinuity, and poor correspondence related to the IVC-based frameworks, likewise on account of the restrictions of the substantial handling and cost of the robust frameworks. The information trade would advance the information area of an individual controller, and possibly improve the standard of control by allowing using higher degrees of knowledge to upgrade the determination of the heuristic assessment capacities basic the nearby choice guidelines. This new framework will be meant during this paper by the unified decentralized framework.

12.3 EXPERTS COOPERATING

A recognizable allegory for a drag solver working during a disseminated processor might be a gathering of human specialists competent at cooperating, attempting to complete an outsized assignment. This analogy has been used in a few AI frameworks [15–18]. Of essential enthusiasm to us in looking at a gaggle of human specialists' activity is that the route during which they interface to disentangle the general issue, the path during which the outstanding task at hand is circulated among them, and how results are coordinated for correspondence outside the gathering.

It is expected that no one master is in all-out control of the others, albeit one master could likewise be at last liable for conveying the appropriate response of the top-level issue to the client outside the gathering. In such a circumstance, every master may invest a large portion of his energy working alone on different subtasks apportioned from the most assignment, stopping periodically to communicate with different individuals from the gathering. For the most part, these communications include demands for help on subtasks or the trading of results.

Singular specialists can help each other in at least two different ways. First, they will separate the remaining task at hand among themselves, and each hub can autonomously tackle some subproblems of the general issue. We call this assignment sharing (as in [16] and [18]). During this collaboration method, we are worried about how specialists conclude who will perform which task. We hypothesize that one intriguing technique for affecting this understanding is using arrangement.

A specialist (El) may demand help since he experiences an undertaking overlarge to deal with alone or an errand that he has no ability. On the off chance that the undertaking is excessively enormous, he will initially segment it into sensible subtasks and endeavor to discover different specialists who have the satisfactory abilities to deal with the new assignments. If the principal task is past his ability, he promptly endeavors to search out another progressively suitable master to deal with it.

In either case, if E1 knows which different specialists have the necessary skill, he can advise them straightforwardly. On the off chance that he doesn't know anybody in particular who could likewise be prepared to help him (or on the off chance that the errand requires no unique ability), at that point, he can depict the assignment to the entire gathering.

If another master (E2) accepts that he is equipped for finishing the errand that El portrayed, he educates El of his accessibility and possibly demonstrates any particularly pertinent aptitudes he may have. El may find a few such volunteers and may pick between them. The picked volunteer demands extra subtleties from El, and consequently, the two take part in further direct correspondence for the span of the errand.

Those with undertakings to be executed and individuals equipped for executing the assignments along these lines connect during a straightforward kind of exchange to disseminate the Exceptional task to hand. They powerfully structure subgroups as they advance toward an answer. The second kind of collaboration is appropriate when sub-problems can't be solved by independent experts working alone. The experts periodically report back to each other during this form the partial results they need to obtain during individual tasks. We call the sharing of findings (as in [17] and [19], for example). In this cooperation model, it is assumed that a priori problem partitioning was carried out in which individual experts worked on sub-problems with a point of commonality (e.g. interpreting data from overlapping image portions). An expert (El) reports to his neighbors (E2 and E3) a partial result for his sub-problem when that result may relate to their processing. (For instance, a partial result could be the best result E1 can obtain by using only the information and knowledge available to it.) E2 and E3 attempt (1) to use El's result to confirm or reject competing results for its sub-problems, or (2) to aggregate partial results of their own with El's outcome to provide a result relevant to El's subproblem and their own, or (3) use El's outcome to point out alternative lines of attack that they may wish to solve their sub-problems. Subgroups offer two advantages. To begin with, correspondence among the individuals doesn't unnecessarily divert the entire gathering. This is frequently significant because correspondence itself can be a genuine wellspring of interruption and trouble in enormous gatherings. Consequently, one of the primary motivations behind an association is to lessen the required amount of correspondence. Second, the subgroup individuals might have the option to talk with each other during a more productive language for their motivation than the language being used by the entire gathering.

12.4 A DISTRIBUTED PROBLEM-SOLVING PERSPECTIVE

This section proposes a model for the stages that an appropriated soluble goes through because it unravels a drag (Figure 12.1). The model comes with a system to stay the two sorts of collaboration that are the principal focal point of this chapter. It empowers us to find out the utility of the two structures, the sorts of issues that they're most appropriate, and along these lines the path during which they're correlative.

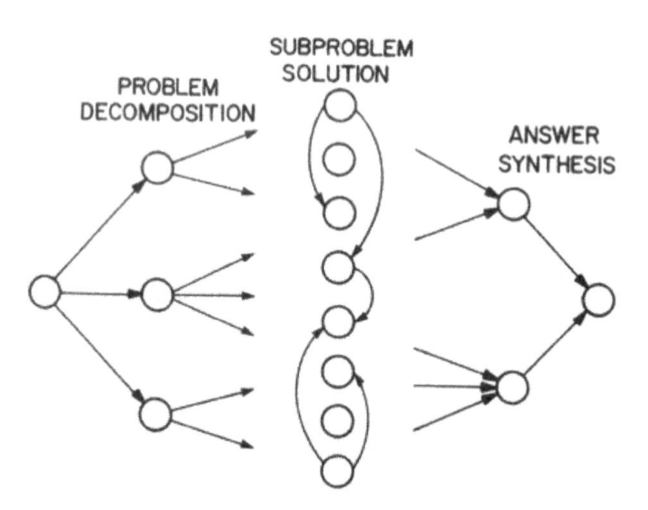

FIGURE 12.1 Distributed problem-solving phases.

In the principal stage, the issue deteriorates into subproblems. As Figure 12.1 shows, the disintegration procedure may include a chain of command of partitionings; also, the strategy may itself be appropriated to keep away from bottlenecks. Disintegration continues until part (nondecomposable) subproblems are created. Consider, for instance of a simple circulated detecting framework (DSS). Inside the difficult decay stage, the subproblems of distinguishing objects in explicit segments of the general region of intrigue are characterized and appropriated among the available sensors.

The subsequent stage includes the arrangement of the part subproblems. As appeared inside the figure, this may require correspondence and participation among the hubs to disentangle the individual subproblems. Inside the DSS model, correspondence is required inside the subproblem arrangement stage (1) if articles can move starting with one territory then onto the next, so it's useful for sensors to advise their neighbors regarding the development of items they need recognizing, or (2) if it's hard for one sensor to dependably distinguish objects without help from different sensors.

Answer amalgamation is performed inside the third stage; that is, a combination of subproblems results in understanding an answer for the general issue. Like issue decay, answer amalgamation could likewise be various leveled and circulated. In the DSS model, the arrangement combination stage includes the age of a guide of the articles inside the general zone of intrigue. For some unexpected issues, the three stages may fluctuate in intricacy and significance. A few stages may either be absent or paltry. For example, inside the traffic-light control issue considered in [19], the issue deterioration stage includes no calculation. Traffic-light controllers are just positioned at every crossing point. For a DSS, the issue disintegration is normally suggested straightforwardly by the spatial circulation of the issue.

There is likewise no answer amalgamation stage for traffic-light control. The response to a piece subproblem might be a smooth traffic progression through the related crossing point. There's no opportunity to combine a general guide of the traffic. Accordingly, the response to the general issue is obtained via the answer to the subproblems. (This is normally valid for control issues; note that it does not mean, in any case, that correspondence among the hubs comprehending individual subproblems is not required.)

Many pursuit issues (like emblematic joining [20]) likewise include an insignificant answer combination stage. When the issue has been decayed into piece subproblems and illuminated, the sole answer amalgamation required is the restatement of the rundown of steps that are followed to find the solution. In any case, for a couple of issues, the arrangement union stage is the predominant stage. A model is that the COGEN program [21]. COGEN is used in atomic structure explanation. It produces every auxiliary isomer that is both per a given equation, which incorporates basic sections known to be available inside the substance (superatoms). Inside the difficult disintegration stage, COGEN produces all structures that are as per the data (by first creating intermediates structures, decaying those structures, at that point on until just structures containing iotas or super-particles remain). The superatoms (like the molecules) are considered by name and valence as it were. Inside the appropriate response combination stage, the superatoms are supplanted by the specific auxiliary pieces they speak to and installed inside the created structures. Since installing can frequently be cleared out somehow or another, this stage represents an enormous bit of the general calculation. It will be clear that the model is furthermore pertinent to bring together critical thinking. Notwithstanding, the unmistakable stages are progressively clear during a dispersed solver, principally because correspondence and participation must be tended to unequivocally during this case.

12.5 CAVEATS FOR COOPERATION

Probably the most point of receiving a disseminated approach is to acknowledge fast critical thinking. To do this, circumstances during which processors "get in one another's way" must be maintained from a strategic distance. This relies upon the issue itself (e.g., there are issues that information or calculation can't be parceled into enough for the most part autonomous pieces to involve the entirety of the processors). The exhibition likewise depends, be that as it may, on the critical thinking design. It is, in this way, suitable to consider structures for participation. Note that the issue solver should, in any case, execute even an absolute disintegration. Hubs should even now go to a concession to which hub deals with which part of the general region. It is regular in AI issue solvers to parcel skill into area explicit KS's, every one of which is master during a specific a piece of the general issue. KS's are normally framed exactly because of assessing contrasting sorts of information that will be conveyed to endure on a particular issue. For instance, in our example, in a discourse on the signal of the discourse itself, from the linguistic structure of the articulations from the semantics of the errand area [22]. The decisions about which KS's are to be shaped are typically made along with the arrangement of a progressive system of information reflection levels for a drag. For example, the sum used in the chain of command of the HEARSAY-II discourse understanding framework was parametric, segmental, phonetic, surface-phonemic, syllabic, lexical, phrasal, and theoretical [22]. KS's are normally picked to deal with information at one degree of deliberation or to connect two levels (see, for example, [22] and [23]).

Associations among the KS's during a conveyed processor are costlier than during a uniprocessor because correspondence during an appropriated engineering is typically much slower than calculation. The system for participation must, along these lines, limit correspondence among processors. The accessible correspondence channels could likewise be soaked all together that hubs are compelled to remain inert while messages are transmitted.

As a simple case of the issue that over the top correspondence can cause, consider a conveyed processor with 100 hubs interconnected with one communication channel. Accept that all of the hubs work at 108 guidelines for every second; the calculation and correspondence load is shared similarly by all hubs. Subsequently, the critical thinking engineering is with the end goal that every hub must impart the slightest bit for each of the ten directions that it executes. With these parameters, it's promptly indicated that the interchanges channel must have a transmission capacity of at any rate 1 Gbit/s (in any event, disregarding the impact of conflict for the channel) [18]. With a little data transfer capacity, processors are compelled to confront inactive anticipating messages.

There are, obviously, numerous structures that don't prompt channel transfer speeds of a similar extent. In any case, the reason remains that unique consideration must be paid to internode correspondence and control in circulated critical thinking if huge quantities of quick processors are to be associated.

The participation system should likewise disperse the preparing load among the hubs to keep away from calculation and correspondence bottlenecks. By and large, execution could likewise be restricted by the grouping of unbalanced measures of calculation or correspondence at a number of processors. It is additionally the situation that the control of preparing must itself be appropriated. Something else, demands for choices about what to attempt to next could in time. The attention here is on speed yet the contrary purposes behind embracing a dispersed methodology likewise are significant—for example, dependability (i.e., the possibility to get over the disappointment of individual segments, with agile corruption in execution) and extensibility (i.e., the ability to change the quantity of processors applied to an issue). amass at a "controller" hub quicker than they may be handled. Dissemination of control does, notwithstanding, cause troubles in accomplishing internationally intelligible conduct since control choices are made by singular hubs without the benefit of a general perspective on the issue.

12.6 TASK SHARING

Assignment sharing is a type of participation where singular hubs help each other by sharing the computational burden for the execution of subtasks of the general issue. Control in frameworks that utilization task-sharing is normally objective coordinated; that is, the handling done by singular hubs is coordinated to accomplish subgoals whose outcomes can be incorporated to take care of the general issue.

Assignment sharing is indicated schematically in Figure 12.2. The individual hubs are spoken to by the undertakings in whose execution they are locked in.

The key issue to be settled in task-sharing is how assignments are circulated among the processor hubs. There must be a method whereby hubs with undertakings to be executed can locate the most suitable inactive hubs to execute those assignments. We call this the association issue. Tackling the association issue is critical to keeping up the focal point of the difficult solver. This is particularly obvious in

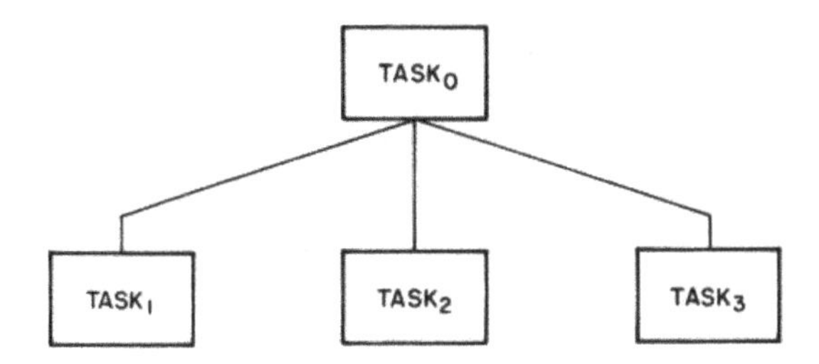

FIGURE 12.2 Sharing the tasks.

AI applications since they don't, for the most part, have completely characterized calculations for their answer. The most fitting KS to conjure for the execution of some random undertaking by and large can't be recognized earlier. There are, for the most part, numerous prospects to attempt every one of them. In the rest of this area, we consider exchange as an instrument that can be utilized to structure hub collaborations and tackle the association issue in task-shared frameworks. Arrangement is recommended by the perception that the association issue can likewise be seen from an inactive hub. It must discover another hub with an appropriate undertaking that is accessible for execution. So as to boost framework simultaneousness, the two hubs with assignments to be executed and hubs prepared to execute errands can continue all the while, connecting each other in a procedure that takes after agreement arrangement to take care of the association issue.

In the agreement net way to deal with arrangement [18, 24], an agreement is a specific understanding between a hub that produces an errand (the director) and a hub ready to execute the assignment (the temporary worker). The administrator is liable for checking the execution of an assignment and preparing the aftereffects of its execution. The temporary worker is at risk for the specific execution of the errand. Singular hubs aren't assigned from the earlier as administrator or temporary worker; these are just jobs, and any hub can battle either job powerfully throughout critical thinking. Hubs are subsequently not statically attached to an effective order.

An agreement is built up by a nearby common determination procedure that upheld a two-path move of information. To sum up, the chief for an undertaking declares the existence of the assignment in various hubs with an errand declaration message (Figure 12.3). Accessible hubs (potential temporary workers) assess task declarations made by a few supervisors (Figure 12.4) and submit offers on those that they're fit (Figure 12.5). A private supervisor assesses the offers and grants contracts for the execution of the assignment to the hubs it decides to be generally suitable (Figure 12.6). Hence, the chief and temporary worker are connected by an agreement (Figure 12.7) and convey secretly while the agreement is being executed.

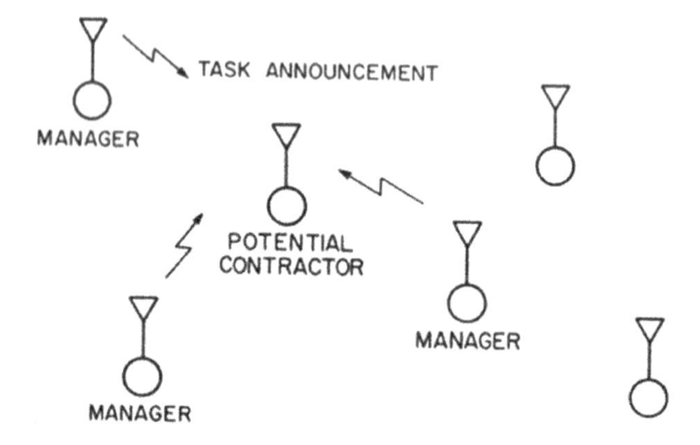

FIGURE 12.3 Submit a job notification.

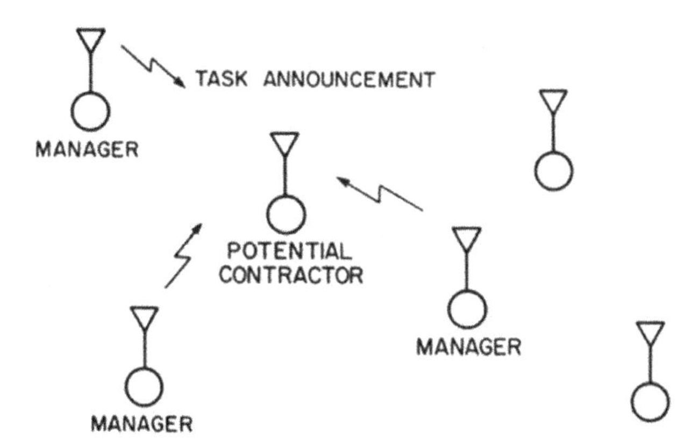

FIGURE 12.4 Receiving mission alerts.

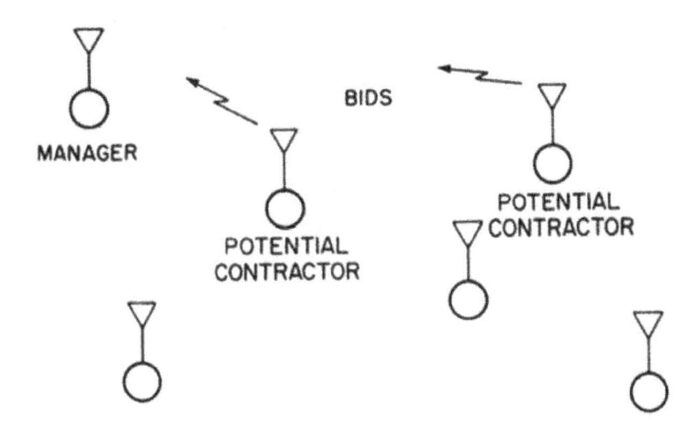

FIGURE 12.5 Qualify.

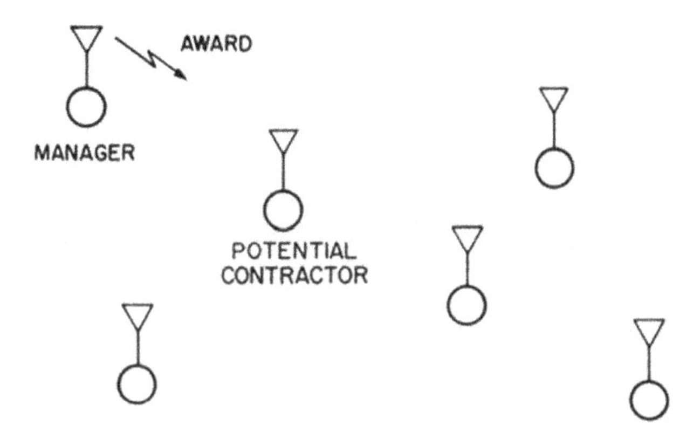

FIGURE 12.6 Making a medal.

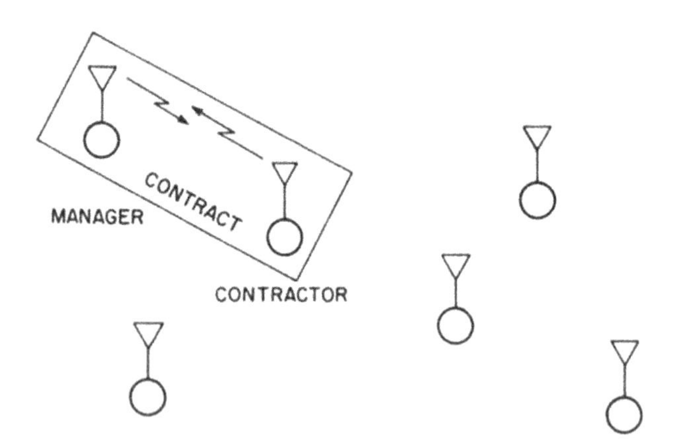

FIGURE 12.7 Linkage between manager and contractor.

The exchange procedure may then repeat. A temporary worker may additionally parcel an assignment and grant agreements to different hubs. It is then the administrator for those agreements. This results in the various leveled control structure that is regular of assignment sharing. Control is appropriated in light of the fact that preparing and correspondence aren't engaged at specific hubs, but instead each hub is fit for tolerating and doling out undertakings. This maintains a strategic distance from bottlenecks that would corrupt execution. It likewise improves dependability and grants elegant debasement of execution inside the instance of personal hub disappointments. There are no hubs whose disappointment can totally hinder the agreement arrangement process.

We have just quickly outlined the exchange procedure. A few inconveniences emerge in its execution. An assortment of augmentations to the basic technique exists that empower productive treatment of particular associations where the total multifaceted nature isn't required (e.g., when straightforward solicitations for data are made). See [24] for work.

Next is a case of arrangement for an undertaking that includes the social event of detected information and extraction of sign highlights. This is taken from a reenactment of an appropriated detecting framework (DSS) [25]. The detecting issue is apportioned into an assortment of errands. We will think about one among these errands, the sign assignment, that emerges during the instatement period of DSS activity.

The supervisors for this errand are hubs that don't have detecting abilities yet have broad handling capacities. They intend to discover a gathering of sensor hubs to gracefully them with signal highlights. On the contrary hand, the sensor hubs have constrained handling abilities and endeavor to search out chiefs, which will additionally process the sign highlights they remove from the crude detected information.

Review that we see hub communication as an understanding between a hub with an undertaking to be performed and a hub fit for playing out that task. In some cases, the disposition on the ideal character of that understanding varies depending on the member's perspective. For example, from the sign assignment administrators' demeanor, the least complex arrangement of temporary workers has a satisfactory spatial appropriation about the including region and sufficient circulation of sensor types. From the motivation behind perspective on the potential sign assignment contractual workers, on the contrary hand, the least difficult administrators are those nearest to them, in order to constrict potential correspondence issues.

Each message type inside the agreement net convention has spaces for task-explicit data. The spaces are picked to catch the sorts of information conveniently passed between hubs to work out fitting associations without excessive correspondence. for example, signal assignment declarations incorporate the resulting openings.

1) An undertaking reflection opening is packed with the assignment type and this way the situation of the administrator. This allows a potential temporary worker to work out the director to which it ought to react.
2) The qualification detail space substance demonstrates that bidders must have detecting capacities and must be situated inside a similar region due to the director. This lessens superfluous message traffic and offers preparing by expressly determining the traits of a contractual worker that are regarded as fundamental by the supervisor.
3) The opening of the offer determination indicates that the bidder must indicate its position and the name and form of its sensors along these lines. This diminishes the length of offer messages by indicating the information that a chief must choose a proper arrangement of temporary workers.

The potential contractual workers hear the undertaking declaration from the shifted administrators. In the event that qualified, they react to the nearest administrator with an offer that contains the data spread out in the errand declaration. The chiefs utilize this data to pick a gathering of bidders at that point grant signal agreements. The honor messages determine the sensors that a temporary worker must use to gracefully flag highlight information to its administrator.

Utilization of the agreement net convention during a DSS makes it feasible for the sensor framework to be arranged progressively, thinking about such factors in light of the quantity of sensor and processor hubs accessible, their areas, and along these lines the simplicity with which correspondence is regularly settled.

Arrangement offers a more remarkable instrument for association than is out there in current critical thinking frameworks. The association that is influenced by the agreement net convention is an augmentation to the example coordinated conjuring used in numerous AI programming dialects (see [26] for an inside and out conversation).

It is most helpful when errands require specific KS's to the point that the satisfactory KS's for a given assignment aren't known from earlier, and when the undertakings are sufficiently enormous to legitimize a more significant exchange of data before conjuring than is typically permitted in problem solvers.

12.7 RESULT-SHARING

Result-sharing might be a kind of collaboration during which singular hubs help each other by sharing fractional outcomes, bolstered to some degree alternate points of view on the general issue. In frameworks that utilization result-sharing, control is normally information coordinated; that is, the calculation done at any moment by a private hub relies upon the data accessible, either locally or from remote hubs. A specific progression of assignment subtask connections doesn't exist between singular hubs. Result-sharing is demonstrated schematically in Figure 12.8. The individual hubs are spoken to by KS's.

A basic case of the usage of result-sharing is the improvement of predictable labelings for "squares world" pictures [27]. A squares world picture might be an outline that shows the sides of a lot of clear articles (e.g., 3D shapes, wedges, and pyramids) during a scene. Each picture is spoken to as a diagram with hubs that relate to the vertices of the articles inside the picture and circular segments that compare to the sides that interface the vertices. The objective is to decide a correspondence among hubs and curves inside the chart and real articles.

A truly feasible vertex is regularly given a gathering of from the earlier potential names upheld the quantity of lines that meet at the vertex and in this manner the points between the lines (e.g., "L," "T," "Bolt," and "FORK") [27]. A vertex is additionally particular by the character of the lines that form it (e.g., a line can characterize an arched limit between surfaces of an item, a limit among light and shadow, at that point on). Such labeling is defined by looking at the vertices of separation. Vagueness exists in view of the fact that for each vertex, for the most part, only one name can be accomplished. Notwithstanding, the quantity of potential marks is regularly decreased (frequently to one name) by considering the limitations forced by the collaborations between vertices that offer edges in an article. just a couple of the gigantic number of combinatorically conceivable vertex types can share a solid footing during a truly feasible item. Along these lines, the way to accomplish predictable picture marking is to coordinate each vertex's name set with those of its neighbors and dispose of conflicting names.

If we parcel the issue all together that a processor hub is obligated for one vertex inside the picture, at that point the basic outcome sharing procedure is clear. Hubs impart their nearby name sets to their neighbors. Every hub utilizes these remote mark sets close by consistency conditions to prune its own name set. It at that point transmits the new name set to its neighbors. the strategy proceeds until special marks

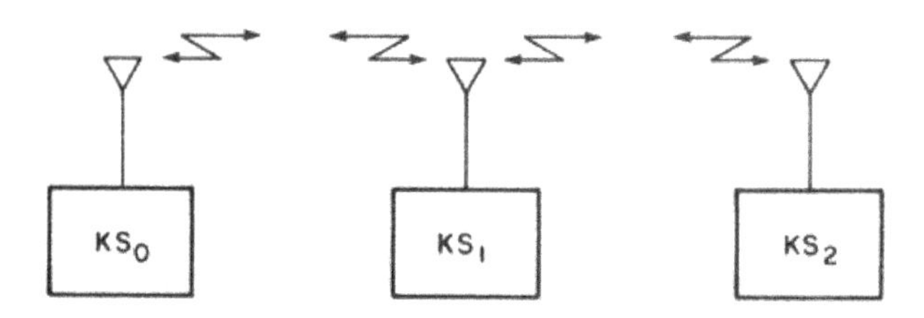

FIGURE 12.8 Outcome-sharing.

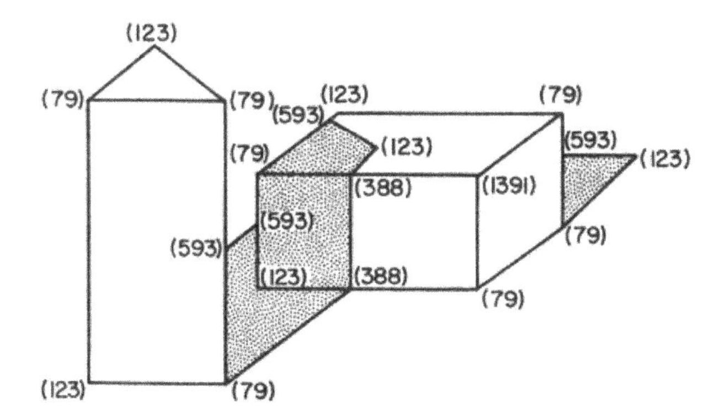

FIGURE 12.9 Sample blocks world issues (from [28]).

are set up for all hubs or no further pruning is doable. (This procedure of iterative refinement of name sets is named unwinding, requirement proliferation, or range restriction.)

Figure 12.9 shows a simple picture considered by Waltz. The numbers appeared in enclosures adjacent to every vertex show the measure of from the earlier labelings workable for that vertex, inside the nonappearance of any inward vertex constraints.

Notwithstanding the inconsistency that emerges in pictures from the squares world, genuine pictures additionally experience the ill effects of the vagueness that emerges because of boisterous information and incorrect element locators. The picture is again viewed as a chart; however, during this case, the hubs relate to little locales of the picture [29, 30]. Test marks during this setting are line sections with indicated direction, or articles (e.g., entryways, seats, and wastebaskets).

Similarly, as with the squares world issue, the point is to decide one of a kind names for each hub by thinking about logical data from adjoining hubs. During this case, no outright requirements are conceivable. Rather, the limitations (or compatibilities as they need likewise been called) express a level of conviction that the marks identified with neighboring hubs are reliable (e.g., a line section with a particular direction recognized at one hub includes a high level of similarity with a different line portion with a proportional direction distinguished at an adjoining hub).

The strategy is instated by partner a gathering of marks with every hub on the possibility of nearby component recognition. A numerical assurance measure is also allotted to each name. As in the past, hubs at that point convey their nearby name sets to their neighbors. Instead of pruning its mark set, every hub utilizes these remote name sets to refresh the information estimates identified with the names in its own name set. Though an elevated level convention has been created to encourage task-sharing [24], no closely resembling convention has risen up out of research on the outcome sharing. We are by and by inspecting the structure of correspondence for result-imparting to a view to stretching out the agreement net convention to raised consolidate it. The three-step dance tackled this issue by utilizing a brought together calculation that thought about only each vertex in turn. The calculation required 80 emphases to gracefully particular marking for this picture

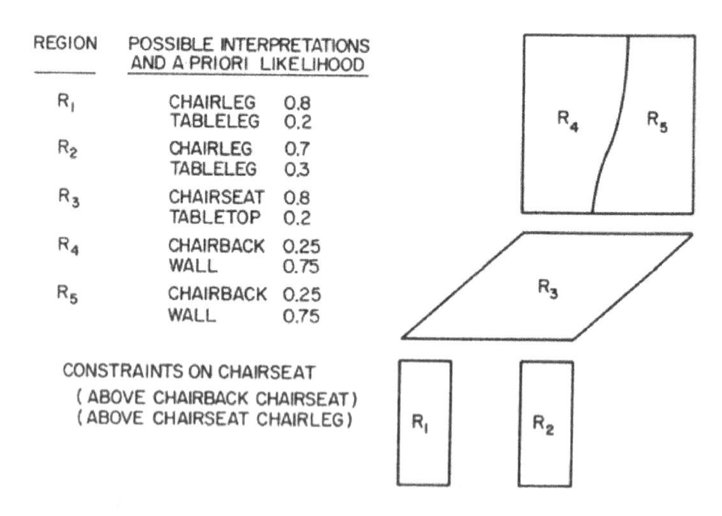

FIGURE 12.10 MSYS: reference debate.

The refreshing is finished on the possibility of the communications among marks portrayed above and reinforces or debilitates the information measure for each name. This procedure proceeds until one-of-a-kind marks are set up for all hubs (i.e., one name at every hub includes an enormous conviction measure regarding those identified with the contrary names for that hub) or no further refreshing is practical.

Figure 12.10 shows an example assortment of districts of the sort considered by MSYS [29]. for each locale, potential understandings and they are from the earlier probabilities are appeared. Additionally, demonstrated are the imperatives set on any area that will be deciphered as a "seat." These limitations increment the information that the social occasion of locales ought to be deciphered as a seat.

Lesser and Erman [31] have explored different avenues regarding a conveyance of the HEARSAY-II discourse understanding framework [17]. Appropriation has been affected by dividing every expression into portions covering in time and doling out each section to a hub. Figure 12.11 shows the structure of the division that has been executed.

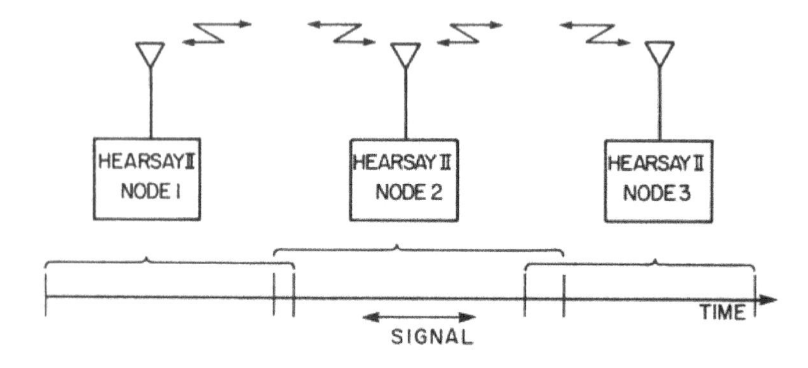

FIGURE 12.11 Interpretation distributed: Segmentation.

Every hub endeavors to build up an understanding regarding the data that it has received. It does this by making halfway translations or theories and testing them for believability at each phase of the preparation. (This is the exemplary AI worldview of theorizes and test.) An answer is developed through the steady collection of commonly compelling or fortifying fractional arrangements while conflicting incomplete arrangements cease to exist.

Provisional choices are made on the possibility of incomplete data at that point reexamined when additional data opens up (either inside the kind of more information or inside the kind of halfway translations got from different hubs). The requirement in the discourse understanding space is reachable by the requirement for consistency of translation of the covering portions and by the syntactic and semantic limitations that one piece of an expression may put on another part.

The techniques utilized by disseminated HEARSAY-II are practically similar to those utilized by the picture naming frameworks. Progressive refinement of theories is affected in a way practically like the refreshing of name sets. Be that as it may, the picture marking frameworks accomplish participation exclusively by shared limitation or limitation on the outcomes accomplished by singular hubs. Disseminated HEARSAY-II adopts an increasingly broad strategy. It accomplishes collaboration by both shared limitations and by common collection of results accomplished by singular hubs (i.e., fractional understandings accomplished at neighboring hubs are joined to make increasingly finish translations).

In beginning tests, the outcome sharing methodology in dispersed HEARSAY-II has exhibited a fascinating capacity to influence equivocalness and vulnerability in information and information. during a minor departure from the quality hunt versus information exchange off, the outcome sharing methodology recommends an accumulation versus information exchange: amassing fractional, vague arrangements can here and there be a lot simpler than endeavoring to flexibly one, complete and definite arrangement, and ought to really end in basically no loss of precision.

12.8 TASK-SHARING AND RESULT-SHARING: A COMPARATIVE ANALYSIS

Errand sharing is utilized to orchestrate disintegration by developing an unequivocal assignment subtask associations between hubs. The resultant pecking order is furthermore helpful as a method of organizing answer amalgamation. Assignment sharing accepts that piece subproblems are regularly understood by singular hubs working autonomously with negligible internode correspondence. The primary concern is the effective coordinating of hubs and undertakings for rapid critical thinking. It is generally valuable for issue spaces during which it is proper to characterize a progressive system of assignments (e.g., heuristic pursuit) or levels of information deliberation (e.g., sound or video signal understanding). Such issues loan themselves to deterioration into a gathering of generally free subtasks with no utilization for worldwide data or synchronization. Individual subtasks are frequently allocated to isolate processor hubs; these hubs would then be able to execute the subtasks with no utilization for correspondence with different hubs. If so, at that point task-sharing might be an adequately amazing kind of collaboration to deal with every one of the three periods of circulated critical thinking. During an outcome sharing framework, hubs are confronted with an association issue similar to thereto portrayed for task-sharing frameworks. inside the outcome sharing case, a hub must choose, from among all outcomes produced, the genuine outcomes to be transmitted, likewise as the contrary hubs to which they're to be transmitted. Endless supply of an outcome, a hub must choose whether or to not acknowledge it, and what activity to require upheld the got outcome. Besides, by and large we can't accept that a hub will discuss just with its neighbors. this can block the probability of tackling issues that include nonlocal cooperation between subproblems (e.g., in spite of the fact that shadow areas in squares world pictures may not be adjoining, they should be as per regard to their connection to the wellspring of enlightenment). the issue is that we should recognize physical contiguousness inside the correspondence organize and causal or data sway nearness. Result-sharing is utilized to encourage subproblem arrangement when part subproblems are such that they can't be illuminated by singular hubs working autonomously without huge correspondence with different hubs. Result-sharing offers no component for issue decay. Consequently, it must be utilized alone as a kind of participation for issues during which issue deterioration and conveyance of subproblems to singular hubs are taken care of by an operator outside to the circulated solver. The outcome sharing offers a negligible system for answer blend. It is helpful during this reference to the degree that a proportionate outcome sharing component is regularly utilized for by and large answer amalgamation likewise as a subproblem arrangement.

Result-sharing is generally helpful in issue spaces during which (1) results accomplished by one hub impact or oblige individuals who are regularly accomplished by another hub (i.e., the outcomes are applicable to each other), (2) sharing of results drives the framework to unite to a response to the issue (i.e., results obtained from remote hubs don't cause wavering), and (3) sharing of results drives the framework to an exact answer for the issue.

Minimization of internode correspondence is significant for both the undertaking sharing and result-sharing sorts of participation because of the calculation/correspondence speed unevenness in conveyed processors. The agreement net convention utilizes components simply like the qualification particular crush task declarations to downsize superfluous offer messages, while disseminated HEARSAY-II utilizes an assortment of intriguing instruments to restrict the quantity of speculations conveyed between hubs. For example, one technique is just to consider the transmission of results that no further refinement or expansion is attainable through nearby preparing. (This kind of result has been classified as "locally complete" [31].)

It has recently been expressed [19] that the significant preferred position of result-sharing is its resilience to vulnerability. Notwithstanding, it's intriguing to note that task-sharing can even be wont to accomplish resilience to vulnerability. Consider, for example, an application during which three hubs attempt to understand a uniform translation of information that is taken from covering parts of an image. during an outcome sharing methodology, they intend to accomplish agreement by imparting incomplete translations of the data. during an errand sharing methodology, the three hubs each process their own piece of the information again, instead of imparting their incomplete translations to each other, they convey them to a fourth hub (an administrator in contract net terms) that has the undertaking of looking at the irregularities. This hub occasionally retasks the three different hubs, utilizing the premier current information and halfway understandings. There still remain the issues of choosing when to end critical thinking movement and choosing which hub will answer the client outside the gathering. inside the disseminated HEARSAY-II framework, all hubs will determine a translation for the whole expression in the long run. this may be satisfactory for a three-hub framework yet will cause an unsuitable measure of correspondence for a greater framework. This short model brings out a genuine contrast between the 2 methodologies, to be specific, that outcome sharing might be a more understood kind of collaboration than task-sharing. Collaboration and assembly are accomplished via cautious plan of individual KS's to utilize results gotten from remote KS's. Assignment sharing, on the contrary hand, makes the participation express by setting up formal lines of correspondence and embeddings hubs whose particular errand is to incorporate the fractional translations from the hubs that work the specific information.

The model additionally outlines one among the primary unsolved issues in disseminated critical thinking—the best approach to accomplish reasonable conduct with a framework during which control is dispersed among an assortment of independent hubs. When the quantity of errands or results that would be prepared surpasses the quantity of realistic hubs, hubs with assignments or results to share must seek the eye and assets of the gathering.

On account of undertaking sharing, instruments must be structured that gives some affirmation that individual subproblems are truly handled, that processors don't get in one another's way in attempting to unwind indistinguishable subproblems while different subproblems are unintentionally overlooked. Also, it's fundamental that the subproblems that in the end influence answers for be prepared in inclination to subproblems that don't cause arrangements. We have proposed arrangement as a system for taking care of these troubles and have structured the agreement net convention in view of them [19]. In any case, it's evident that much work remains to be done.

On account of result-sharing, there must be some affirmation that hubs impact each other in such how to merge to an exact arrangement. indeed, even as incomplete outcomes got from an outside hub can recommend productive new lines of assault for a drag, they will even be diverting. In ongoing work on result-sharing frameworks, it has been seen that conviction measures produced at various hubs are often especially hard to incorporate. In conveyed HEARSAY-II, for example, it had been discovered that conviction estimates used in an incorporated methodology aren't really proper for a dispersed definition. The impact was that remotely created outcomes sometimes made hubs seek after lines of assault that weren't as productive because the ones that they had been seeking after before receipt of those outcomes. Some proof of this wonder can be induced from the analyses that were performed during which a few outcomes were lost in transmission. Sometimes, framework execution really improved, a sign that hubs a few times occupy their neighbors. again, much work stays to be cleared out of this region. In association hypothetical terms, the fourth hubs complete an "incorporating job." Hierarchical control of this sort might be a standard component utilized by human associations to influence vulnerability [11, 12]. inside the agreement net methodology, the chiefs for undertakings are inside the best situation to perform such obligations.

12.9 CONCLUSION

Two correlative sorts of participation in conveyed critical thinking are examined: task-sharing and result-sharing. These structures are valuable for different sorts of issues and for different periods of circulated critical thinking. Undertaking sharing is advantageous inside the difficult decay and answers combination periods of circulated critical thinking. It accepts that the subproblem arrangement can be accomplished with negligible correspondence between hubs. Result-sharing helps the subproblem arrangement stage when part subproblems can't be understood by hubs working autonomously without correspondence with different hubs. It is additionally useful somewhat inside the appropriate response union stage—particularly for issues during which the arrangement blend stage is a continuation of the subproblem arrangement stage. In the long run, we hope to determine frameworks during which the two sorts of collaboration are utilized, drawing upon their individual qualities to tackle issues or which neither one of the forms is adequately ground-breaking without anyone else.

REFERENCES

1. J. L. Baer, "A survey of some theoretical aspects of multiprocessing," *Computing Surveys*, Vol. 5, 1973, pp. 31–80.
2. E. K. Bowdon, Sr., and W. J. Barr, "Cost-effective priority assignment in network computers," in *FJCC Proceedings*, Vol. 41, R. Davis and R. G. Smith, Eds. Montvale, NJ: AFIPS Press, 1972, pp. 755–763.
3. C. R. D'Olivera, *An Analysis of Computer Decentralization*, Cambridge: Massachusetts Institute of Technology, MIT LCS-TM-90, October 1977.
4. L. Chen, et al., "VGITS: ITS based on intervehicle communication networks and grid technology," *Journal of Network and Computer Applications*, Vol. 31, No. 3, 2007, pp. 285–302. doi:10.1016/j.jnca.2006.11.002
5. Y. E. Hawas, "A microscopic simulation model for incident modeling in urban networks," *Transportation Planning and Technology*, Vol. 30, No. 2, 2007, pp. 289–309. doi:10.1080/03081060701398117
6. Y. E. Hawas, et al., "Comparative assessment of Inter Vehicular Communication (IVC) algorithms for real-time traffic route guidance," *Journal of Intelligent Transportation Systems: Technology Planning and Operations*, Vol. 13, No. 4, 2009, pp. 199–217. doi:10.1080/15472450903323107
7. Y. E. Hawas, and M. A. Hameed, "A multi-stage procedure for validating microscopic traffic simulation models," *Transportation Planning and Technology*, Vol. 32, No. 1, 2009, pp. 71–91. doi:10.1080/03081060902750686
8. Y. E. Hawas, and H. S. Mahmassani, "Comparative analysis of robustness of centralized and distributed network route control systems in incident situations," *Transportation Research Record*, Vol. 1537, 1996, pp. 83–90. doi:10.3141/1537-12

9. J. Anda, et al., "VGrid: vehicular ad hoc networking and computing grid for intelligent traffic control," *Proceedings of the IEEE 61st Vehicular Technology Conference*, Vol. 61, No. 5, 2005, pp. 2905–2909.
10. J. L. Adler, et al., "A multi-agent approach to cooperative traffic management and route guidance," *Transportation Research Part B*, Vol. 39, No. 4, 2005, pp. 297–318. doi:10.1016/j.trb.2004.03.005
11. J. Jeremy, et al., "Challenges of intervehicle ad hoc networks," *IEEE Transactions on Intelligent Transportation Systems*, Vol. 5, No. 4, 2004, pp. 347–351. doi:10.1109/TITS.2004.838218
12. C. J. Jiang, et al., "Research on traffic information grid," *Journal of Computer Search and Development*, Vol. 40, No. 12, 2003, pp. 1676–1681.
13. H. S. Mahmassani, and S. Peeta, *System Optimal Dynamic Assignment for Electronic Route Guidance in a Congested Traffic Network*. Heidelberg: Springer-Verlag, 1995.
14. H. S. Mahmassani, and Y. E. Hawas, *Experiments with a Rolling Horizon Dynamic Route Guidance Algorithm: Robustness under Stochastic Demands*. Paper presented at the INFORMS, Atlanta, 1996.
15. C. Hewitt, "Viewing control structures as patterns of passing messages," *Artificial Intelligence*, Vol. 8, 1977, pp. 323–364.
16. D. B. Lenat, "Beings: knowledge as interacting experts," in *Proceeding of the 4th International Joint Conference on Artificial Intelligence*, September 1975, pp. 126–133. https://dl.acm.org/doi/10.5555/1624626.1624646
17. V. R. Lesser, R. D. Fennell, L. D. Erman, and D. R. Reddy, "Organization of the HEARSAY II speech understanding system," *IEEE Transactions Acoust Speech and Signal Process*, Vol. ASSP-23, 1975, pp. 11–24.
18. R. G. Smith, *A Framework for Problem-Solving in a Distributed Processing Environment*. Stanford, CA: Department of Computer Science, Stanford University, STAN-CS-78-700 (HPP-78-28), December 1978.
19. V. R. Lesser, et al., *Working Papers in Distributed Computation I: Cooperative Distributed Problem Solving*. Amherst: Department of Computer and Information Science, University of Massachusetts, July 1979.
20. N. J. Nilsson, *Problem Solving Methods in Artificial Intelligence*. New York: McGraw-Hill, 1971.
21. R. E. Carhart, D. H. Smith, H. Brown, and C. Djerassi, "Applications of artificial intelligence for chemical inference—XVII: an approach to computer-assisted elucidation of molecular structure," *Journal of the American Chemical Society*, Vol. 97, 1975, pp. 5755–5762.
22. L. D. Erman, and V. R. Lesser, "A multi-level organization for problem-solving using many, diverse, cooperating sources of knowledge," in *Proceeding of the 4th International Joint Conference on Artificial Intelligence*, September 1975, pp. 483–490. https://dl.acm.org/doi/10.5555/1624626.1624702
23. H. P. Nii, and E. A. Feigenbaum, "Rule-based understanding of signals," in *Pattern-Directed Inference Systems*, D. A. Waterman and F. Hayes-Roth, Eds. New York: Academic, 1978, pp. 483–501.
24. R. G. Smith, "The contract net protocol: high-level communication and control in a distributed problem solver," *IEEE Transactions on Computers*, Vol. C-29, 1980, pp. 1104–1113.
25. R. G. Smith, and R. Davis, "Applications of the contract net framework: distributed sensing," in *Proceedings of the ARPA Distributed Sensor Net Symposium*. Pittsburgh, PA, December 1978, pp. 12–20.
26. R. Davis, and R. G. Smith, "Negotiation as a metaphor for distributed problem solving," *Artificial Intelligence*, Vol. 20, No. 1, 1983, pp. 63–109. doi: 10.1016/0004-3702(83)90015-2

27. D. L. Waltz, *Generating Semantic Descriptions from Drawings of Scenes With Shadows*. Cambridge: Massachusetts Institute of Technology, MIT AI-TR.-271, November 1972.
28. P. H. Winston, *Artificial Intelligence*. Reading, MA: Addison- Wesley, 1977.
29. H. G. Barrow, and J. M. Tenenbaum, *MSYS: A System for Reasoning About Scenes*. Menlo Park, CA: SRI International, SRI AIC TN 121, April 1976.
30. S. W. Zucker, R. A. Hummel, and A. Rosenfeld, "An application of relaxation labeling to the line and curve enhancement," *IEEE Transactions on Computers*, Vol. C-26, 1977, pp. 394–403.
31. V. R. Lesser and L. D. Erman, "Distributed interpretation: a model and experiment," *IEEE Transactions on Computers*, Vol. C-29, 1980, pp. 1144–1163.

13 Comparative Studied Based on Attack Resilient and Efficient Protocol with Intrusion Detection System Based on Deep Neural Network for Vehicular System Security

Naziya Hussain and Preeti Rani

CONTENTS

13.1 INTRODUCTION

In recent years, both industry and academia have devoted attention to wireless networks capable of supporting high-mobility broadband connectivity [1–3]. The definition of linked automobiles or vehicle networks, in particular, as shown in Figure 13.1. It has gained tremendous traction in modern vehicle connectivity. Along with the latest onboard computing and recognizing methods, it's the important enabler of keen conveyance systems and smart cities [4–6]. Ultimately, this new network generation should have a major effect on society, making everyday journeys healthier, greener, more effective, and more relaxed. In addition to advancements in a variety of artificial intelligence (AI) technologies, it helps pave the way for self-directed driving during the fifth generation's advent of cellular networks (5G).

Over the years, many coordination protocols such as ad hoc networks of vehicles (VANETs) Over the years, many coordination protocols for ad hoc vehicle networks (DSRC) in the United States [7] and the ITS-G5 in Europe [8], Both built on the latest IEEE 802.11p [9]. AI and V2X can support unconventional applications, including real-time prediction and management of traffic flows, locational applications, free transport facilities, vehicle platoons, vehicle data storage, and congestion control. The use and adaptation of AI technology methods to address the demands of vehicle networks, however, remains an area of research in its infancy.

However, recent studies have shown these technologies [10] suffered from various problems, including indefinite delays in access for networks, lack of facility quality assurances, and short-term vehicle-to-infrastructure (V2I). The 3GPP has begun to explore associate vehicle-to-all services (V2X in Figure 13.2) in the LTE network

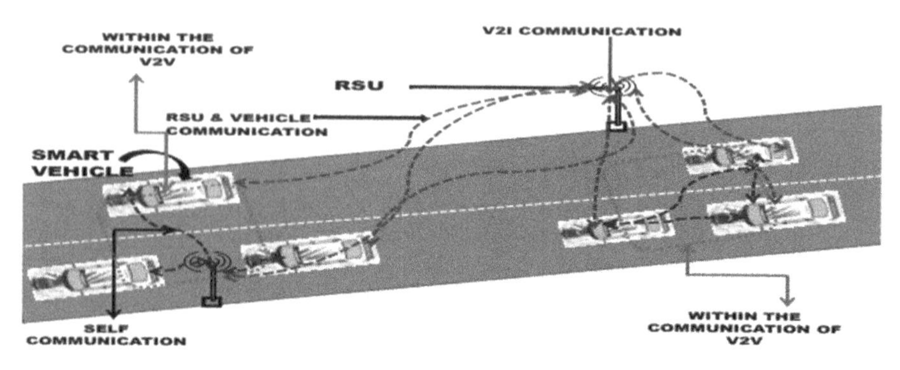

FIGURE 13.1 Communication system of VANETS.

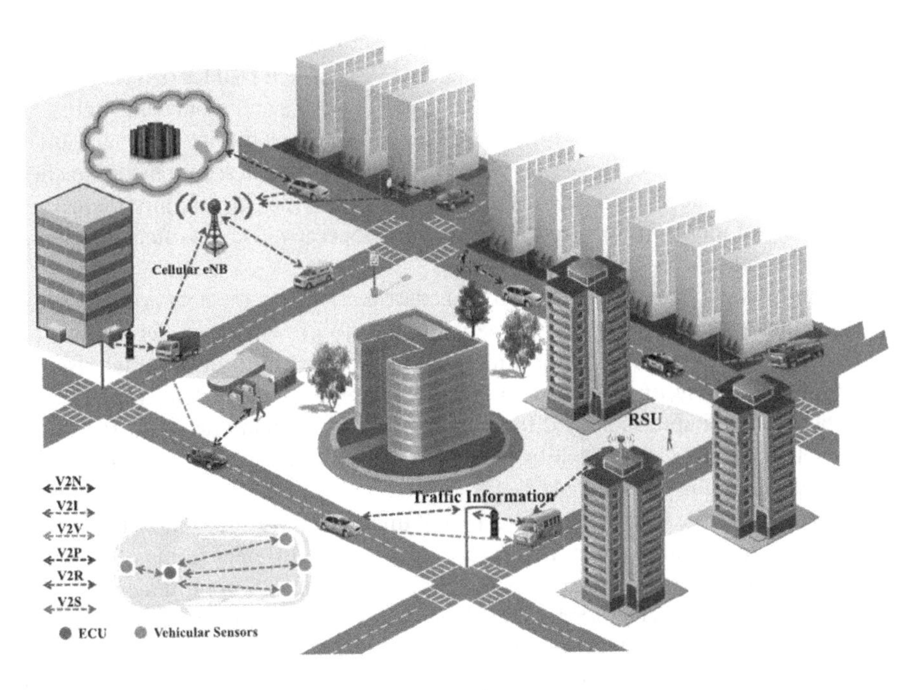

FIGURE 13.2 An overview of the V2X scenario.

and the potential of 5G cellular systems to reduce the limitations of IEEE 802.11p related skills and exploit high rate of cellular network penetration [11,12].

Some recent works can be found in this line of effort [13,14], which studies the efficient allocation of radio resources in-vehicle networks using the communication technology device-to-device (D2D) enable V2V communication into mobile systems.

As graph theory techniques extensively reviewed in-vehicle network resource allocation design [15] in current years. The rigorous, heterogeneous QoS specifications of vehicle applications and strong inherent dynamics in the vehicle environment pose major tests in developing wireless networks for effective and efficient support for highly mobile environments. Four types of vehicle ad hoc communication are closely related to the vehicle ad hoc device mechanisms as described above and describe the main functions of every communication type as depicted in Figure 13.1 [16,17].

In VANETs science, also referred to as the in-vehicle domain, it is becoming more important to import in-vehicle communication. The efficiency of a car and, in particular, exhaustion and driver drowsiness is important to drivers and public safety [18,19].

The V2V contact will provide drivers with a data exchange channel for sharing information and warning messages. A further useful area of research in VANETs is contact on the vehicle-to-road infrastructure (V2I), enabling drivers to track traffic and weather in real-time and provides environmental singing and monitoring [20,21]. The V2V contact will provide drivers with a data exchange channel for sharing information and warning messages. A further useful area of research in VANETs is contact on the vehicle-to-road infrastructure (V2I), which enables drivers to track traffic and weather in real-time and provides traffic details and monitoring data.

This misconduct contributes to several negative effects on the network, such as (1) reducing the reliability of the network and (2) growing the interruptions of bunches that increase the need for a monitoring mechanism to detect this misconducting MPR. Many of the surviving processes are non-cooperative, which often contribute to undependable choices.

This misconduct contributes to several negative effects on the network [22], such as (1) reducing the reliability of the network and (2) growing the interruptions of clusters that increase the need for a monitoring mechanism to detect this misconducting MPR. Many of the surviving processes are non-cooperative, which often contribute to unreliable choices [23]. Random Mobility System Waypoint (RWMM) and Reference Point Group Mobility (RPGM) Gauss-Markov, Manhattan for simulation and evaluating tests [23,24].

To detect a hostile vehicle in Figure 13.3 in this case. It illustrates how the assailant node interacts with the car. This figure shows three different forms of communication: malicious communication with vehicles (M2V), V2V communication, and V2I communication. When an actual vehicle sends a data packet to another actual vehicle and hires a route discovery process, When a malicious vehicle believes that it has the active route, it pretends to know the target vehicle and its received RREQ

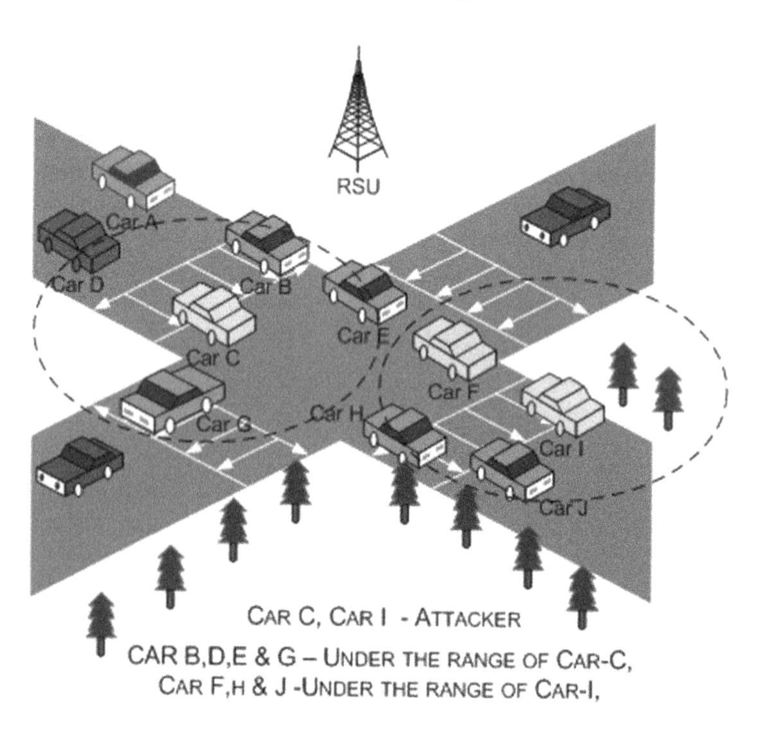

FIGURE 13.3 Attack scenario with network topology.

packets. Vehicle communications happen between the actual and malicious vehicles and a confirmation is sent to the actual original vehicle before the malicious vehicles receives the answer. It was based on this initial vehicle, which is active and has fully discovered the current path. The initial node rejects all responses and data messages from other railway vehicles and missing data packets.

Here is a suitable method for attack detection using ANN. ANN are computer plans that method data to simulate the neural fundamental of the human brain. It consists of hundreds of neurons or components in the input, output level, and various hidden levels of processing. In the framework of the issue that we are speaking about, the ANN identifies the misbehaving vehicles. It analyzes the recorded findings from both the GPCR-MA and GPCR-ARE c. The task of each vehicle is to route the GPCR-MA to overhear its one or several hops. A network training algorithm is used to change neuronal weights before you use the Neural Network.

In summary, recommending a supportive detection method based on ANN to detect the mischievous GPCR-MA routing. Our method can:

- Comment on a final joint decision.
- Take advantage of the previous knowledge of continuous learning identification.
- Increase the likelihood of identification and reduce false alarms.

The research outcome indicates that the ANN has the detection of attack probability and a low false alarms rate (FAR) compared to the GPCR-ARE model.

13.2 RELATED WORK

This section analyzes the previous work with a short description of identifying the attack vehicle in the VANET network.

If this VANET node comes within a transmission range, VANET nodes can be communicated with each other directly; the sender nodes transmit packets to the destination side. This research concentrates on dismembering and developing the Greedy Perimeter Coordinator Routing (GPCR) most frequently used VANET protocol. Because of the total number of spells in the networks, IDS is built on the network. The networks will distribute the high pace of the present-day lengthways with a broad network structure. The research focuses on the dispersed system of intrusion detection and performs circulated and scattered capture of data and analysis of data. This distributed structure helps our software to operate efficiently within both heavy traffic and large-scale structures. This technique design and develop an NS2 environment. The team subsequently compared our predictable scheme with the GPCR-MA PDR protocols listed with network latency, throughput, and overall power usage of VANET nodes can be directly connected while this VANET node is in the transmission series, source nodes are the transmission nodes that transmit packets to recipients. The sender nodes are the transmission nodes. This chapter concentrates on dismembering and developing the Greedy Perimeter Coordinator Routing (GPCR), the most frequently used VANET protocol.

VANET is vulnerable to many threats to security. A major attack is the Sybil attack, in which a malicious node forges many fake disruptive identities VANET applications' proper functioning. A distributed and flexible strategy to protect against Sybil's attack is provided. The suggested scheme detects suspicious vehicles' false identities by examining the clear similarity of the neighborhood details. Both vehicles exchange beacon packets periodically to signal their presence and to be aware of neighboring nodes. Every node periodically keeps a record of its neighboring nodes. In this process, each node exchanges its neighboring nodes and performs their intersections. If any vehicles observe that they have similar neighbors for a long time, this is known as a Sybil attack.

Without secret information sharing and special hardware assistance, the proposed approach will easily identify Sybil nodes and test the updated approach on a practical traffic scenario. Experimental tests show that the detection rate increases when the intruder forges full Sybil nodes [25].

New threats and vulnerabilities are emerging as more software modules and external interfaces are installed on vehicles. Researchers have shown how to compromise and control car maneuver in-vehicle electronic control units (ECUs). Various types of security mechanisms have been proposed to resolve these vulnerabilities. Still, they do not satisfy the need to provide effective protection for vital safety ECUs against in-vehicle network attacks. We suggest a framework of anomaly-based intrusion detection (IDS) called Clock-based IDS (CIDS) to alleviate this deficiency. It tests and then uses periodic in-car messages intervals for fingerprint ECUs. The resulting fingerprints are then used to construct a baseline of clock behavior of ECUs using the Recursive Least Squares (RLS) algorithm.

On that basis, CIDS identifies irregular shifts in recognition errors with the Accumulated Description (CUSUM)-a simple indication of intrusion. It easily detects the intrusions of the vehicle network at a 0.055% fake-positive pace. Contrary to state-of-the-art IDSs, fingerprinting CIDS of ECUs also facilitates the study of the root cause, the ECU, which mounted the attack. Our tests on a prototype CAN bus and on genuine vehicles have shown that CIDS can detect a wide range of Sybil nodes' in-vehicle network attack number by the attacker [Cho, K. T., & Shin, K. G. (2016)] [26].

To enhance the safety of the vehicle network, a revolutionary (IDS) utilizing DNN is proposed. Feature vectors have trained the DNN structure building parameters based on probability derived network in-vehicle packets. The DNN supplies the chance of every class distinguishing between usual and bout packets for a given packet so that the device can detect any malicious vehicle attack. In contrast with the current neural artificial network used by IDS, the suggested technique draws on recent progress in profound learning science, such as the initialization of the limits over the unattended pretraining of profound confidence networks (DBN). Experimental findings show that the technology proposed can respond to the attack in real-time by increasing the detection ratio of the CAN bus [27].

The emerging trend for automakers is to build seamless communication between car management and the fleet, providing remote diagnostics and airborne software updates. It is necessary to connect the previously isolated car network to an external network and, therefore, to be a completely new class of threats recognized as cyberattacks are revealed. In this article, we are debating the applicability of an approach to identifying cyberattack requirements in the vehicle network. We collect information to create security details for communication and ECU behavior in the open May draft of standard 3.01 communication protocol and directory objects pages. We also include a collection of examples, suggest an effective position for the bout detector, and test the finding using a variety of attack measures [28].

Vanessa has been worthy of tremendous recognition for many years. The implementation of wireless communication in VANET ensures that security requirements are taken into account. Within VANET, various attacks are based on a multiple identity generation by the attacker to mimic several nodes called the Sybil attack. In this chapter, it is proposed that the effect of different assumptions (a type of antenna, signal strength of transmission) on the effectiveness of a Sybil attack be quantified precisely [26].

Vehicle communications play a major role in ensuring secure travel through the exchange of health posts. Several approaches were suggested by researchers to secure messages of safety. Fixed key infrastructure protocols have greater deployment efficiency and greater security compared to dynamic systems. The resolution of this chapter is to present a method of detecting impersonation based on a fixed key infrastructure in an ad hoc vehicle network; in other words, Sybil attack. This attack has a significant effect on the network's output. The suggested solution is to detect Sybil by using a cryptography method. Finally, the effects of this approach are checked using the Mat laboratory simulator, as this system has the Certification Authority to perform a short detection time for a Sybil attack [29].

They propose a security protocol for the Sybil attacks detection in privacy vehicle ad hoc networks (VANETs) for location-based applications. In our protocol, vehicles define Sybil attacks locally by analyzing the logic of the vehicles' behavior against their neighbors cooperatively. The attack detection uses communications characteristics and GPS locations of vehicles included in the security messages regularly transmitted. No new hardware and limited connectivity and overhead processing will be implemented for automobiles. Our protocol is, therefore, very light weighted and suitable for real applications. Also, a clever intruder scenario in which a malicious vehicle will change its communications range to avoid detection and the collusion scenario of malicious vehicles is considered. NS2-based simulation results demonstrate the protocol's performance [30].

To fix problems of identifying misbehaved ad hoc road network vehicles (VANET), using the QoS-OLSR protocol. Under this protocol, vehicles may include misbehavior during the creation of clusters by demanding false information or forming clusters. If a vehicle over-speeds or below speed, the minimum speed limit is considered selfish or misbehaving, where such a behavior contributes to a disconnected network. We are proposing a two-phase model that motivates nodes to cooperate during cluster formation and identifies misbehavioral nodes following clusters' formation. Reputation opportunities and network infrastructure are offered to enable vehicles to operate cooperatively during the first process.

Misbehaving vehicles will still benefit from the infrastructure of the network by acting normally in the creation and misbehavior of clusters. A Dempster–Shafer cooperative monitoring model is modeled where information is aggregated, and cooperative decisions are taken to identify misbehaving cars. Simulation results show that the model proposed of detection will increase detection probability, decrease false negatives, and the percentage of selfish nodes in the vehicle network while preserving service quality and stability [31].

13.3 BACKGROUND

ANN are computer systems that model the processing of information in a human mind. An ANN is generated from many input layers, the outcome layer, and several hidden neuron coatings or elements processing (PE). Each PE is equipped with weight, output, and transfers. The weights of dispensation elements are enhanced while training pending the error is minimalized, and the network achieves the required level of correctness.

13.3.1 Processing Phase

The multilayer perceptron (MLP) is, as its name implies, an ANN model consisting of multiple layers. The contribution layer is conventionally 0, so a multilayer perceptron in three layers [32] proceeds the method shown in Figure 13.4. Every neuron's outcome is the transmission function F(x) in the x, where the variable x equals the number of inputs reproduced by the weight of the neuron.

13.3.2 Training Phase

We will train the ANN, pending the error is reduced. MLP uses a reverse spread method of network training [33]. A series of vectors from original data collected or created by a suitable simulant is used in the training process. Learning occurs by

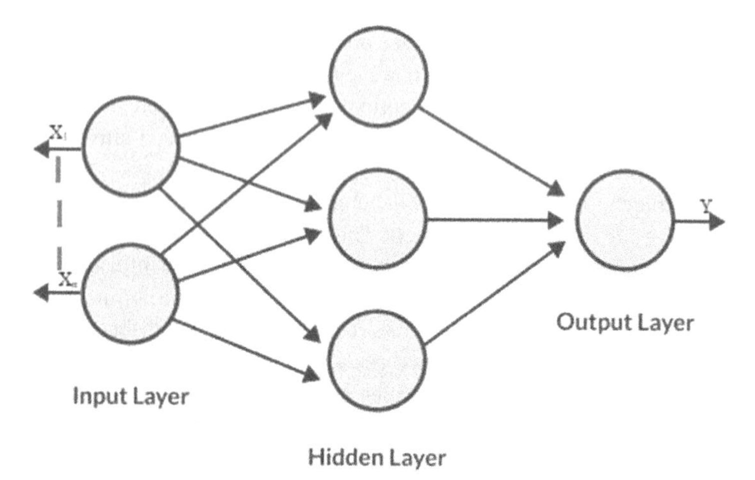

FIGURE 13.4 Architecture of three-layer MLP.

adjusting the weights of output error reduction relations compared to the expected outcome by exercise vector.

13.4 INTRUSION DETECTION SYSTEM

An ideal IDS should meet various necessities for VANETs:

- The use of VANETs does not present new flaws and vulnerabilities to the current network.
- It necessitates a few resources and should not deteriorate the entire network's performance by presenting pointless overheads—no trade-off among detection rate and detection overhead.
- No changes/upgrades to current facilities.
- Automatic real-time security against intruders without any human intervention should be given. It should be insubstantial and less composite.
- It should run patiently and reliably without any knowledge of its operation. Provides a nominal false positive rate (FPR) and false-negative rate (FNR).
- It should be able to self-defense and monitor itself to verify whether the attacker has compromised it. It should be able to communicate with other IDSs.
- High rate of detection and short detection time.
- IDS should not only detect the attack but should also be able to identify the source and type of attack.
- The detection rate and network output should be as sparse as dense as the network is not affected.
- The node mobility does not affect IDS detection.
- Primary distributions such as variables do not influence the output of the IDS.
- A single IDS could detect multiple attacks equally efficiently.

Some studies [34,35] use the IDS system for GPCR-MA with many parameters such as high mobility, large area and road lane segments, and measure the variation QoS parameter for performance verification. The IDS detects the attack vehicle and passes that ID into a modified protocol of GPCR-MA is known as GPCR-ARE, as mention in the flowchart (Figure 13.5).

13.5 IDS WITH MACHINE LEARNING

Methods of intrusion detection were extensively the traditional network has been studied to help cope with malicious attacks. With various literary techniques for intrusion detection, the attack packets' patterns are different from those of the normal packets based on machine learning techniques. In some studies [36–38], the intrusion detection uses statistical modeling of a packet data to apply (ANN) and support vector machine (SVM).

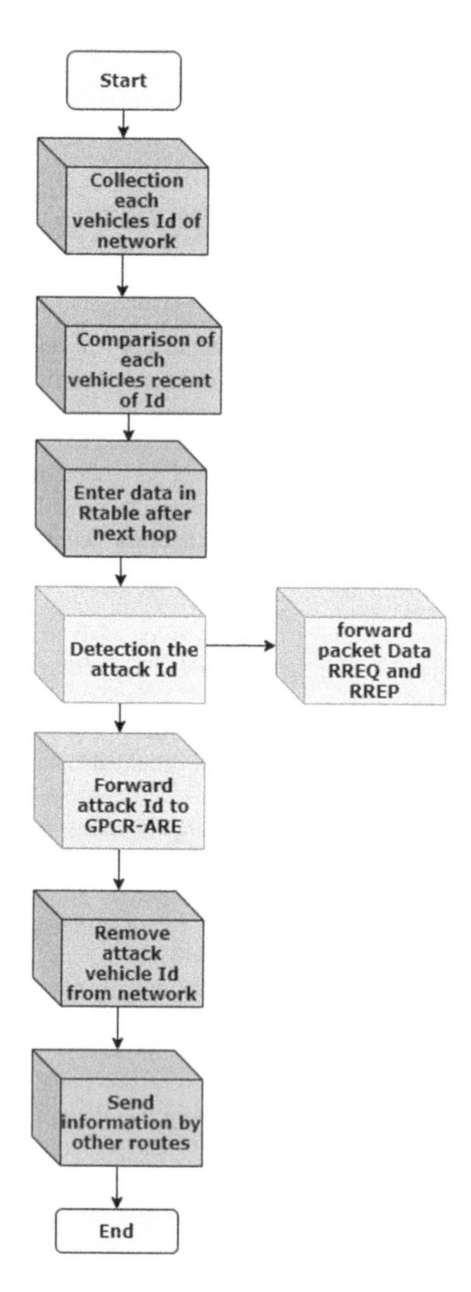

FIGURE 13.5 Proposed Flowchart of Secure IDS (GPCR-ARE).

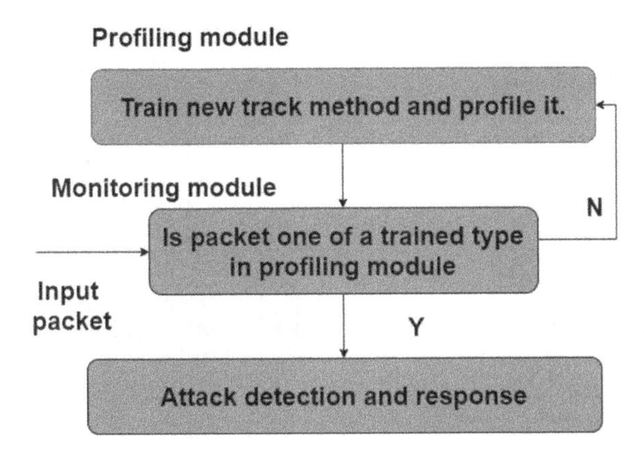

FIGURE 13.6 IDS architecture based on machine learning.

In one study [39], an encoding approach based on frequency is used in both ANN and SVM for a packet dimension. The above work is focused on supervised computer teaching, and therefore several labeled document sets are necessary for the training. In contrast, Kayak et al. employ unattended machine knowledge techniques, such as a self-organized map (SOM) for the detection of network interruption. Figure 13.6 shows an increasing IDS design focused on computer education. The IDS contains different modules for the collection and storage of a large number of data packets. After extraction, the monitoring module usually identifies an input packet part. Includes the off-line profiling module functionality. If a new type of attack is detected in the monitoring module, the profiling module can update the module database.

13.6 PROPOSED TECHNIQUE

13.6.1 Proposed Deep Neural Network Intrusion Detection System

The suggested ID architecture is a general example of targeting malicious data packets in the car on a CAN bus. Mobile phone networks [40], for example, 3G, 4G, and WiFi, can be reached from car networks or Self-diagnostic applications like OBD and mobile driver [41]. The proposed system for intrusion detection tracks the transmission of CAN packets are in the bus and decides attack.

The IDS designations two main stages, namely preparation and detection, as shown in Figure 13.7. The planning takes place off-line, as it takes time to plan. During processing, a CAN packet is analyzed to retrieve a function that reflects a network analytical performance. May packet is labeled with a binary, i.e., a normal packet or a controlled attack packet via training. The corresponding characteristics must, therefore, reflect the specifics of the mark. We follow the DNN Learning System weight parameters of the node edges. The method of detection is also shown in Figure 13.7. The same function is extracted via a CAN bus, and the DNN structure from an incoming packet checks the qualifying binary limits.

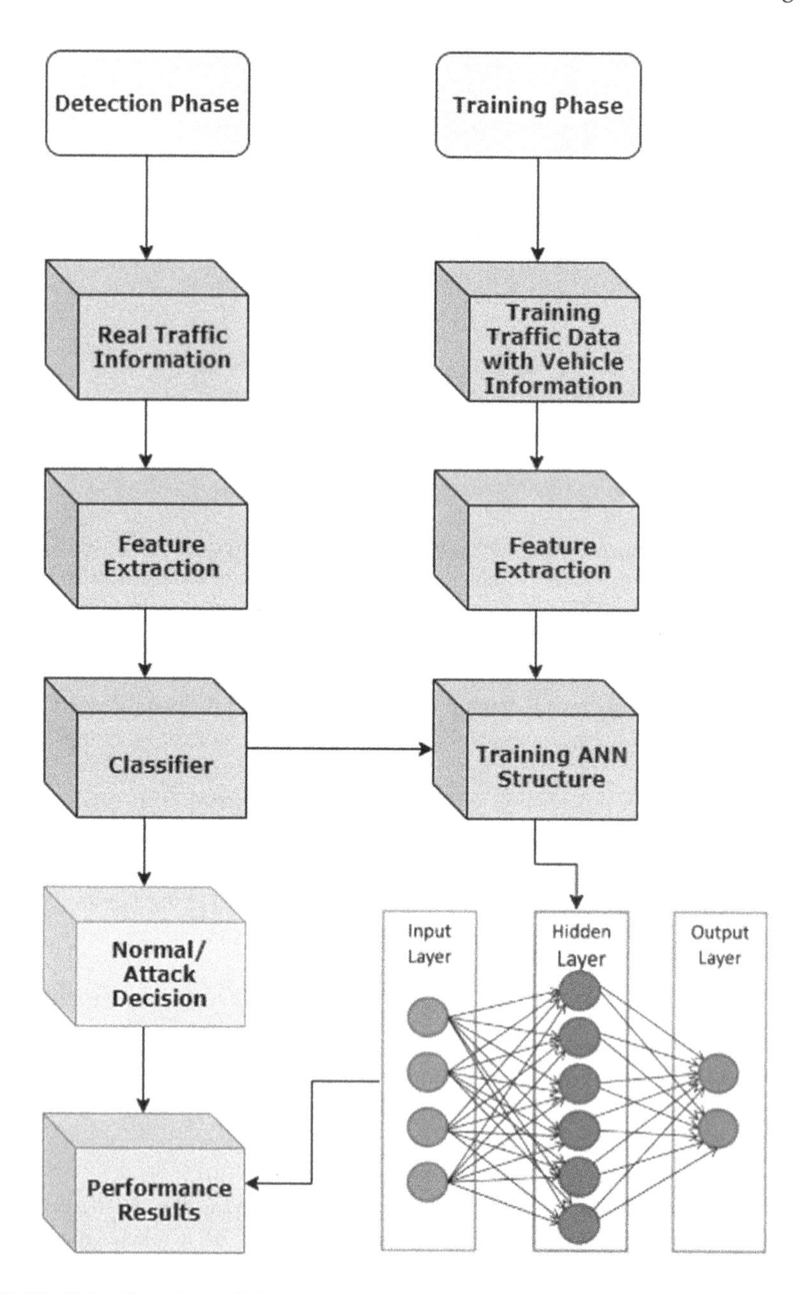

FIGURE 13.7 Overview of the proposed intrusion detection system.

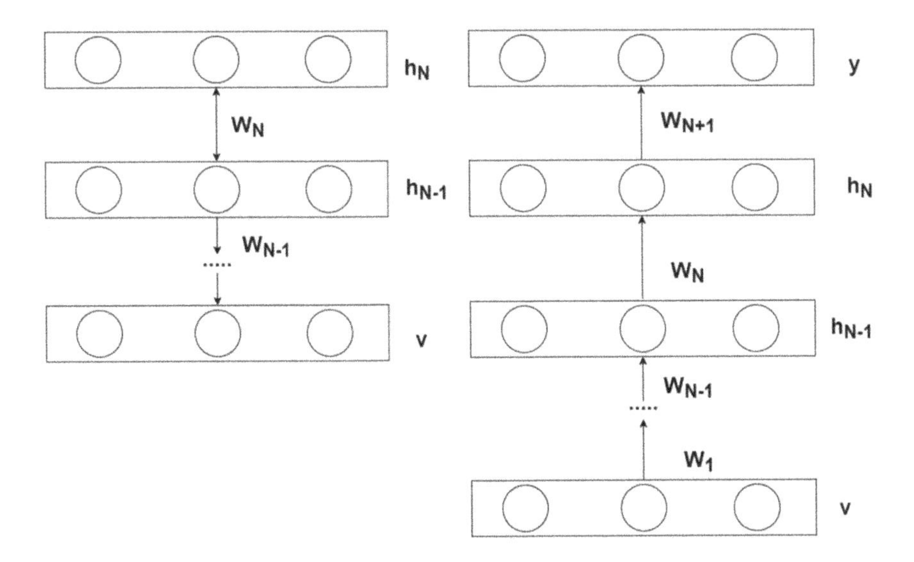

FIGURE 13.8 DBN and DNN structure involving the pretrained weight parameters in n hidden layers built with a bottom-up manner.

The learning framework for supervised learning should be optimized as an unattended learning mechanism is offered by the DBN model in Figure 13.8. To create a deeper framework for discrimination, the final classification layer with label information is applied to the top layer of the DBN model. The layout has been changed into a deep ANN feed system, where the mechanism is upwardly trained, based on the mark. Figure 13.8, the weight of the hidden nodes of the DBN system is first attained from uncontrolled pretraining. However, the parameters are only used to initialize the weights and are finished later on with the deep feedforward ANN gradient descent method.

13.6.2 Training the Deep Neural Network Structure

The study mechanism of the proposed DNN is to differentiate between a normal packet and an attack packet. Figure 13.7 displays a data layer, several hidden layers, and a layer of output. The function vector is entered into the structure's input nodes. Figure 13.9 node calculates an exit with a rectified linear unit (ReLU) function. The proposed DNN learning mechanism differentiates between the standard packet and the attack packet. Figure 13.9 displays a data layer, several hidden layers, and an outcome layer. The function vector is entered into the input nodes of the structure.

13.6.2.1 ANN Parameters

Let us define:

N_i = No. of input layer's neurons
N_h = No. of hidden layer's neurons

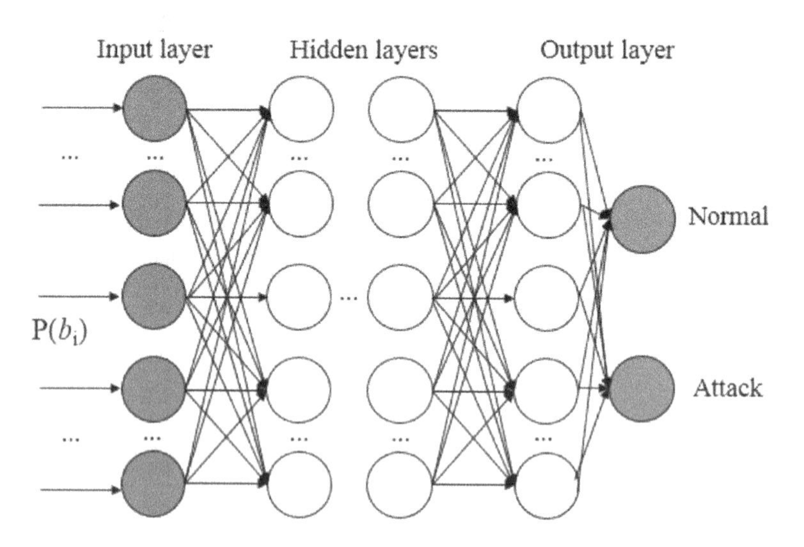

FIGURE 13.9 Proposed technique DNN system.

N_o = No. of output layer's neurons
F(x) = Transfer Function

13.6.2.2 Input Layer's Neurons

1. Ni: The total neurons in the input layer is equal to the total vehicles needed.

13.6.2.3 Hidden Layer's Neurons

2. N_h: The total neurons in the hidden layer is set according to the empirical equation:

$$Nh = \_\left(Ni + No\right) + l, l \text{ between 1 and 10}$$

13.6.2.4 Output Layer's Neurons

3. N_o: The total neurons in the output layer is equal to one and indicates the final judgment.

13.6.2.5 Transfer Function

4. F(x) = tansig(x) = 2/(1 + exp(−2x)) − 1

13.7 SIMULATION PARAMETERS

The 900/600 m^2 simulation surface area comprises a 20 to 100 nodes vehicle collection, 300 m transmission range, and multi-lane highway topology. The Network

TABLE 13.1
Simulation Parameters

Parameters	Values
MAC layer protocol (Wireless)	IEEE 802.11
Network Simulator	NS2.35
Application Layer	TCP
Routing protocols	GPCR-MA and GPCR-ARE
Channel mode	Wireless channel
No. of Vehicles	20; 40; 60; 80; and 100
Simulation time	100; 200; 300; 400; and 500 seconds
Traffic type	CBR
Mobility paradigm	Freeway, Manhattan, Gauss-Markov, RPGM, and Random Way Point
Simulation district	900 × 900
Performance Metrics	Packet Delivery Ratio, Average Network Throughput, Average Energy, and Average Delay
Interface queue type	Tail/ Drop/PriQueue
Antenna type	Omni-directional
Network interface mode	Phy/WirelessPhy
Radio-propagation pattern	Two-Ray Ground

Simulator 2 creates a file with parameters to simulate the MATLAB method. Traffic model parameters to the vehicle position and vehicle mobility are summarized in Table 13.1.

Based on the simulation data, we can calculate the end-to-end average, average energy and average network throughput, and PDR (packet deliver ratio) using the following formulae, which are mentioned below in Equations 13.1, 13.2, and 13.3.

13.7.1 Average End-to-End Delay

Here, Equation (13.1) represents the Average end-to-end delay formula as "D" while the total number of successfully delivered packets is "n."

$$\text{Average delay} = \frac{\sum_{i=1}^{n}(\text{Received Packet Time} - \text{Send Packet Time}) * 1000(\text{ms})}{\text{Total Number of Packets Delivery Successfully}} \tag{13.1}$$

13.7.2 Average Energy Consumption

The total energy consumption is the sum of total use up the energy of general vehicles network and energy spending of the vehicle is the summing up of energy used for communication.

13.7.3 Average Network Throughput

Here, Equation (13.2) represents the mathematical formula of throughput, the sum of data the destination receives. Packet size represents the size of the ith the packet entering the destination; the last packet arrival time is known as a packet arrival while the first packet's arrival time is marked as a packet launch. Box scale the amount of packed used in the device layer. The average throughput is the throughput per unit of time.

$$\text{Average Throughput} = \frac{\text{Total number of packet received by destination} * \text{Packet Size}}{(\text{Packet Arrival} - \text{Packet Start})} \tag{13.2}$$

13.7.4 Packet Delivery Ratio (PDR)

The mathematical formula to calculate PDR is stated as below.

$$\text{Packet delivery ratio} = \frac{\text{received packets}}{\text{generated packets}}$$

$$\text{Packet Delivery Ratio} = \frac{\sum \text{Total packets received by all destination node}}{\sum \text{Total packets send by all source node}} \tag{13.3}$$

13.8 CONCLUSION

We briefly presented the fundamentals of artificial intelligence and provided examples of how to learn dynamics using these tools and make smart decisions in the vehicle networks. We also highlighted some open issues and areas of concern. In this research, we address the detection of faulty nodes in ad hoc vehicle networks. A vehicle is known to be misbehaving when it overspeeds or underspeeds. Most of today's strategies focus on collective decisions. Others use a helpful scheme and methods based on an indication to measure the different outcomes.

We have proposed an efficient (IDS) to protect the car network based on a deep neural network (DNN). We have trained DNN parameters with probability dependent functional vectors derived with unregulated pretraining from vehicle network packages methods of profound creed networks and the traditional stochastic downward gradient approach. The DNN gives each class the chance to distinguish between regular and hacking packets, which means that the device can detect a malicious attack on the vehicle. We have proposed a new feature vector, which includes network packet mode information and value information efficiently used in training and testing.

REFERENCES

1. Peng, H., Liang, L., Shen, X. and Li, G.Y., 2018. Vehicular communications: a network layer perspective. *IEEE Transactions on Vehicular Technology*, *68*(2), pp. 1064–1078.
2. Araniti, G., Campolo, C., Condoluci, M., Iera, A. and Molinaro, A., 2013. LTE for vehicular networking: a survey. *IEEE communications magazine*, *51*(5), pp. 148–157.
3. Cheng, X., Chen, C., Zhang, W. and Yang, Y., 2017. 5G-enabled cooperative intelligent vehicular (5GenCIV) framework: when Benz meets Marconi. *IEEE Intelligent Systems*, *32*(3), pp. 53–59.
4. Hussain, N., Singh, A. and Shukla, P.K., 2016. In-depth analysis of attacks & countermeasures in vehicular ad hoc network. *International Journal of Software Engineering and Its Applications*, *10*(12), pp. 329–368.
5. Sun, W., Ström, E.G., Brännström, F., Sou, K.C. and Sui, Y., 2015. Radio resource management for D2D-based V2V communication. *IEEE Transactions on Vehicular Technology*, *65*(8), pp. 6636–6650.
6. Zhang, R., Cheng, X., Yang, L., Shen, X. and Jiao, B., 2014. A novel centralized TDMA-based scheduling protocol for vehicular networks. *IEEE Transactions on Intelligent Transportation Systems*, *16*(1), pp. 411–416.
7. Kenney, J.B., 2011. Dedicated short-range communications (DSRC) standards in the United States. *Proceedings of the IEEE*, *99*(7), pp. 1162–1182.
8. Eckhoff, D., Sofra, N. and German, R., 2013, March. A performance study of cooperative awareness in ETSI ITS G5 and IEEE WAVE. In 2013 10th Annual Conference on Wireless On-demand Network Systems and Services (WONS) (pp. 196–200). IEEE, Banff, AB, Canada.
9. Eichler, S., 2007, September. Performance evaluation of the IEEE 802.11 p WAVE communication standard. In 2007 *IEEE* 66th Vehicular Technology Conference (pp. 2199–2203). IEEE, Baltimore, Maryland.
10. Hassan, M.I., Vu, H.L. and Sakurai, T., 2011. Performance analysis of the IEEE 802.11 MAC protocol for DSRC safety applications. *IEEE Transactions on Vehicular Technology*, *60*(8), pp. 3882–3896.
11. Seo, H., Lee, K.D., Yaskawa, S., Peng, Y. and Sartori, P., 2016. LTE evolution for vehicle-to-everything services. *IEEE Communications Magazine*, *54*(6), pp. 22–28.
12. Wang, J., Shao, Y., Ge, Y. and Yu, R., 2019. A survey of vehicle to everything (v2x) testing. *Sensors*, *19*(2), p. 334.
13. Boston, M., Klügel, M., Kellerer, W. and Fertl, P., 2014, April. Location dependent resource allocation for mobile device-to-device communications. In *2014 IEEE Wireless Communications and Networking Conference (WCNC)* (pp. 1679–1684). IEEE, Atlanta, Georgia.
14. Liang, L., Xie, S., Li, G.Y., Ding, Z. and Yu, X., 2018. Graph-based resource sharing in vehicular communication. *IEEE Transactions on Wireless Communications*, *17*(7), pp. 4579–4592.
15. Liang, L., Ye, H. and Li, G.Y., 2018. Toward intelligent vehicular networks: a machine learning framework. *IEEE Internet of Things Journal*, *6*(1), pp. 124–135.
16. Faezipour, M., Nourani, M., Saeed, A. and Addepalli, S., 2012. Progress and challenges in intelligent vehicle area networks. *Communications of the ACM*, *55*(2), pp. 90–100.
17. Liang, W., Li, Z., Zhang, H., Wang, S. and Bie, R., 2015. Vehicular ad hoc networks: architectures, research issues, methodologies, challenges, and trends. *International Journal of Distributed Sensor Networks*, *11*(8), p. 745303.
18. Yousefi, S., Mousavi, M.S. and Fathy, M., 2006, June. Vehicular ad hoc networks (VANETs): challenges and perspectives. In 2006 6th International Conference on ITS Telecommunications (pp. 761–766). IEEE, Chengdu, China.

19. Paul, B. and Islam, M.J., 2012. Survey over VANET routing protocols for vehicle to vehicle communication. *IOSR Journal of Computer Engineering (IOSRJCE)*, *7*(5), pp. 1–9.
20. Belanovic, P., Valerio, D., Paier, A., Zemen, T., Ricciato, F. and Mecklenbrauker, C.F., 2009. On wireless links for vehicle-to-infrastructure communications. *IEEE Transactions on Vehicular Technology*, *59*(1), pp. 269–282.
21. Miller, J., 2008, June. Vehicle-to-vehicle-to-infrastructure (V2V2I) intelligent transportation system architecture. In 2008 IEEE Intelligent Vehicles Symposium (pp. 715–720). IEEE, Eindhoven, Netherlands.
22. Hussain, N., Maheshwary, P., Shukla, P.K. and Singh, A., 2018. Attack resilient & efficient QoS based GPCR-ARE protocol for VANET. *International Journal of Applied Engineering Research*, *13*(3), pp. 1613–1622.
23. Hussain, N., Maheshwary, P., Shukla, P.K. and Singh, A., 2018. Mobility-aware GPCR-MA for vehicular ad hoc routing protocol for highways scenario. *International Journal of Organizational and Collective Intelligence (IJOCI)*, *8*(4), pp. 47–65.
24. Hussain, N., Maheshwary, P., Shukla, P.K. and Singh, A., 2019. Manhattan & RPGM parameters based mobility aware GPCR-MA routing protocol for highway scenario in VANET. In *Proceedings of Recent Advances in Interdisciplinary Trends in Engineering & Applications* (RAITEA).
25. Grover, J., Gaur, M.S., Laxmi, V. and Prajapati, N.K., 2011, November. A Sybil attack detection approach using neighboring vehicles in VANET. In *Proceedings of the 4th International Conference on Security of Information and Networks* (pp. 151–158).
26. Cho, K.T., and Shin, K.G., 2016, October. Error handling of in-vehicle networks makes them vulnerable. In *Proceedings of the 2016 ACM SIGSAC Conference on Computer and Communications Security* (pp. 1044–1055).
27. Kang, M.J. and Kang, J.W., 2016. Intrusion detection system using deep neural network for in-vehicle network security. *PloS one*, 11(6), p. e0155781.
28. Larson, U.E., Nilsson, D.K. and Jonsson, E., 2008, June. An approach to specification-based attack detection for in-vehicle networks. In *2008 IEEE Intelligent Vehicles Symposium* (pp. 220–225). IEEE.
29. Rahbari, M. and Jamali, M.A.J., 2011. Efficient detection of sybil attack based on cryptography in VANET. arXiv preprint arXiv:1112.2257.
30. Hao, Y., Tang, J. and Cheng, Y., 2011, December. Cooperative Sybil attack detection for position based applications in privacy preserved VANETs. In *2011 IEEE Global Telecommunications Conference-GLOBECOM 2011* (pp. 1–5). IEEE.
31. Wahab, O.A., Otrok, H. and Mourad, A., 2014. A cooperative watchdog model based on Dempster–Shafer for detecting misbehaving vehicles. *Computer Communications*, 41, pp. 43–54.
32. Chidzonga, R.F., 1997, September. Matrix realizations of multilayer perceptron ann. In *Proceedings of the 1997 South African Symposium on Communications and Signal Processing* COMSIG'97 (pp. 181–186). IEEE, Rhodes University, Grahamstown.
33. Fu, L., Hsu, H.H. and Principe, J.C., 1996. Incremental backpropagation learning networks. *IEEE Transactions on Neural Networks*, *7*(3), pp. 757–761.
34. Hussain, N., Maheshwary, P., Shukla, P.K. and Singh, A., 2019. Detection of black hole attack in GPCR VANET on road network. In International Conference on Advanced Computing Networking and Informatics (pp. 183–191). Singapore: Springer.
35. Hussain, N., Shukla, P.M.D.P.K. and Singh, A., 2019. Detection of Sybil attack in vehicular network based on GPCR-MA routing protocol. *Current Science*, *20*(1), pp. 1–11.
36. Golovko, V. and Kochurko, P., 2005, September. Intrusion recognition using neural networks. In *2005 IEEE Intelligent Data Acquisition and Advanced Computing Systems: Technology and Applications*, Sofia, Bulgaria (pp. 108–111). IEEE.

37. Hu, W., Liao, Y. and Vemuri, V.R., 2003, June. Robust anomaly detection using support vector machines. In *Proceedings of the International Conference on Machine Learning, Washington, DC* (pp. 282–289).
38. Zhang, Z., Li, J., Manikopoulos, C.N., Jorgenson, J. and Ucles, J., 2001, June. HIDE: a hierarchical network intrusion detection system using statistical preprocessing and neural network classification. In *Proc. IEEE Workshop on Information Assurance and Security*, United States (pp. 85–90).
39. Chen, W.H., Hsu, S.H. and Shen, H.P., 2005. Application of SVM and ANN for intrusion detection. *Computers and Operations Research*, 32(10), pp. 2617–2634.
40. Koscher, K., Czeskis, A., Roesner, F., Patel, S., Kohno, T., Checkoway, S., McCoy, D., Kantor, B., Anderson, D., Shacham, H. and Savage, S., 2010, May. Experimental security analysis of a modern automobile. In *2010 IEEE Symposium on Security and Privacy*, Berleley/Oakland (pp. 447–462). IEEE.
41. Woo, S., Jo, H.J. and Lee, D.H., 2015. A practical wireless attack on the connected car and security protocol for in-vehicle CAN. *IEEE Transactions on Intelligent Transportation Systems*, 16(2), pp. 1–14.

14 A Secure Electronic Voting System Using Decentralized Computing

Dhruv Rawat, Amit Kumar, and Suresh Kumar

CONTENTS

14.1 INTRODUCTION

In every single democracy, the voting system is of topmost priority, as a threat to it is a threat to the nation's security. Using blockchain technology in the field of voting, the main goal is to minimize the cost of the whole election process. While making this voting procedure reliable and efficient, we can replace the traditional voting system, which is viewed as flawed. Electronic voting and online surveys are a few ways to learn the opinions of the people.

Computer technology has come a long way in previous decades. Electronic voting machines (EVMs) are being viewed as flawed because of the physical concern that poses a threat to the votes being cast and also to the secrecy of voting [1]. Electronic voting systems are facing a threat from various hackers. It is difficult to remove scams from happening to e-voting and it is also very difficult to perform the high cost of data cleanup after the system has been hacked. Hacking even a single system will be viewed as the failure of the entire voting system and also the citizens will worry about their anonymity being compromised, thereby questioning the integrity of the system.

An electronic voting system should be capable of performing authentication, it should provide transparency, it should protect the anonymity of the people casting vote, and in the end it should provide us with correct and accurate results.

There is a great deal of research going on regarding the adoption of blockchain technology into electronic voting. Japan is the first country to lead this research to implement their country's voting system and transform it into a blockchain electronic voting system. Other countries are leading the way, too.

A blockchain is a ledger that stores transaction details/data and is public, distributed, immutable, and peer-to-peer. It has a few main features:

(i) As the ledger is peer-to-peer, there is no single point, i.e., the ledger is present in dissimilar locations so there is no doubt of failing of the process at a single point and maintenance is easy.
(ii) The control is distributed.
(iii) Whenever we create a new block, it is the citation of the foregoing version of the ledger that creates an unchangeable chain. This prevents tampering with blockchain integrity.
(iv) When adding the new blocks to the blockchain, consensus algorithms must be followed to make the proposed new block a permanent part of the ledger.

14.2 BACKGROUND AND MOTIVATION

There are a few basic principles of voting:

- **Secret ballot:** Your vote is secret. No one should be ready to connect your vote back to your private profile.
- **One vote per person**: each elector votes once and therefore the election commission should be ready to adapt the heterogeneity of voters, the votes, and people who failed to vote [2].
- **Elector eligibility:** Only the citizens who have been verified as the citizens of a country and have the right to vote are eligible.
- **Transparency**: Tallying of votes must be transparent, i.e., government should show who has won and how many votes acquired on their behalf so that citizens are well aware of this.
- **Votes are accurately recorded:** Vote tally is congruous. The total tally of votes is revealed to the public and the result, once stated, cannot be changed.

- **Reliability:** The system has to be reliable and should keep fraud from occurring as well as stop various security violations.

We have to keep this in mind that these are the main principles that we have followed up to now in our mainstream e-voting systems. We also need to implement these principles in case we make a new one.

Blockchain will solve the numerous issues discovered in these early attempts to make an e-voting system. A blockchain DApp does not concern the safety of the net affiliation, as any hacker with access to the end node will not be able to have an effect on different nodes. Voters can submit their vote without revealing their name or political preferences. Admins will count votes with complete confidence, knowing that every ID attribute to at least one vote, and no more than one that means meddling is not possible.

Now, if we use blockchain, we will have to make a ballot that implements proper voting procedures. This should adhere to the following points.

14.2.1 Secret Ballot

In the ballot contract, only pocketbook addresses and names are noted. Each elector is known by their MetaMask pocketbook address. Apart from the chairman, the one that created the new instance of the ballot contract, nobody else will be able to tell how they voted.

The votes array is asserted non-publicly; therefore, nobody will be able to browse the contents of the votes array.

The reality could be a very little a lot of difficult than this. On a blockchain, even non-public variables are unit clear if you are attempting to read clearly enough. For this to really be non-public, users of the contract can have to be compelled to code the values before writing them to the good contract (encryption).

14.2.2 One Man, One Vote

The voter array stores an inventory of voters that have voted. It ensures that nobody will vote a second time. Once an elector votes, his standing changes to "voted" and the ballot contract checks to confirm that they do not vote again.

14.2.3 Voter Eligibility

Voter eligibility is decided by collecting an array of pocketbook addresses before the proof of voting begins.

The voter's address must match the one that the chairman registers before the process begins. However, if a voter's MetaMask pocketbook is taken, then someone else can vote on their behalf. However, on a blockchain, the important factor is the wallet's non-public key.

But how will the chairman make sure that that it is the correct person who is casting the vote? Facial recognition? Yes, to vote as long as you pass a fingerprint or biometric authentication check, this can work.

14.2.4 Transparency

Transparency relates to the things that blockchain always does by default. Each action taken and each record saved on the blockchain is changeless, i.e., can't be changed whatsoever if the contract is deployed.

14.2.5 Votes Accurately Recorded and Counted

In Smart contract, the ballot goes through many states, from creation of the ballot, to where we cast the votes, to the closing of the ballot and the final counting of votes.

In each of these states, the contract dictates what the chairman and voters are allowed or not allowed to perform.

14.2.6 Reliability

There is no single location of failure on a blockchain as each node within the chain participates to keep the blockchain running. If Smart contract is deployed on the blockchain is changeless. The business logic of the good contract once deployed is forged in stone. There's no way the chairperson will amend the principles, say, from one-man-one-vote to one-man-two-vote once the contract is deployed.

14.3 LITERATURE SURVEY

Zhang, Wenbin, IBM Center for Blockchain Innovation IBM Research, Singapore

This author has presented a brief survey of blockchain technology and its use over an ecosystem in the 2018 IEEE Conference. He discussed that all of the prerequisites of an e-voting platform and algorithmic schema to append the process of encryption of the voting process and to make it more secure and inexpensive.

He has stated a native mechanism on blockchain to facilitate the process of deciding peers of a blockchain network. The planned protocol protects voters' privacy and allows detection and correction against cheating with none sure party. His implementation on hyperledger material shows that the protocol is possible and can be implemented on a small to medium scale.

Wu, Hsin-Te, National Penghu University of Science and Technology

These authors have developed a blockchain-based network security mechanism for balloting systems. providing blockchains feature a distributed design and solely permit addition however not modification of knowledge, this effectively permits any user to conduct knowledge authentication. They extensively utilized linear pairing in establishing the network security mechanism that has the advantage of low encryption/decryption knowledge volume, which reduces the volume of hold on knowledge in nodes.

14.4 MAIN CONTRIBUTIONS

The contributions of this chapter are as follows:

(i) We have proposed an electronic voting system that uses ether for the implementation of the voting system.

(ii) It is versatile, as it can also be used to perform a survey.
(iii) The process has been made very easy to implement, yet it is very secure.

We have proposed a smart contract that has pseudo-codes, as discussed in this section.

14.4.1 Variables of the Contract

- *struct vote{*
  address voterAddress;
  bool choice;
  }

The vote contains two elements: the address of the voter, and the choice that voter makes. True or False.

- *struct voter{*
  string voterName;
  bool voted;
  }

A citizen has three attributes: his name, and whether or not he has voted.

- *mapping(uint => vote) private votes;*
  mapping(address => voter) public voterRegister;

Votes are kept in mapping named votes and voters are kept in a mapping known as voterRegister.

- *uint private countResult = 0;*

countResult is a non-public variable that no one can see.

- *enum State {Created, Voting, Ended}*
  State public state;

The ballot has three states: the time when it was created, confirmation of the process of voting, confirmation that the votes have been cast. At that time, the contract is closed.

14.4.2 Preparing the Ballot

- *constructor(*
  string memory _ballotOfficialName,
  string memory _proposal) public {
  ballotOfficialAddress = msg.sender;
  ballotOfficialName = _ballotOfficialName;
  proposal = _proposal;

state = State.Created;
}

The chairman initializes the contract with the func^{n} by providing his _ballotOfficialName and _proposal. The func^{n} reads his pocketbook address and update the ballotOfficialAddress.

- ***function addVoter***(*address _voterAddress, string memory _voterName)*
  public
  inState(*State.Created)*
  onlyOfficial
  {
  voter memory v;
  v.voterName = _voterName;
  v.voted = false;
  voterRegister[_voterAddress] = v;
  totalVoter++;
  emit voterAdded(_voterAddress);
  }

Next, the chairman adds voters to the voterRegister mapping. This involves entering the voter's pocketbook address and name into the mapping.

14.4.3 Vote Counting

- *function endVote()*
  public
  inState(State.Voting)
  onlyOfficial
  {
  state = State.Ended;
  finalResult = countResult;//move result from private countResult to public finalResult
  emit voteEnded(finalResult);
  }

The chairmain runs endVote() to finish the ballot. the specified state to run this perform is "Voting" and solelyOfficial ensures that only the chairman will run this fun^{n} The fun^{n} changes the contract state to "Ended" so no voters will continue balloting. It then updates finalResult (a public variable that everybody will read) with the merit of countResult (a non-public variable that I wont to accumulate vote counts as voters were voting). This concludes the ballot method.

The main procedure we need to follow is:

- Set few accounts on meta mask
  - Admin: That runs the smart contract and manages the process of voting.

  - Dhruv, Amit, Ronaldo: Voters.
- Open ganache and give 2 ETH to each voter so that they can run transactions
- The DApp starts by putting in the worldwide variables which will be used throughout the balloting life-cycle, namely:
  - accountaddress: this is often the Ethereum account of the chairman that manages the balloting method. This address is used once when the contract is deployed.
  - ballotContract: this stores the address of the ballot contract which will be generated once the ballot contract is deployed.
  - ballotByteCode: here, the computer memory code of the ballot contract is kept. We are going to populate the code on this variable shortly.
  - Ballot: this variable stores the Ballot object as we run and monitor it.
  - ballotABI: this is often the ABI for the Ballot contract.
  - voterTable: this may store a data table which keeps track of the voters, their addresses, and their balloting standing.

14.5 E-VOTING AND BLOCKCHAIN

An electronic voting system is just another balloting system which uses electronic means to implement the process of voting.

Depending upon the various needs of a democracy, a EVM can be designed accordingly. It can use the internet for interconnecting all the nodes all around the country at the time of voting. And so on, there may be an infinite number of ways of implementing and formulating the use of EVM in a vote.

Coming on to the blockchain, a blockchain is a dispensed log of data. There is no single or centralized governing authority. Each of the nodes is connected to the others (p2p connection).

In the process of e-voting, there are:

- Admin
- Voter
- District node
- Boot node (Figure 14.1)

There are many roles for an election to be conducted.

There are:

- **Administrators:** They are responsible for the running of the election. Many companies and government agencies are given this role.
- **Voters:** There are voters, the eligible citizens who have been given a clean sheet by the government i.e., the government has run a background check of the individual and it is a viable citizen and can vote.
- **District nodes:** They come into action when the main admin creates an election, every ballot is present at a different place in the country, i.e., a

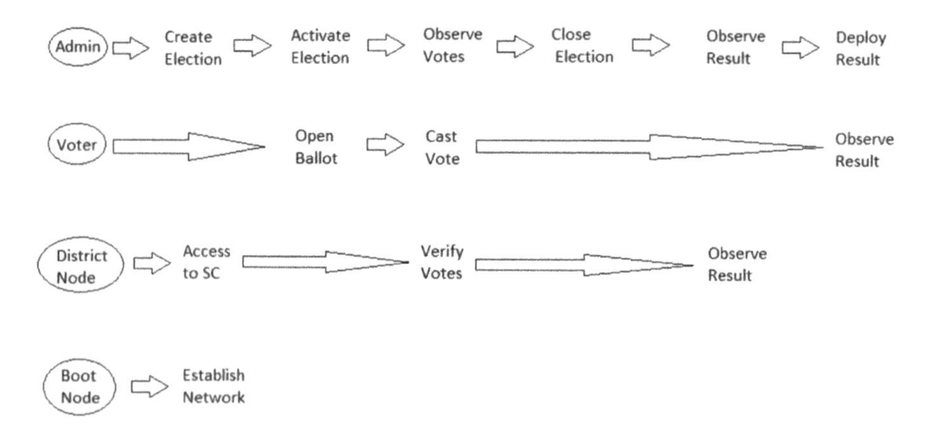

FIGURE 14.1 Election process and roles.

node is present at every district and is responsible for the effortless running of the smart contract and the maintenance of the same.
- **Boot Node:** A boot node is nothing but a helping hand that helps the district nodes find and communicate with one another.

The data is processed and then it has to be mined by a computer on the chain network to verify the authenticity of the data. There is no medial authority present to authenticate and substantiate the transactions, but every transaction in a blockchain is considered to be completely safe and corroborated. This is feasible because there are various consensus algorithms that help maintain the authenticity of a blockchain. Thus, to make this concrete, there are various consensus algorithms, including:

- Proof of work
- Proof of stake
- Proof of capacity

(i) Proof of Work

It is a consensus algorithm that uses the proof for the work done to appoint a miner to mine the next block and is also rewarded accordingly. PoW is the consensus algorithm that is used by bitcoin. The working of this algorithm is based on the fact that each miner in the blockchain is given a mathematical problem and has to give a solution. There may be thousands or millions of miners but only one gets the opportunity to mine the next block so any miner who solves the puzzle in the least amount of time will mine the next block, i.e., they have a very powerful hardware to pull out the solution to the puzzle in the minimum amount of time. There was a trend for building a mining rig just to get some cryptocurrency as reward after mining of a block.

(ii) Proof of Stake

Proof of Stake is again the most usual alternative to PoW Consensus algorithm. Ethereum used Proof of Work, the same as Bitcoin, but now has shifted from Pow to PoS. Now, this algorithm is a little bit different from the Pow algorithm. In PoW, the main focus was to make expensive hardware that can perform solving problems at faster rates but here, the main point is stake, i.e., all the miners have to invest some crypto coins to be able to mine a new block, i.e., stake. It is on the face that no hacker or a fraud person will ever invest their money as a stake because they must go through a background check to be able to do so. Now, after the stakes are placed, all the miners will mine the block together and after mining the block, all the miners will get their subsequent reward according to the bets they have placed, i.e., proportional.

(iii) Proof of Capacity
In the PoC algorithm, miners have to invest in hard drives. As in PoW, the miner with the fastest hardware to be able to perform the puzzle in the least amount of time gets the reward, here the miner will the maximum hard drive space will have chances of being selected to mine the next block.

14.5.1 Cryptography

Blockchain uses cryptography. It is a process of making and using codes to secure the transmission of information.

- **Plaintext:** It is the data to be protected during transmission.
- Encryption algorithm. It is a mathematical process that produces a ciphertext for any given plaintext and encryption key. It is a cryptographic algorithm that takes plaintext and an encryption key as input and produces a ciphertext.
- **Ciphertext:** It is the scrambled version of the plaintext produced by the encryption algorithm using a specific encryption key. The ciphertext is not guarded. It flows on a public channel. It can be intercepted or compromised by anyone who has access to the communication channel.
- **Decryption algorithm:** It is a mathematical process, that produces a unique plaintext for any given ciphertext and decryption key. It is a cryptographic algorithm that takes a ciphertext and a decryption key as input, and outputs a plaintext. The decryption algorithm essentially reverses the encryption algorithm and is thus closely related to it.
- **Encryption key:** It is a value that is known to the sender. The sender inputs the encryption key into the encryption algorithm along with the plaintext in order to compute the ciphertext.
- **Decryption key:** It is a value that is known to the receiver. The decryption key is related to the encryption key, but is not always identical to it. The receiver inputs the decryption key into the decryption algorithm along with the ciphertext in order to compute the plaintext.

There are two types of cryptosystems:

- Symmetric Key
- Asymmetric Key

The main difference between symmetric key encryption and asymmetric key encryption is the interconnection between the decryption and encryption key.

In symmetric key encryption, both of the keys used for encrypting the plain text and then decrypting the cipher text are the same.

In asymmetric key encryption, both of the keys used for encrypting the plain text and then decrypting the cipher text are different, i.e., there is one key for encryption and one for decryption.

Also, blockchain uses a digital signature; it used MD5 Hashing before but now uses SHA-256.

The following is the depiction of the process of a digital signature process (Figure 14.2).

The data is passed on to a hashing function which creates a hash code of (MD5-128 bit, SHA-256 -256 bits) respectively.

Now there is a signer's private key that only the signer has. Now he uses this private key to sign the hash code generated by the hashing function. Then this signature is used to sign and lock the data.

When it reaches the reader's side, the data is put onto a hashing function, which generates a hash code. Then the signature part of the message is verified by a verification algorithm with the help of a public key of the signers that is publicly available.

After this process, it also yields a hash code now this hash code is cross-checked with the previous hash code generated by hashing of the data. If equal it is verified and no tampering is done to the data.

Smart contracts: It is a self-executing contract that is traceable but immutable if once executed. This helps maintain the integrity of the contract.

In other words, smart contracts are lines of code that are stockpiled on a blockchain and they are being performed spontaneously when conditions are met.

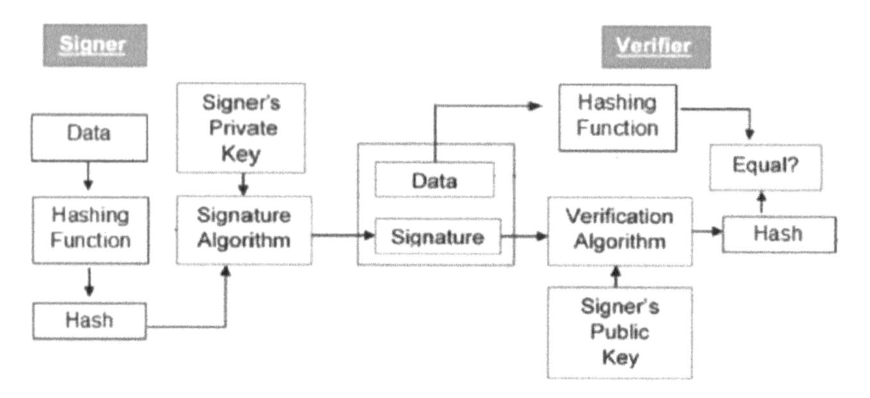

FIGURE 14.2 Digital signature process.

It is also a crypto contract, that directly controls the movement of information between two nodes. And also they work a lot like the vending machines.

14.6 USE OF BLOCKCHAIN IN VOTING SYSTEM

E-voting is not a new process. There have been many attempts in the past decade to make an e-ballot system which is secure and unhackable. All of the previous attempts have not been successful. Now, coming on to the introduction of blockchain in this environment of an e-voting system, it is very appealing as it is very secure even on the internet and also since it follows so many consensus algorithms, the use of hacking and also the use of digital signature, hacking into the one node will not compromise the whole system and also the unethical node will be known of as all the nodes are in a p2p network. The voters will be able to cast their votes anonymously and without revealing to whom they have voted for. Since tampering is not possible, there is no question about fake votes.

Governments around the world have been trying to implement secure and reliable e-voting (location-independent, individualized voting over the Internet) for a long time, to no avail: e-voting remains fundamentally insecure.

This can change with the *blockchain technology*, an open, transparent, and distributed digital ledger. However, blockchain-based e-voting will only work if the blockchain-based e-voting infrastructure is truly distributed and no one entity, not even the government, controls a majority of it.

DApp: decentralized application. It is a computer application that runs on a distributed computing system.

It stores the data in a decentralized database and used decentralized computing resources to work.

Ganache: software that allows us to make a personal blockchain with many accounts and let us customize the ether values as well.

MetaMask: an extension that eases the way to communicate with the Ethereum blockchain, i.e., it helps us use the Ethereum wallet easily.

The process of election can be explained in five points:

(i) Creation of the Election:

The election has to be created by the administrator using DApp (decentralized application). This interacts with the smart contract, in which there is a list of all the voters and all the district nodes and each and every information about the voting system.

All of the names of the candidates has to be there and along with their election sign. When this is deployed, the names of the candidates should appear on the screen along with the election sign.

Also to generate the election smart contract some ether would be required to make the election live so as there is proof that the creator is not fake.

(ii) Adding the List of Voters:

When the contract has been made, now the admin needs to make the list of the citizens that are eligible to vote. This may be done before the process of voting. The committee must run a background check on the citizens to see if they are more than 18 years of age and are the citizen of our own country and so grant them permission to vote and add their name on the voter list. Now adding them will again require some ether. After adding all the voters to the list the contract is all set to be deployed.

(iii) Vote Transaction:

This is the process after the contract has been deployed. When each individual is voting at their respective district node, they interact with the main ballot smart contract. But they can only interact with the ballot smart contract they have been registered to, i.e., only one district node all over the country.

Before the election each voter is given 1 ether, and it would cost 1 ether to vote. Hence, the voter can only vote once.

Whenever an individual casts their vote, it will cost them an amount of ether and after doing so they can cast their vote (1 ether in this case). After casting of the votes, the admin will be notified that the given candidates has voted, but for whom they voted would be never known. That is one of the main features of voting, i.e., anonymity. Each vote will be attached to the blockchain and so on till the voting ends. Only the voter would know for whom they voted and no one else.

(iv) Closing the Election:

When the time is up, the administrator will end the voting smart contract and this will in turn cost him some ether so as to only authentical admin (person) will end the voting.

(v) Counting Votes:

Each smart contract will tally their own results and this saves the result in their own dedicated storage. After the election is finished, final results of all smart contracts will be announced individually.

14.7 RESULT AND ANALYSIS

There are several ways in which blockchains will create government a great deal more responsible, clear, economical, and fraud-proof. Some of the fields involved include contract management, e-voting, and health care. There are already many developments in numerous countries concerning the employment of blockchain technology in e-health, e-resident systems, elections, and particularly land and property registration. One country that has already implemented many applications of blockchain technology is the Republic of Estonia. Other countries that have used blockchain technology include Sweden, Hong Kong, Ghana, Kenya, the Federal Republic of Nigeria, and Georgia. However, despite this, blockchain technology continues to be in its infancy, as there are still unknown factors and vulnerabilities.

The blockchain electronic voting system possesses many plus points over the other alternate systems,

- **Anonymity:** Each person in this e-voting system uses a specific ID rather than his real name. Thus, the privacy of the users is protected.
- **Security:** As the blockchain system has many consensus algorithms to follow as well as uses hashing techniques, decentralized ledger, and many more features, hence making it very secure.
- **Withdrawable:** We style a transactional paradigm that permits voters to change their vote before an arranged period. It meets the vital needs of an e-voting method.

Recommended actions to deploy blockchain technology in today's environment:

- To offer a balance between privacy and confidentiality on the one aspect and transparency on the opposite aspect
- Resolve challenges like group action speed, the verification method, and information limits.
- Provide superior, low-latency operations
- Ensure that distributed ledgers are climbable, secure and supply proof of correctness of their contents
- Energy potency
- Ensure a high level of cryptography

14.8 CONCLUSION

There are many places in which blockchain technology is being used. In this chapter, I have described an e-voting system using blockchain technology and have implemented it successfully. Since blockchain is highly encrypted and is peer-to-peer connected to the nodes and has various consensus algorithms, it is efficient and cost-effective. Many countries have already adopted a blockchain voting system and, in the future, many more will follow.

The idea is to move to a digital vote system to make the general public voting method cheap, quick, and easy. Creating a low-cost electoral method that also is quick will remove hurdles between officials and voters.

REFERENCES

1. Science Direct. *Democracy Online: An Assessment of New Zealand Government Web Sites*. Accessed: August 1, 2018. Available: www.sciencedirect.com/science/article/pii/S0740624X00000332
2. M. Volkamer, O. Spycher, and E. Dubuis. Measures to establish trust in internet voting. In *Proc. 5th Int. Conf. Theory Pract. Electron. Governance*, Tallinn, Estonia, 2011, pp. 1–6.

15 DAI for Document Retrieval

Anuj Kumar, Satya Prakash Yadav, and Sugandha Mittal

CONTENTS

15.1 INTRODUCTION

The record retrieval administration accepts the correct data is moved through the proper ways to the right area. Utilizing a few techniques for looking, removing, separating, perusing, etc., the applicable information for a given area is gathered from archives. A "site" here refers to an individual, a specialist, or a machine who has requested data through a specific arrangement of inquiries. A "data" unit is from "archives," and a record is a surface of data that mirrors the report [2]. A unit of data from an assortment of reports is gathered under a specific limited review. A "dream" of a spot (or creator) makes the introduction of the data interesting. Since each writer has clashing perspectives, all content is deciphered in an alternate path as the writer's composed item, although a similar material for data is introduced. Similar data can be separated from many records and each one has distinctive surface value [3].

Finding the most productive and briefest course from the area to the data (or report) or the other way around is the correct data bearing. At the point when we are at a "data" perspective, data trips through one content to the next, and then shows up at the right spot. From this perspective, we call it "document extraction (or retrieval)" or "retrieval of information," and record individually. Then again, standing apart from a "site," we change starting with one documentation then onto the next through them, looking through them until an informational index contains the right data [4]. That would be classified "search data" (or report space) or "exploring data space." "Datastream" is a term that portrays information recuperation and route. "Savvy" record retrieval depends on some "cognizance" of reports to get the productive streaming data method. A shrewd method of recovering reports isn't a consequence of an absence of attention to such records.

A wise method of getting to data doesn't originate from an absence of information on certain data. If the area is an individual, an operator, or a machine, we can recognize the right way to correct the reports for the area question given.

"Understanding" refers both to records and to areas. On the off chance that a content is supposed to be justifiable, it is written in very much framed sentences, pleasing sorts, all around organized information, all around communicated circumstance content dependent on the information on accepted perusers, perfectly changed in accordance with the perspectives on expected perusers, incredibly acclimated to the circumstance of accepted perusers, and thinking abilities. "Appreciation" alludes both to records and to areas. If a content is supposed to be justifiable, it is figured in very much framed words, first-rate types, all around organized information, potentially the best-communicated situation material dependent on the experience of assumed perusers, balanced to the perspectives on expected perusers, composed to the state of accepted clients, and thinking capacity. (We didn't determine what "level of information" here is) Look at another way, there's a distinction in data among them places. Presently, we'd like back to our "proficient report retrieval" issue. An audit of what is next we've been talking about up until this point. Accepted effective record retrieval that "understanding" states expected the supposed "total report express." An entire book might be drawn up by consolidating both the archive and understandability of the setting. The following is a summary of what we have been debating So far. Restoration Success of documents supposed "understanding" statements, which assumed a so-called "complete document state." A complete text may be drawn up by combining both the completeness of the document and the skills and technologies of the venue. It comes out our question is what constitutes a complete document. In the next part, a design document structure defining a "knowledge" of the documents in question will be addressed.

By way of cooperation or competition, several locations are required in the restoration process. We call this the Architecture of Information. Moreover, multi-lingual processing (or translation) on the information architecture is simply the process of restoration. Ultimately, the multiple language (or translation) processing of the information architecture of the document has to reflect a full understanding of the document's condition. A comprehensive report is a consequence of the compensatory mix of record as expressed above [5,6]. The procedure of reestablishment to full record requires a few procedures and assets. This recovery cycle is achieved by assuming the uniform specification of each of its layers is not restricted to one location implementing document architecture. Here, the "information interchange format" means the communication specification among every stratum of paper architecture and information architecture.

15.2 ARTIFICIAL INTELLIGENCE

Artificial Intelligence is the improvement of PC frameworks that can perform assignments that include intelligence. Discourse acknowledgment, visual discernment, discourse acknowledgment, complement, and choice—to-language correspondence are instances of such exercises. A computer with strong AI might benefit from observations. It also can think and behave just like a human. A usage of Artificial Intelligence that enables PCs to learn and advance without human help or current programming. "Making computational model of human behavior" is one thing. Since we believe that humans are smart, models of intelligent behavior must therefore be AI. Turing wrote a great paper that set up this concept of an AI as creating models of human behavior [7].

15.2.1 Some Real-Life Examples of AI

- Self-driving cars
- Boston dynamics
- Navigation systems
- Chatbots
- Human vs. computer games
- Many more!

15.2.2 Advantages of AI

The primary objective is to reduce injury in:

- Wars
- Dangerous workspaces
- Natural disasters
- Transportation
- Cleaning
- Many more!

15.2.3 Information Retrieval

The software engineering division's primary reason for existing is to store and recover an enormous amount of literary or interactive media data explicit to a client, for

TABLE 15.1
Divergences between the Systems IR and DB.

	Information Retrieval	Database/Data Retrieval
Matching	Partial match	Exact match
Inference	Induction	Deduction
Model	Probabilistic	Deterministic
Classification	Polythetic	Monothetic
Query Language	Natural	Artificial
Query Specification	Incomplete	Complete
Items Wanted	Relevant	Matching
Error Response	Insensitive	Sensitive

instance, message, pictures, discourse, and so forth. The articles dealt with by an IR application are generally called reports, and the product device, which consequently deals with these records, is called Information Retrieval System (IRS). The assignment of an IR framework is to enable a client to discover those containing the data the client is searching for in an assortment of archives or give help with satisfying the client requirement for data. IR is set up innovation that offers customers answers for more than four decades and will be a functioning zone of study [8]. This means that while there is a lot of work to be completed, plenty needs to be done. In reality, a confounding variety of indexing and text retrieval technologies have been proposed and tested over forty years. These techniques developed slowly and were strengthened by many minor refinements, rather than every big development (Table 15.1).

The significantly more fundamental distinction between IR and DB is that IR needs to get archives that apply to the client's information that must be communicated using a question, rather than simply coordinating this equivalent inquiry, much like DB. The differentiation is key, since centrality is a convoluted idea that goes past just coordinating certain inquiry terms and terms found in the record for an increasingly point by point conversation of the effects [1]. One of the ramifications of this is that the association between the requirement for information on a client and the substance of a report is viewed as vague, and the IR framework can endeavor to estimate it, leaving the client with a definitive judgment regarding its real significance. A couple of the ramifications of this is the association between a client's requirement for data and the substance of an archive is viewed as flighty, so the IR framework can just look to get it, leaving the client with a definitive judgment about its real importance. Along these lines, an IRS recuperates records that are probably going to be esteemed significant by the client, that is, liable to meet the client's data needs. As such, DB realities acquired in light of an inquiry are typically viewed as a genuine and full response to the inquiry. In IR, the apparent significance of an archive contrasts altogether among clients and on different occasions, just for a solitary individual. This IR function, of course, has some effects. Secondly, user requests sent to an IR device are usually vaguer.

15.2.4 Information Retrieval Assessment

In this brief, one significant research field worthy of mention IR review is concerned with evaluating IR programs. A lot of work and effort has gone into the analysis of the question assessment in IR. Considering the method suggested by Mr Rijsbergen in attempting to evaluate an IRS, at least a couple must be attempted to respond to questions:

1. What to assess?
2. How to assess?

In light of the primary inquiry, one should note that a scientist might be keen on deciding the speed of the mending procedure or the IR's level of collaboration. The framework takes into consideration, for instance. There are different aspects of IR that an analyst may have been engaged with looking at them, as should be obvious a few times in this book. All things being equal, the key perspective we should talk about here is the viability of the restoration procedure. Rather the subsequent inquiry requires an increasingly nitty gritty answer. This book discusses numerous approaches to assess an IR framework, especially if the framework is intended for a particular type of correspondence and important archives. In any case, care and precision are the two most popular adequacy checks generally applied in IR. On the other hand, the subsequent inquiry requests an increasingly specialized answer. In this book, we will see a few different ways to assess an IR framework, particularly if the framework is worked for a particular type of collaboration and various sorts of archives [2] (Table 15.2).

15.3 DISTRIBUTED ARTIFICIAL INTELLIGENCE

Artificial intelligence, distributed in 1975, an area of interest of artificial intelligence that deals with intelligent communications between agents emerged. Distributed artificial intelligence frameworks were considered as a gathering of savvy individuals, called specialists, who convey through coordinated effort, conjunction, or competition. It rose as a field of investigation of artificial intelligence that manages savvy cooperation's between specialists. Distributed artificial intelligence frameworks

TABLE 15.2
Precision Discernment and Evaluation Values Are Recalled

Documents	Relevant	Not Relevant	
B	$A \cap B$	$\neg A \cap B$	B
Not Retrieved	$A \cap \neg B$	$\neg A \cap \neg B$	$\neg B$
	A	$\neg A$	

were developed as a general public of keen individuals, called operators, conveying through joint effort, conjunction, or competition.

15.3.1 Introduction to Distributed Artificial Intelligence

Distributed Artificial Intelligence (DAI) is called Decentralized Artificial, a subfield of artificial intelligence examine committed to creating distributed critical thinking arrangements. DAI has close relations to multi-specialist frameworks and a field forerunner. Distributed Artificial Intelligence (DAI) is an answer for profound thought, arranging, and choice taking. It is embarrassingly equal, so it can grow huge scope registering and the spatial circulation of PC assets. These highlights permit it to take care of issues that include the preparation of exceptionally huge informational collections. DAI frameworks are self-governing learning intelligence hubs (operators), regularly broadly distributed. DAI frameworks are self-ruling learning intelligence hubs (specialists), regularly broadly distributed.

The DAI hubs may work freely, and incomplete arrangements are joined through contact between hubs, frequently no concurrently. Because of its scale, DAI frameworks are powerful and versatile, and essentially inexactly coupled. DAI frameworks are intended to acclimate to changes in the difficult portrayal or hidden informational indexes because of the size and multifaceted nature of redeployment. Dissimilar to homogenous or brought together Artificial Intelligence frameworks that have firmly associated and topographically close preparing hubs, DAI frameworks don't permit the assortment of every material datum in one single area. Subsequently, DAI frameworks regularly deal with subsamples or hashed encounters from huge datasets.

What's more, the source informational collection might be changed or altered during the execution of a DAI program. DAI creating and maintaining the knowledge base, expert programs, and learning processes. Research moves into another software engineering field, the distributed artificial intelligence (DAI). The primary goal of a DAI program is to follow various master participation to take care of a confounding issue. This kind of framework is commonly characterized on two levels. At the smaller scale level the master is seen as an information autonomous individual fit for performing such explicit obligations. The commitment of each master could be very straightforward: The program's quality stems from the commitment between the different specialists. The large scale level characterizes the master society, which takes care of the difficult given. A multi-master engineering causes the program to follow the participation of specialists. The benefits of such a program intrigue: Problem investigation is less unpredictable, as the issue isolates into independent sub-issues. The engineering is extremely secluded, bringing about improved execution. Moreover, framework development and re-utilization of modules inside different frameworks will be simpler. At long last, the chance of utilizing an equal machine truly suits such a design.

DAI points are simply to explain the issue of thinking, arranging, learning, and discernment by appropriating the issue to self-governing handling hubs (specialists), especially when they require huge information. To accomplish this point, DAI requires:

- A distributed system with reliable and elastic computing of unreliable and unsuccessful resources which are loosely linked
- Coordination of behavior and node communication
- Larger data sets and online machine learning tests

There are several reasons to want multi-agent systems to spread information or cope with such. Mainstream issues in research at DAI include:

- Parallel problem-solving: primarily addresses how to adjust traditional principles of artificial intelligence so that multiprocessor systems and clusters of computers can be used to speed up calculation.
- Distributed problem-solving (DPS): The entity concept, autonomous entities that can communicate with each other, was created to serve as an example for DPS implementation. For more details, please see below.
- Multi-agent based simulation (MABS): a DAI branch that provides the basis for simulations that need to analyze not only events at the macro-level but also at the micro-level.

15.3.2 Distributed Artificial Intelligence Tools

It is a data recuperation framework that utilizes DAI instruments comprising two slate-based modules: a semantic parser, and a reformulation module. The Linguistic Parser extricates the syntactic structure from an application and makes a system of formats which incorporates the solicitation's key information components. Semantic parser, and a reformulating module. The Linguistic Parser extricates the syntactic structure from an application and creates a system of layouts containing the data recovery framework which utilizes DAI programming comprising of two writing board-based modules: a key solicitation data work. The reformulation module uses this system to get to certain pieces of the applicable reports. In this part we characterize the capacity and design of every module in more prominent accuracy.

15.3.3 Complete Document and Document Interchange Format

"Complete record" requires a portrayal of the degree of information inside archives under the idea of report engineering. At whatever point correspondences summon, the two sides of records accept that they recuperate the total archive from the report's surface structure, that is, from the writer's composition. Recovery of the full archive is the presumption of legitimate correspondence. We name framework engineering detail "web exchange design."

A full record is gotten from the data of the two gatherings. Since the two sides participate for the comprehension of the per user area, they compensate for one another. Furthermore, they go after fulfillment as well. The following area will introduce an idea of "data design" to explain the correspondence information on the substances and procedures in the two sides (Figure 15.1).

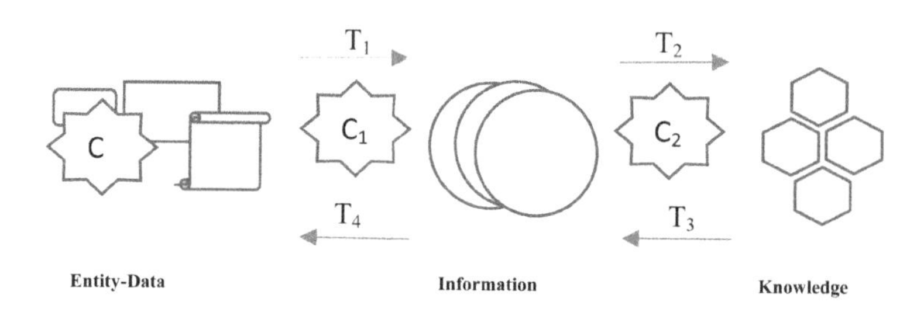

* Transformation: *Constraint:

-T1: Extraction of the information -Cl: View, C2: Situation

-T1: Extraction of Knowledge

-T1: Creating the Concept

-T1: Data Processing

FIGURE 15.1 Information architecture system.

15.3.4 Data Network Architecture for Distributed Information Retrieval

The creator has total information to comprehend the creator's report, as found in the past segments, yet the creator doesn't give the full archive. A full archive is essential for shrewd record recovery. A definitive point of recovering records is to accomplish proficient contact among authors and per users. The issue is the place the total report can be recouped. It is asserted that one methodology is the data engineering (Figure 15.2).

15.3.5 Types of DAI

- In multi-agent structures, agents organize their expertise and actions and the reasoning for the communication processes. Agents are physical or virtual entities capable of behaving, perceiving its environment, and interacting

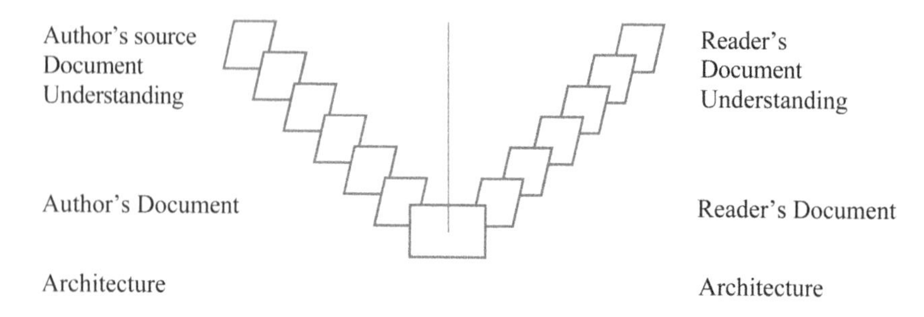

FIGURE 15.2 Distributed information architecture.

with others. The agent is independent and has the ability to achieve goals. Through their actions, the agents change the state of their environment. There are a variety of different forms of teamwork.

- The work is divided among nodes, and the knowledge is shared in distributed problem-solving. The main concerns are the task of breaking down and synthesizing the knowledge and solutions.

15.3.6 Challenges in Distributed AI

1. When to execute agent communication and interaction, and which language or protocols of communication should be used
2. How to ensure agents are reliable
3. How to synthesize the results between groups of "intelligent agents" by wording, definition, decomposition and allocation

DAI frameworks don't require all the pertinent information to be collected in a solitary area, as opposed to solid or incorporated Artificial Intelligence frameworks which have firmly coupled and topographically close handling hubs. In this manner, DAI frameworks frequently work on sub-tests or hashed impressions of enormous datasets. Moreover, the source dataset may change or be refreshed throughout the execution of a DAI framework.

15.3.7 The Objectives of Distributed Artificial Intelligence

To fathom the thinking, arranging, learning, and discernment issues of artificial intelligence, particularly on the off chance that they require huge information, by dispersing the issue to self-governing handling hubs (specialists). To arrive at the target, DAI requires:

- A distributed system with robust and elastic calculation of unreliable and unsuccessful resources which are loosely linked.
- Coordination of actions and coordination between nodes.
- Online machine learning and samples of large data sets.

There are several reasons to want multi-agent systems to spread information or cope with such. Mainstream issues in research at DAI include:

- Parallel problem-solving: primarily discusses how to change traditional artificial intelligence principles so that multiprocessor systems and computer clusters can be used to speed up calculation.
- Distributed problem-solving (DPS): the agent definition has been developed to serve as an interface for creating DPS, autonomous entities that can interact with one another.
- Multi-agent based simulation (MABS): a DAI branch that provides the basis for simulations that need to analyze not only macro-level phenomena but also micro-level phenomena.

- Operators facilitate their insight and exercises in multi-specialist frameworks and the thinking for the correspondence forms. Agents are real or virtual beings able to function, perceive their surroundings, and interact with others. The operator is independent and is fit for accomplishing destinations. The operators adjust the condition of their condition through their conduct. There are a considerable number of various procedures for cooperation.
- The agent is confident and can accomplish objectives. The specialists modify the condition of their reality through their conduct. There are a considerable number of various techniques for synchronization.

DAI may apply a base up way to deal with AI, like the subsumption engineering, just like the traditional top-down way to deal with AI. What's more, DAI can likewise be a vehicle for crises.

Similar to the subsumption architecture and the conventional top-down approach to AI, DAI can apply a bottom-up approach to AI. Additionally, DAI may also be an emergency vehicle.

15.3.8 Areas in Which DAI Is Implemented

- Electronic exchange, such as the DAI method, learns from the subsamples of very broad samples of financial data for trading strategies
- The DAI program controls networks in the WLAN network http:/dair.uncc.edu/projects/past-projects/wlan-resource, e.g., in telecommunications, shared resources
- Routing, for example model vehicle flows within transportation networks
- Scheduling, e.g., flow shop scheduling, where local optimisation and coordination efficiency and local continuity are ensured by the resource management agency
- Multi-agent frameworks like artificial computers, computer simulation analyses
- Electric power stations, e.g., Multi-Agent Condition Monitoring System (COMMAS) and IntelliTEAM II Automatic Restoration System applied to track transformer conditions

15.3.9 Software Agents

The key idea utilized in DPS and MABS is the program called machine operators for reflection. A specialist is a self-sufficient virtual (or physical) object that comprehends its condition and follows up on it. In a similar procedure, a specialist is generally ready to team up with different operators to accomplish a typical target that one operator couldn't accomplish alone. This strategy for correspondence utilizes a language for speaking with the beneficiary. A valuable first grouping is to isolate the specialists into:

- Reactive agent: not much more than an automaton that receives input, processes it, and generates output.

- Deliberative agent: by contrast, the deliberative agent should have an unbiased view of his environment and be in a position to execute his plans.
- Reactive agent: not much more than a car that receives input, processes it, and generates output.
- Deliberative agent: a deliberative agent will, on the contrary, have an internal perception of its world and be able to implement its plans.
- Hybrid agent: a hybrid agent is a mixture of reactive and deliberative, implementing their plans but also reacting directly without deliberation to external events.

Well-recognized architectures of agents that define how an agent is organized internally include the following:

- ASMO (emergence of distributed modules)
- BDI (Believe Desire Intention, a general architecture that describes how plans are made)
- InterRAP (A three-layer architecture, with a reactive, a deliberative, and a social layer)
- PECS (Physics, Emotion, Cognition, Social, describes how those four parts influence the agent's behavior)
- Soar (a rule-based approach)

15.4 CONCLUSION

The fair progression of data is exemplified in the design of the archive and its normalization of the configuration for record trade. Under the data engineering worldview, the general picture is drawn. One idea of data engineering is the mono-lingual record recovery, and the multi-lingual interpretation administration. Archive recovery's "intelligence" accept full understanding, and that ends up being a finished report.

The normalization issue in an activity of this sort is one of the handy curios for viable correspondence. Accepting such guidelines, the information to recuperate the whole record can be situated in a "distributed" organize, independent of whether it is epitomized in an essentially or truly. Under this worldview, reasonable collaboration is conceivable in a distributed domain. The procedure starts from the creation stage. This is an intriguing methodology since every module will boost the advantages of each kind of control Linguistic parser effectiveness, adaptability of the particular, and working reformulation module perspective. These practical focuses bolster this recovery.

REFERENCES

1. M. Braunhofer, M. Kaminskas, and F. Ricci. Location-aware music recommendation. *International Journal of Multimedia Information Retrieval*, 2(1), 31–44, 2013.
2. Y.-S.Choi, J. Lee, J.-X. Huang, and K.-S. Choi. Cross-language information retrieval system for Korean-Chinese-Japanese-English languages. *Proceedings of ACL2000 (Demonstration)*, Hong Kong, 2000.

3. G. Grefenstette, ed. *Cross-Language Information Retrieval.* Kluwer Academic Publishers, 1998. www.springer.com/gp/book/9780792381228
4. K. Sparck Jones, and P. Willett, eds. *Readings in Information Retrieval.* Morgan Kaufmann, 1997. https://dl.acm.org/doi/book/10.5555/275537
5. G. Schecter.*Information Retrieval: A Critical View.* Academic Press, London, 1967.
6. W. Goffman. A searching procedure for information retrieval. *Information Storage and Retrieval*, 2, 294–304, 1977.
7. Swets, J.A. *Effectiveness of Information Retrieval Methods.* Bolt, Beranek and Newman, Cambridge, 1967.
8. S. Mizzaro. How many relevances in information retrieval? *Interacting with Computers*, 10(3), 303, 1998.

16 A Distributed Artificial Intelligence

The Future of AI

Pushpa Singh, Rajnesh Singh, Narendra Singh, and Murari Kumar Singh

CONTENTS

16.1 INTRODUCTION

Distributed artificial intelligence (DAI) is a subfield of AI that is used for learning, planning, decision capability, and remote processing. DAI techniques are particularly valuable for those problems that are intrinsically distributed in nature or required to process vast volumes of data computation quickly (Kokkinaki & Valavanis, 1995). There is a huge dataset used in AI-based applications for learning and prediction. These datasets are coming from a wide range of computational resources and coming from different areas. In order to analyze large amounts of data quickly, IoT and 5G network offers best-connected channel model (Singh & Agrawal, 2018a) as an internet infrastructure for AI- and IoT-based applications. These applications need scalability, fault tolerance, and fast processing that is very tough in centralized systems (Patel, 2019). Distributed computing environment is an ultimate solution to implement AI and IoT-based applications. Modern AI-based applications should be designed in such a way that can support and implement on distributed nodes. This type of AI that supports a distributed environment is calledDAI and a distributed node is termed an "agent." DAI is also called as Decentralized Artificial Intelligence devoted to the development of distributed solutions for real-world problems. DAI is based on assignment, controlling, coordination, and integration of multiple agents (multi-agent). Current approaches to DAI consist of the design and development of multi-agent and distributed AI systems. The objectives of DAI is to design multiple independent node or agents that can achieve a common goal when run together, specifically if they involve vast data set, by distributing the task to the independent processing agent.

AI states to a wider knowledge where machines can execute tasks "smartly" and think like a human being. Machine learning and deep learning are subsets of AI that are used to solve real-world problems. Machine learning can be defined as a system that can learn on its own without explicit programming. Deep learning is a subset of machine learning applied to large data sets and inspired by the processing pattern of the human mind in terms of the network of neurons. In a deep network there are a number of layers and each layer contains a number of neurons to process, learning algorithms to build a model; for example CNN, RNN, etc. (Alom et al., 2019). Applications based on this type of model need several machines and must exploit parallelism and distributed environment. DAI must entail a framework that comprises the performance benefits of an optimized system without requiring the user to reason about scheduling, data transfers, and machine failures. "Ray," a distributed framework for evolving AI applications, employs a distributed, fault-tolerant, single dynamic execution engine targeted at large-scale machine learning and reinforcement learning applications (Moritz et al., 2018). DAI overcome the challenges of traditional AI and offers to learn constantly and adapt to the dataset changes according to the distributed approach. Distributing the processing to a set of autonomous processing nodes helps in speeding up computation. DAI is concerned with synchronized intelligent behavior: intelligent agents, coordinating their knowledge, skills, and plans to solve problems and working toward specific goals that interact (Bond & Gasser, 2014). DAI is based on Agent-Based Modeling (ABM) and Multi-Agent Systems (MAS) to run a multiprocessor system and clusters of computer or node or entity (Simmonds et al., 2019).

In this chapter, the author proposes an AI-enabled node for distributed environments. AI-enabled node is actually referred to as an agent in a multi-agent system of DAI. This chapter aims to elaborate on real-time applications of healthcare systems where the system requires complex learning, prediction, and decision-making due to huge-scale calculations and spatial distribution of computing nodes. DAI offers real-time processing, consumption, storage, and distribution in various applications ranging from healthcare (Datta et al., 2019), agriculture (Eli-Chukwu, 2019), and waste management (Dubey et al., 2020). The medical and healthcare applications will offer several health evaluations to patients, decision support for prescribing medicines, and suggestion. AI in agriculture has important potential to leverage new technologies such as satellite imaging, distributed sensor, cloud, IoT, etc., to process a system that forms a distributed environment with a different agent like soil, farmer, irrigation, etc.

The remainder of the chapter is structured as follows: Section 16.2 discusses the historical background and challenges of traditional AI. Section 16.3 introduces the foundation of distributed environment and the objective of DAI. DAI-based real-time applications are given in Section 16.4. Section 16.5 includes future work directions and Section 16.6 concludes the chapter.

16.2 BACKGROUND AND CHALLENGES OF AI

AI is the area of computer science that simulates the human brain. AI is not a new phenomenon or concept. The term “Artificial Intelligence” was coined for the first time in 1956, and it took a couple of decades to make significant progress toward making it a technological reality. A brief development history of AI is represented in Table 16.1.

AI researchers functioned on traditional AI and strategies are mainly dependent on heuristic algorithms. For example, Deep Blue, the first chess-playing framework to beat a world chess champion, utilizes heuristic to discover an improved play during the game. However, as it may, Deep Blue could scarcely deal with large-scale data (e.g., the game of Go) because of the high complexity of its heuristic algorithms and its powerlessness to be applied to any situations other than chess (Zheng et al., 2019).

16.2.1 Hardware for AI

In 1965, Gordon Moore detected that the number of transistors, in a thick integrated circuit. Since 2005, researchers were no longer getting faster computers and the hardware was designed in multicore fashion. So for taking full advantage of new hardware implementation, the software has to be developed in a multi-threaded environment also. Current machines consolidate incredible multicore CPUs with devoted hardware intended to tackle parallel processing. GPU and FPGA are the most famous devoted hardware ordinarily accessible in workstations creating AI frameworks. A GPU is a chip envisioned to amplify the processing of multi-dimensional data such as images.

16.2.2 Platform and Programming Languages for AI

AI platform is characterized as a combination of hardware architecture or software framework with an application structure that permits a program to be executed. The AI platform also includes the utilization of machines to play out quicker than human beings. The platform simulates cognitive functions that the human brain can perform such as critical thinking, reasoning, learning, planning, social knowledge as well

TABLE 16.1
Development History of Artificial Intelligence

Year	Description
1943–1954	McCulloch and Pitts proposed that neurons could be combined to build a Turing machine in 1943. In 1950, Alan Turing proposed "Computing Machinery and Intelligence" also known as the "Turing Test." Marvin Minsky and Dean Edmunds in 1951, formulated SNARC as the first artificial neural network. In 1952, Arthur Samuel builds up the first PC checkers-playing program and the main Computer program to learn all alone.
1955–1961	The first artificial intelligence program by Herbert Simon and Allen Newell. Rosenblatt perceptron model an ANN model for pattern recognition (1957). Lisp Programming was introduced by John McCarthy for AI in 1958. James Slagle develops SAINT (Symbolic Automatic INTegrator) program in 1961.
1962–1969	A natural language understanding Computer program introduced by Daniel Bobrow. Edward Feigenbaum et al. took a shot at DENDRAL at Stanford University in 1965. Terry Winograd creates SHRDLU, an early natural language understanding Computer program in 1968. In 1969, Minsky and Pappert indicated that perceptron could not learn to evaluate the logical function of exclusive-or (XOR).
1970–1979	The WABOT-1 robot was working at Waseda University in Japan. MYCIN in 1972 for identifying bacteria causing serious contaminations, XCON (Master CONfigurer) program in 1978.
1980–1990	The powerful learning rules had been developed which enabled multiple-layered networks to be trained. The Japanese Service of Global Exchange and Industry, designed a PC that could carry on discussions, decipher dialects, decipher pictures, and reason like people in 1981. A back-propagation networks was proposed by David Rumelhart, Geoffrey Hinton, and Ronald Williams in 1986. In 1988, Marvin Minsky and Seymour published an extended release of their 1969 book Perceptrons. However, from 1969 to 1993 AI experienced major winter due to collapse of the market.
1991–2000	In 1993, Vernor Vinge publishes "The Coming Technological Singularity," where he predicts that "inside thirty years, we will have the mechanical way to make superhuman insight. Not long after, the human period will be finished." In 1993, research began to rise again, and in 1997, IBM's Deep Blue beat a chess champion.
2001–2010	The first DARPA Fabulous Test, a prize rivalry for self-sufficient vehicles, was held in the Mojave Desert. None of the self-sufficient vehicles completed the 150-mile route in 2004. Fei Li (2007) and associated with ImageNet.
2011–2020	A CNN planned by specialists at the University of Toronto accomplishes an error rate of just 16% in the ILSVCR in 2011. In 2016, a computer Go program developed by Google DeepMind to "solve intelligence." An OpenAI-machine bot played and won the tournament in 2017. Alibaba's AI Outguns Humans in Reading Test in 2018. The service of Google Duplex permits an AI assistant to book appointments on phone in 2018. In 2020, Robots Can Help Combat COVID-19 and AI as Mediator: Communication During Pandemic is some recent development of AI.

as general insight. The top AI platforms are Google AI, TensorFlow, Vital AI, Wit, Receptiviti, Microsoft Azure, Rainbird, Infosys Nia, Wipro HOLMES, Dialogflow, Premonition, Ayasdi, MindMeld, Meya, KAI, Watson Studio, Lumiata, Infrrd, etc.

Programming languages play a vital role in the development of AI. Various researchers carried out their research projects in the area of AI, and found special necessities; for example, the ability to easily manipulate symbols instead of preparing numbers or series of characters. However, a researcher from MIT, John Maccarthy, built an ad-hoc programming language for a rational framework called LISP. From that point, several programming languages have been developed, such as IPL, LISP, C, Algo 68, C++, C#, GO, Java, Julia, Matlab, Python, R, Prolog, Clojure, Scala, etc.

16.2.3 Challenges of AI

Computer-based AI is likewise called Machine Intelligence, having the expectation to build human insight by having the option to obtain and apply information and aptitudes (He et al., 2019). The traditional architectures were very limited and only capable of handling a simplified version of the application. Whenever a machine ends tasks based on a set of required rules that solve problems, such "intelligent" behavior is called artificial intelligence. This intelligent behavior is possible by "learning" and learning is possible with large-volume datasets, rules, and mathematical algorithms. In order to learn, predict, and analyze, traditional architectures of AI framework are not suitable. There are the following challenges of the traditional AI system:

- Traditional AI framework lacks specialized software to take input as unstructured data, and process and display output.
- A traditional AI framework must require training and human interaction to understand the result.
- Risk of hardware failure.
- Slow processing capability due to large dataset and complex computation.
- Lack of a specific processing unit such as CPU to process deep learning.
- The dataset needs higher RAM, we had no way to rent a bigger machine.
- Poor network infrastructure.
- Lack of infrastructure to integrate with the latest technological advancement.

16.3 COMPONENTS AND PROPOSED ENVIRONMENT OF DISTRIBUTED AI

The current AI architecture framework must include different components such as software that are used to design a multi-tenant data science environment, the distributed computing, training and inference environment, database management requirements, and highspeed network environment. MapReduce of Hadoop was fabricated to permit automatic parallelization and distribution of large-scale computations to process huge amounts of raw data, such as crawled documents or weblogs, and calculate numerous types of derived data. Figure 16.1 indicates the theoretical component representation of a single AI entity.

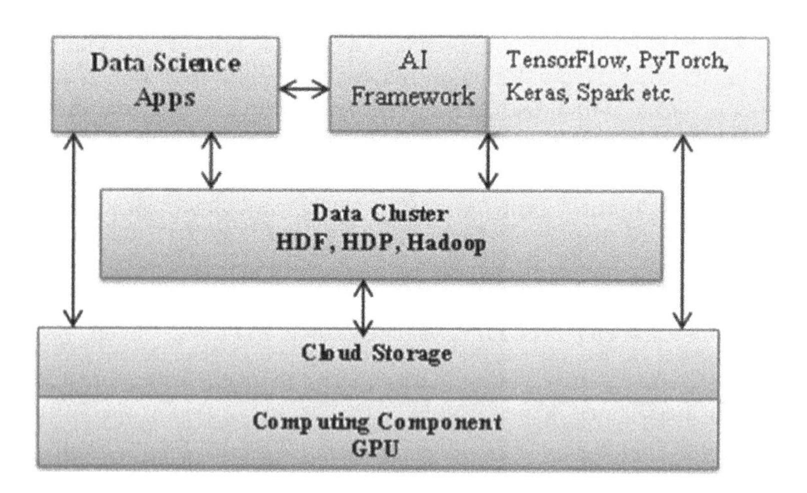

FIGURE 16.1 Component in AI-enabled node.

The important components used to evolve DAI that offers the capability to meet requirements with a distributed environment are listed in the following sections.

16.3.1 Graphical Processing Unit (GPU)

The primary resource of AI is the GPU that is acting as accelerators and offers storage solutions and networking. The GPU is on the IBM Power system, server nodes. NVIDIA Tesla V100 GPU model offers high processing computing (HPC) and graphics. It provides performance of 100 CPUs in single GPUs. GPUs are suitable for the matrix and vector math involved in machine learning/deep learning and has the ability to enhance the speed of deep-learning systems more than 100 times, reducing running times from weeks to days. If this is processed on different computing nodes or as a separate agent and run parallel, reducing running time from hours to seconds.

16.3.2 Storage

In order to support a variety of data formats and velocity of data produced by AI, the storage system must distribute on the cloud. The cloud provides storage, archives, and backups. Distributed snapshot protocols are required for effective AI computation in cloud computing environments. In order to provision the demand for AI applications and promise the service level agreement, cloud computing should deliver not only computing resources, but also essential mechanisms for effective computing (Lim et al., 2018).

Hadoop Distributed File System (HDFS) is a distributed and scalable file system that was written for the Hadoop framework. This file system attains consistency by duplicating the data across multiple node sin workflows of AI, the data growth can be massive and sudden. HDFS is used to maintain multiple copies of large datasets for the performance of both random and sequential I/O.

NetApp enterprise storage for Hadoop Distributed File System (HDFS) offers solutions that permit seamless data management and deployment options across the edge environments, on-premise data centers, and the cloud (Ranganathan, 2018).

16.3.3 High-Speed Reliable Network

5G networks offer the fastest network to resolve any network performance problems (Varga et al., 2020). 5G services offer reliable and unlimited network facilities to geographically distributed systems in order to support distributed AI. In IoT, every object has the ability to sense, transmit information, and provide feedback through the application of AI process in a distributed architecture with processing, intelligence, and connectivity at the cutting edge of AI application (Vermesan et al., 2018).

16.3.4 Proposed Distributed Environment of DAI

DAI is the science of distributing, managing, and predicting the performance of tasks, goals, or decisions in a multiple agent environment (Deshmukh, 1992). DAI is a class of technologies based on a multi-agent system to distribute problem-solving over a set of AI-enabled nodes. These nodes are distributed in different geographical locations and connected with high-speed networks to process AI-based applications, for example driverless cars, healthcare systems (from one hospital to other), etc. DAI is based on the multi-agent system. An agent can be said as a means by which something is accomplished. An agent may be intelligent, person, mobile, and laptop, desktop or any other electronic device. This distributed environment offers DAI as a high-speed network where each coordinator agent can divide the task into different distributed locations. A coordinator agent keeps track of all tasks assigned to other agents and control and manages the result processed by distributed agents. The agent-coordinator (head, the central body) decomposes the original problem into distinct tasks. These tasks are distributed between the agent-executant for processing and execution. Each agent-executant solves the task. Then agent-head can monitor the composition and integration of corresponding to the result of the selected task. Agent-integrator is responsible for the overall result (often, this is the same agent-coordinator). An environment of DAI is represented in Figure 16.2.

DAI systems are defined by three main characteristics (Ponomarev & Voronkov, 2017):

1. Task distribution between the agents.
2. Power distribution between agents.
3. Communication between the agents.

The distributed system entails the following featured to present a DAI environment:

- Load sharing, communication, and coordination of the nodes.
- Distribute samples of huge data sets on different nodes and online machine learning.

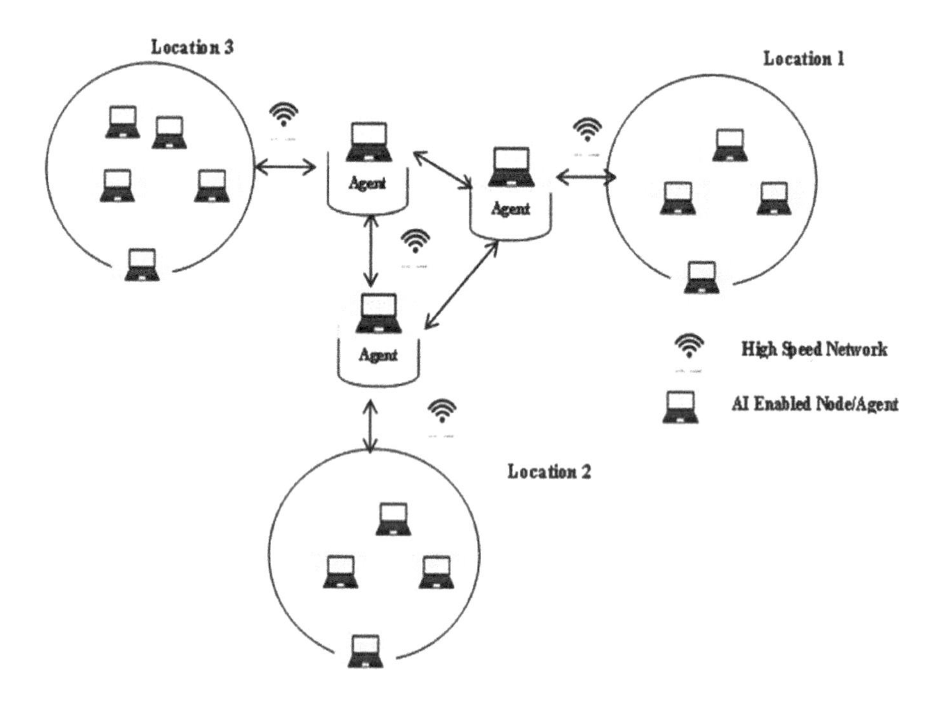

FIGURE 16.2 Agent-based distributed environment for AI.

- Parallel execution of complex set of queries in case of deep learning.
- Distributed machine learning algorithms are fragments of extensive learning and due to its capability to distribute learning methods onto several other nodes. Computing performance of AI and distributed learning algorithms have been scaled up.
- Help in mitigate many of the security problems that one may encounter in such environments (Verma et al., 2018).
- Provides fault-tolerant and robust environment for AI applications.
- Reduced cost of process as well as enhancement in overall performance.

Moreover, a distributed computing environment offers integration of current technology to successfully implement an AI platform. DAI denotes a subfield of AI that is based on multiple processors. There is a group of nodes or agents (logically distinct processing elements) which are trying to solve a problem. The most significant advantage of distributed computing is that it provides unrestricted scalability, without a single point of failure. In case of failure of one node, other nodes can take their responsibility without any data loss that has not happened in the central system.

The BOINC (Berkeley Open Infrastructure for Network Computing) is an open software platform for an organization of distributed computing. BOINC architecture is based on the concept of a distributed AI. It has a BOINC client and BOINC server. BOINC client can communicate with HTTP communications protocol to work. The client can run and control applications. BOINC server entails a set of individual subsystems (agents), each of which is responsible for its own well-defined task, such as performing calculations, file transfer, task scheduler, etc.

16.4 APPLICATION OF DISTRIBUTED AI

A DAI system comprises distributed nodes which act autonomously during communication. Moreover, the different DAI application claims to utilize AI to take the problem-solving capacities of doctors. We trust that AI innovations are developing quickly and may help the capacity of people to give social insurance (Kermany et al., 2018; Esteva et al., 2017). DAI has the prospective of detecting important connections in a dataset and also it is broadly used in numerous applications. The most common applications are healthcare, agriculture, waste management, e-commerce, intelligent networks, etc.

16.4.1 Healthcare Systems

AI is being used diversely for healthcare and research purposes like disease prediction and identification, association between co-disease, provision of health services, and medicine discovery (Tran et al., 2019). DAI is an attractive method for incorporating and evolving distributed healthcare systems. DAI is capable to solve the problem of heterogeneity, supports intelligent and distributed storage, allows an optimal personalized e-health environment, and improves the overall system performance. The agent may function as a part of a distributed system environment and assure interoperability between diverse systems that are to be combined for a working heterogeneous healthcare systems (Al-Sakran, 2015).

A typical healthcare ecosystem with distributed node and their diversified data sources is perceived in Figure 16.3. Each node denotes a different computing agent such as patient agent, doctor agent, insurer agent, pharmaceutical agent, research agent, etc. This agent can communicate with other agents. The agent-coordinator can gather the information from one hospital and provide it to other hospitals for better accuracy and prediction.

Distributed AI can help in finding fast and effective methods and environments that can efficiently combat the COVID-19 disease. The data can be gathered through agents which is distributed location-wise where people are connected to send real-time information about symptoms and diagnosis. An AI-based system offers important information and recommendations for the experts and research communities (Pham et al., 2020). The capabilities of bioinformatics also contribute in infectious disease detection and drug discovery in order to stop epidemic and pandemic situations in the future (Singh & Singh, 2020a).

16.4.2 Agriculture Systems

Agriculture automation is the foremost area and emergent topic for all countries. Automation of agricultural systems has enhanced the gain from the soil and also strengthened the fertility of the soil (Al-Sakran, 2015). AI-based agriculture systems,

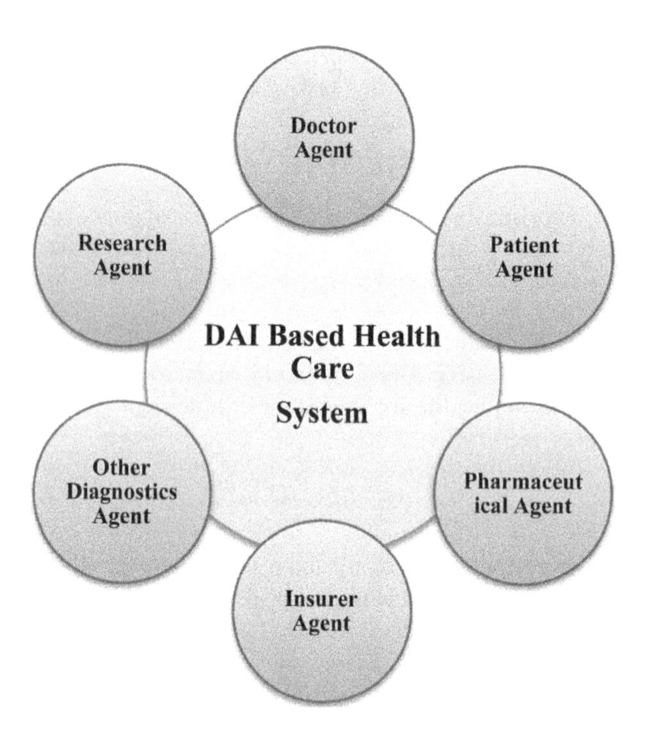

FIGURE 16.3 DAI environment for healthcare systems.

as shown in Figure 16.4, have various nodes such as soil, farmer, weather, irrigation, fertilizer, and crop management. These nodes or agents are distributed nodes that are connected through IoT, stored as big data, and analyzed and predicted by using AI. The various sensors are used for monitoring agricultural systems. For example, DHT11 is a low-cost digital temperature and humidity sensor applied in agricultural systems, and a soil moisture sensor is applied in determining the soil moisture level. Data from each node or agent are collected and analyzed. Deep learning and machine learning are certain domains which help expand a more advanced technology based on AI. The machine or agent can communicate to decide which crop is appropriate for harvesting and which fertilizers can produce the maximum growth of crops (Jha et al., 2019). The weather is continuously analyzed and provides alerts to the farmers. Irrigation departments and government update their policies to the farmer. The irrigation system transports the water to distinct fields within the farm at the right time and right quantity. These distributed agents are connected and facilitate smart farming. The overall objective is to increase productivity, prediction, and estimation of farming parameters to enhance economic efficiency. These data are used to carry out the machine learning processes, store data, and communicate with cloud services to represent different value like soil moisture, soil temperature, etc., that let the farmer decide to use irrigation and design new rules to enhance production (Ferrández-Pastor et al., 2018).

Agent in a distributed system can be physically located at the same place and connected via a local network or connected by a Wide Area Network to other location

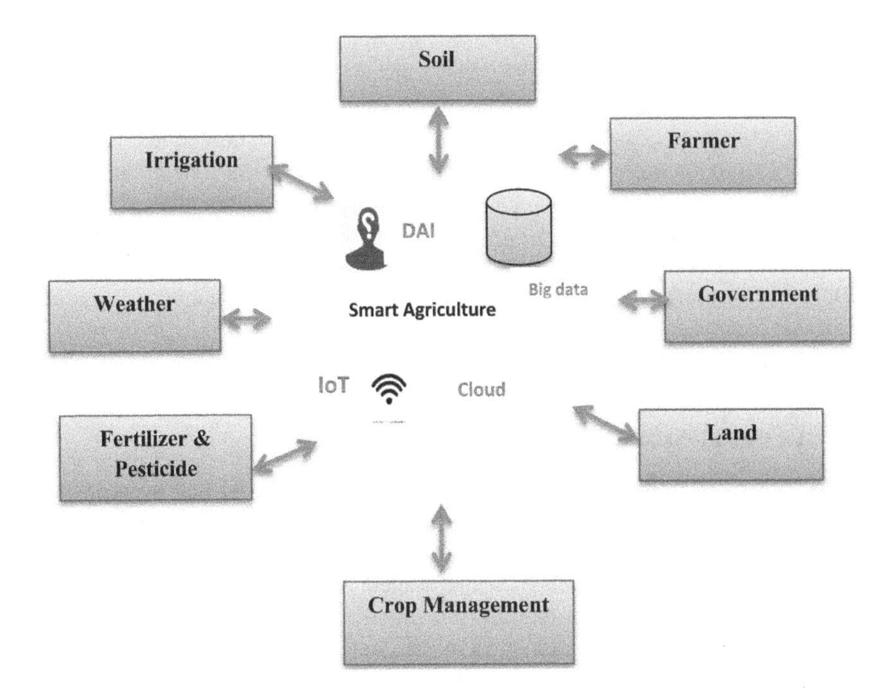

FIGURE 16.4 DAI-based agriculture systems.

to resulting in low latency and better performance. DAI systems permit to allocate complex problems/data into smaller parts and assign on multiple computers (agent) to work in parallel manner, which can reduce the processing time needed to predict, identification, classification and automation.

16.4.3 E-Commerce

The DAI system can learn financial transaction rules and policies from subsamples of very large samples of financial data in E-Commerce. AI with E-Commerce provides various services such as intelligent logistics, recommendation systems, electronic assistant in product searching, and optimal pricing facility (Song et al., 2019). Experiments on large-scale real online banking data, customer data and electronic shops that form a distributed environment can attain significantly higher accuracy and also identify a loyal customer (Singh & Agrawal, 2018b; Singh & Agrawal, 2019) consumer behavior, customer segmentation (Singh et al., 2020), and product preference in the digital market. A significant advancement in Blockchain and the ability to integrate with AI and IoT provides a transparent transaction system to business (Briggs & Buchhols, 2019).

16.5 FUTURE SCOPE

Distributed AI is the need of the hour since future data analytics will be primarily done in a distributed environment. Integration and coordination with heterogeneous nature of agent, component, protocol, etc. are major challenges that will be tackled near future. Furthermore, security and transparency are remaining challenging issues in the distributed environment of AI. A clear procedure must be required to accept the associated complexities and legal difficulties to guarantee that the society will benefit from the improvement of AI. AI-based decisions may lead to completely new dynamics and technologies fostering explainability, authenticity, and user-centricity. Blockchain technology can be combined with AI to support transparent and distributed decision-making processes of multi-agent systems (Singh & Singh, 2020b)). The multiple-agent system can be used for the management and organization of fog computing and blockchain (Calvaresi et al., 2019). DAI can be realizing to expect an autonomous car on the road in the next five years.

16.6 CONCLUSION

In this chapter, we introduced a distributed environment for AI that overcomes the challenges of traditional AI systems. Distributed system with AI termed as DAI, which offers allocation, integration, communication among distributed node that is termed as an agent. Hence, DAI is a multi-agent system that leverages the latest technology such as IoT, cloud computing, etc. in order to solve real-time applications. AI applications such as healthcare, agriculture, and e-commerce are based on large data set and complex algorithm. An agent-based distributed environment offers ease of computation and quick prediction and classification to these applications. In other words, DAI would be able to solve absolutely any problems with virtually unlimited storage, power, connectivity, and maximum efficiency.

However, DAI often makes the system very difficult to correctly and completely specify behavioral dynamics of agents. Leveraging the agent with recent technology such as IoT, cloud, blockchain, etc. enabled the DAI more robust, secure, and fast.

REFERENCES

Al-Sakran, H. O. (2015). Framework architecture for improving healthcare information systems using agent technology. *International Journal of Managing Information Technology*, *7*(1), 17.

Alom, M. Z., Taha, T. M., Yakopcic, C., Westberg, S., Sidike, P., Nasrin, M. S., ... Asari, V. K. (2019). A state-of-the-art survey on deep learning theory and architectures. *Electronics*, *8*(3), 292.

Bond, A. H., & Gasser, L. (Eds.). (2014). *Readings in Distributed Artificial Intelligence*. Morgan Kaufmann Publishers Inc., San Francisco, CA.

Briggs, B., & Buchhols, S. (2019). Tech trends 2019: Beyond the digital frontier. *Deloitte Insights*, *90*, 19–26.

Calvaresi, D., Mualla, Y., Najjar, A., Galland, S., & Schumacher, M. (2019, May). Explainable multi-agent systems through blockchain technology. In *International Workshop on Explainable, Transparent Autonomous Agents and Multi-Agent Systems* (pp. 41–58). Springer, Cham.

Dubey, S., Singh, P., Yadav, P., & Singh, K. K. (2020). Household waste management system using IoT and machine learning. *Procedia Computer Science*, *167*, 1950–1959.

Datta, S., Barua, R., & Das, J. (2019). Application of artificial intelligence in modern healthcare system. In *Alginates—Recent Uses of This Natural Polymer*. IntechOpen. https://doi.org/10.5772/intechopen.90454

Deshmukh, A. V. (1992). Distributed artificial intelligence. *Knowledge Management Intelligence Learning Complex, 1*, 1–6.

Eli-Chukwu, N. C. (2019). Applications of artificial intelligence in agriculture: A review. *Engineering, Technology & Applied Science Research*, *9*(4), 4377–4383.

Esteva, A., Kuprel, B., Novoa, R. A., Ko, J., Swetter, S. M., Blau, H. M., & Thrun, S. (2017). Dermatologist-level classification of skin cancer with deep neural networks. *Nature*, *542*(7639), 115–118.

Ferrández-Pastor, F. J., García-Chamizo, J. M., Nieto-Hidalgo, M., & Mora-Martínez, J. (2018). Precision agriculture design method using a distributed computing architecture on internet of things context. *Sensors*, *18*(6), 1731.

He, J., et al. (2019). The practical implementation of artificial intelligence technologies in medicine. *Nature Medicine*, *25*(1), 30–36.

Jha, K., Doshi, A., Patel, P., & Shah, M. (2019). A comprehensive review on automation in agriculture using artificial intelligence. *Artificial Intelligence in Agriculture, 2*, 1–12.

Kermany, D. S., Goldbaum, M., Cai, W., Valentim, C. C., Liang, H., Baxter, S. L., ... Dong, J. (2018). Identifying medical diagnoses and treatable diseases by image-based deep learning. *Cell*, *172*(5), 1122–1131.

Kokkinaki, A. I., & Valavanis, K. P. (1995). On the comparison of AI and DAI based planning techniques for automated manufacturing systems. *Journal of Intelligent and Robotic Systems*, *13*, 201–245. https://doi.org/10.1007/BF01424008

Lim, J., Gil, J. M., & Yu, H. (2018). A distributed snapshot protocol for efficient artificial intelligence computation in cloud computing environments. *Symmetry*, *10*(1), 30.

Moritz, P., Nishihara, R., Wang, S., Tumanov, A., Liaw, R., Liang, E., ... Stoica, I. (2018). Ray: A distributed framework for emerging {AI} applications. In *13th {USENIX} Symposium on Operating Systems Design and Implementation ({OSDI} 18)* (pp. 561–577). Carlsbad, CA. ISBN 978-1-939133-08-3

Patel, D. T. (2019). Distributed computing for internet of things (IoT). In *Computational Intelligence in the Internet of Things* (pp. 84–109). IGI Global. https://doi.org/10.4018/978-1-5225-7955-7.ch004

Pham, Q. V., Nguyen, D. C., Hwang, W. J., & Pathirana, P. N. (2020). Artificial intelligence (AI) and big data for Coronavirus (COVID-19) pandemic: A survey on the state-of-the-arts. 2020040383 (doi: 10.20944/preprints202004.0383.v1).

Ponomarev, S., & Voronkov, A. E. (2017). Multi-agent systems and decentralized artificial superintelligence. https://arxiv.org/abs/1702.08529

Simmonds, J., Gómez, J. A., & Ledezma, A. (2019). The role of agent-based modeling and multi-agent systems in flood-based hydrological problems: A brief review. *Journal of Water and Climate Change*. https://doi.org/10.2166/wcc.2019.108

Singh, P., & Agrawal, R. (2018a). A customer centric best connected channel model for heterogeneous and IoT networks. *Journal of Organizational and End User Computing (JOEUC)*, *30*(4), 32–50.

Singh, P., & Agrawal, R. (2018b). Prospects of open source software for maximizing the user expectations in heterogeneous network. *International Journal of Open Source Software and Processes (IJOSSP)*, *9*(3), 1–14.

Singh, P., & Agrawal, V. (2019). A collaborative model for customer retention on user service experience. In Bhatia S., Tiwari S., Mishra K., Trivedi M. (eds), *Advances in Computer Communication and Computational Sciences* (pp. 55–64). Springer, Singapore. https://doi.org/10.1007/978-981-13-6861-5_5

Singh, P., & Singh, N., (2020a). Role of data mining techniques in bioinformatics. *International Journal of Applied Research in Bioinformatics (IJARB)*, *11*(6), in press.

Singh, N., Singh, P., Singh, K. K., & Singh, A. (2020). Machine learning based classification and segmentation techniques for CRM: a customer analytics. *International Journal of Business Forecasting and Marketing Intelligence*, *6*(2), 99–117.

Singh, P., & Singh, N. (2020b). Blockchain with IoT & AI: A Review of Agriculture and Healthcare. *IJAEC*, *11*(4), 13–27.

Song, X., Yang, S., Huang, Z., & Huang, T. (2019, August). The application of artificial intelligence in electronic commerce. *Journal of Physics: Conference Series*, *1302*(3), 032030. IOP Publishing.

Sundar Ranganathan, NetApp, (2018). Edge to core to cloud architecture for AI key considerations for the development of deep learning environments. Technical White Paper, NetApp, WP-7271.

Tran, B. X., Vu, G. T., Ha, G. H., Vuong, Q. H., Ho, M. T., Vuong, T. T., ... Latkin, C. A. (2019). Global evolution of research in artificial intelligence in health and medicine: A bibliometric study. *Journal of Clinical Medicine*, *8*(3), 360.

Varga, P., Peto, J., Franko, A., Balla, D., Haja, D., Janky, F., ... Toka, L. (2020). 5G support for industrial IoT applications–challenges, solutions, and research gaps. *Sensors*, *20*(3), 828.

Verma, D., Calo, S., & Cirincione, G. (2018, January). Distributed AI and security issues in federated environments. In *Proceedings of the Workshop Program of the 19th International Conference on Distributed Computing and Networking* (pp. 1–6).

Vermesan, O., Eisenhauer, M., Serrano, M., Guillemin, P., Sundmaeker, H., Tragos, E. Z., ... Bahr, R. (2018). The next generation Internet of things–hyperconnectivity and embedded intelligence at the edge. In: Ovidiu Vermesan (ed.), *Next Generation Internet of Things. Distributed Intelligence at the Edge and Human Machine-to-Machine Cooperation*. SINTEF, Norway. Joël Bacquet, European Commission, Belgium. ISBN: 9788770220088, https://doi.org/10.13052/rp-9788770220071

Zheng, X. L., Zhu, M. Y., Li, Q. B., Chen, C. C., & Tan, Y. C. (2019). FinBrain: When finance meets AI 2.0. *Frontiers of Information Technology & Electronic Engineering*, *20*(7), 914–924.

17 Analysis of Hybrid Deep Neural Networks with Mobile Agents for Traffic Management in Vehicular Adhoc Networks

G. Kiruthiga, G. Uma Devi, N. Yuvaraj, R. Arshath Raja, and N.V. Kousik

CONTENTS

17.1 INTRODUCTION

In recent times, Vehicular Adhoc Networks (VANETs) have offered safety management and data management to users, and they are designed with suitable control methods to operate on any circumstances to emphasize the network dynamics [1]. However, with increased vehicles, traffic management becomes complex with the use of both centralized [1] and distributed [2] algorithms. This indirectly affects urban traffic and increases intersection delays, fuel consumption, traffic accidents, and emission values [3].

Real-time and accurate traffic flow predictions often play a vital feature for traffic management systems. In such scenarios, the Intelligent Transportation System (ITS) offers reliable traffic management services that makes the users aware of the environment by coordinating the network in an optimal and safer way. Adoption of various information and communication technologies in ITS further enhances traffic and mobility management [1]. It highly supports communication in VANETs between the road units (vehicles) and roadside units (RSUs) for effective and safer transportation. Various other challenges lead to time delays and traffic congestion, which entirely affects the VANET performance.

To resolve such complexity in traffic management, research has been carried out in existing literature [4–6]. These methods are adopted to resolve the traffic congestion in VANETs that includes scalability, performance, and management difficulties. Various optimization tasks are carried out that offer a major role in regulating dynamicity of traffic flows in urban scenarios [7].

However, the effects on congestion in existing methods are limited by offering accurate and timely information for traffic predictions. Most recently, the congestion effects are treated as a classification problem and wide-ranging solutions are offered, but most systems lack with limited usage [8,9]. To support both limitations, the ITS systems supports the prediction of traffic conditions by analyzing various network parameters [10]. This prediction involves preplanning of routes and rescheduling that reduces the congestion [11,12]. The stochastic characteristics [13] and non-linear traffic flow further poses a serious challenge to accurate traffic prediction [14,15]. VANET methods use linear and machine learning models for traffic flow prediction based on network density that fails to read the nonlinear uncertainty of the system [16–18].

Traffic management analysis gets complex with exponential traffic growth and with high computational resources [19]. Hence, considering the uncertainty, exponential growth, and high computational resources, high-end intelligent systems are required for offering flexible and open architecture for smoother transmission of vehicles. The high-end intelligent system adopting deep learning models has the ability to manage network-level abstraction and optimization of resources.

With such motivation, we introduce mobile agents (MA) [20] as a vital factor to sense the environment and provide inputs to offer optimal decision-making to the DNN routing algorithm. The main contributions of this chapter include:

1. The author devised a scheme that senses the VANET environment in real time and updates the routing table at rapid instances to reduce the vehicle collision rate.
2. The author combined the Deep Neural Network (DNN) [21–25] algorithm with MA to improve the routing capability of VANETs in a real-time scenario. This ensures optimal routing decisions based on the inputs from MA at regular instances, thereby ensuring better delivery of services across MANETs.
3. The author evaluates the DNN-MA model with existing deep and machine learning models in terms of various collision avoidance metrics like packet delivery rate, latency, error rate, and throughput in a real-time traffic scenario.

The outline of this chapter is as follows: Section 17.2 discusses the network model. Section 17.3 provides the details of traffic management systems. Section 17.4 confers the details of DNN-IA architecture for VANET routing. Section 17.5 concludes the entire work.

17.2 NETWORK MODEL

Figure 17.1 shows the VANET network model with two different entities: vehicle-to-vehicle (V2V) and vehicle-to-infrastructure (V2I).

- *Vehicle units*: The vehicle units, or vehicle in simple terms, is responsible for transportation purpose and is utilized by VANETs for communicating with nearby vehicles along the road segment or with the Road Safety Units (RSUs) at the edges/corners. In this chapter, we use elliptical curve cryptography as an encryption algorithm to generate cryptographic credentials and storing in the vehicle. The global positioning system (GPS) can find the vehicle location. The total number of vehicles in a road segment can be determined by RSUs using a GPS unit.
- *Roadside units* (RSUs): The RSU acts as an access point which is deployed along with the roads, and it holds the details of vehicle units along with the

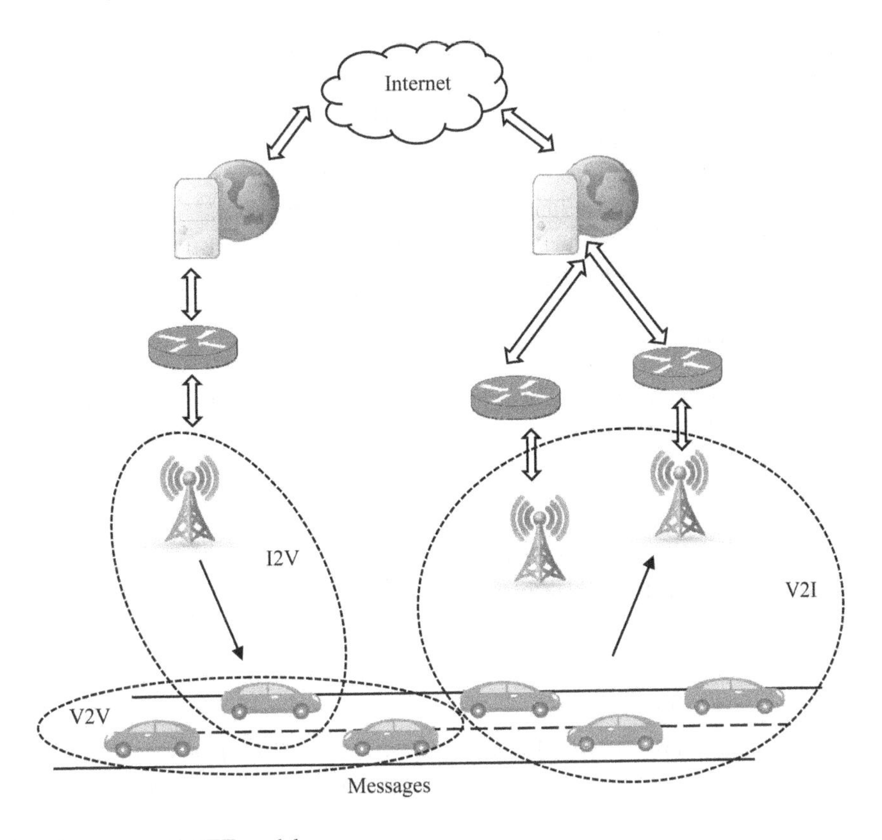

FIGURE 17.1 VANET model.

road segment. The encrypted road segments by the vehicle units are considered as a relay for a traffic message channel (TMC) or an IA. An MA is connected with RSU via a faster communication medium. The RSU functions to run and store the cryptographic credentials to decode the vehicle information and the DNN-IA algorithm.

- *TMC or IA*: The MA uses road segments to calculate the traffic density. The direction connection of TMC with RSUs and also with other MA obtains information about traffic on the road segment. The MA sends the information collected via MA to DNN, which makes optimal routing decisions to avoid congestions on road segments. The difference between a TMC and an MA is that the former is a fixed segment and the latter is the mobile segments that traverse the network.

The proposed method is validated in an urban scenario, and is designed in the form of grids with road segments and edge intersections. The road between any two intersections is considered as the road segment $s(i,j)$ and the intersection $I(i)$ connects two or more road segments. $s(i,j)$ is a collection of various features that includes width and length of road between two intersections, density of vehicular traffic, and number of lanes.

The vehicle unit is designed with a unique ID (UID) that consists of its location, direction, and vehicle velocity. The vehicle accesses the RSU automatically to know the total intersections, road segments along the path, and the traffic congestion level. In such cases, the study uses MA to provide precise information about vehicles rather than existing data communication between the vehicles. In the proposed method, the DNN makes proper decision-making by forming a graph $G = < V, E >$ with V as the intersections between the source and the destination vehicle and E as the road segments connected with intersections in V.

17.3 TRAFFIC MANAGEMENT MODEL

The DNN-IA traffic management operates as a driverless assistant system that uses MA to collect the traffic information along the road segments and intersections and acts as a forwarder of data to RSU. The DNN-IA traffic management is a distributed graph-based model with a set of vehicles (V) and set of edges (E). The MA sends the number of vehicles present in the road segment and the traffic congestion along the intersections and road segments. In the proposed method, the VANETs works with MA that assists the infrastructure unit and then connects with vehicle unit for routing operations.

17.3.1 Mobile Agent Unit

The MA in the proposed VANET architecture is a network module that dynamically moves inside VANETs and it accesses the vehicle units to get connected with the infrastructure unit. The MA consists of four segments, namely the identification unit, execution code unit, routing path unit, and data space unit. The MA uploads the vehicle information to DNN that offers faster computation to distribute the data packets from source to destination node via cooperative MAs by the selective routing paths. Since each MA has varied information; therefore, it is provided with a unique ID. The data of current vehicle passing a road segment is stored at the executive code and the routing path established by DNN is the core that indicates the traversal of the packet. Finally, the data space entirely stores the received data from the vehicle units. Here, the routing path using DNN unit lies in RSU infrastructure unit [21].

17.3.2 Infrastructure Unit

The infrastructure unit lies at the application layer and calculates the routing path via DNN (the architecture of DNN is given in Figure 17.2) using the vehicle location, positions, and velocity in a robust way. The workflow of infrastructure unit is specified in Figure 17.3.

The infrastructure unit is designed to collect the control information like position, location, velocity, and timestamp of vehicle unit. The collection of control information is carried out in an iterative way such that the information collected is precise. The information is collected by the MAs and it is not directly sent by the vehicle units to infrastructure units i.e. RSU. The data from MAs is sent as input to the

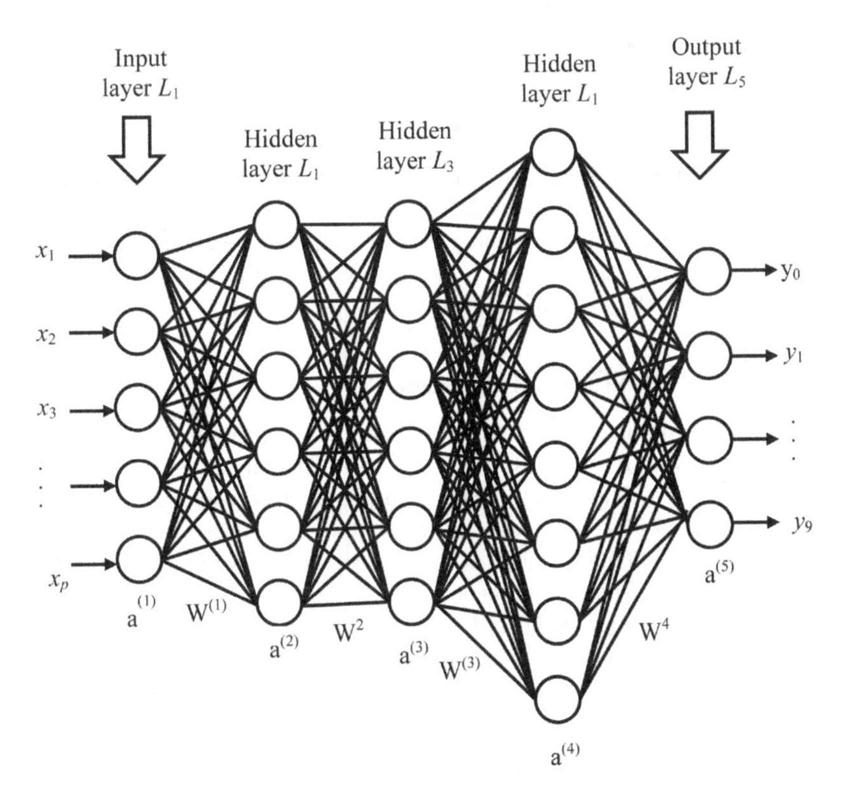

FIGURE 17.2 DNN architecture.

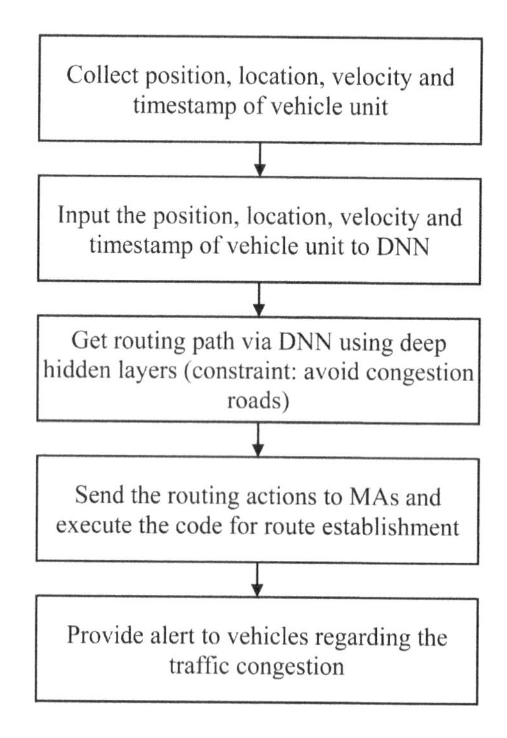

FIGURE 17.3 Infrastructure unit workflow.

DNN model and it processes the route establishment (the routing process is the same as that carried out in [22]) and forwards the routing details to MAs. The MAs then make the routing actions to MAs using code execution unit and then establishes the route to avoid congestion in VANETs. In this way, the entire operation is carried out such that the major challenge of congestion is avoided strictly

17.4 PERFORMANCE EVALUATION

The DNN-MA is simulated in a Python simulator and the parameters for simulating the DNN under VANETs are given in Table 17.1. The simulation is carried out in an area of 1500×1500 m^2 in a two-lane road of fixed road width. The study covers 200 road edges and 250 road segments, where 170 road edges have traffic signals. The simulation covers 500 vehicles with a speed varying between 0 and 50 kmph (kilometers per hour) in an urban scenario.

The validation of proposed DNN-IA is carried out against existing deep learning model: DNN [20] and a machine learning model: artificial neural network (ANN) [21]. The validation is carried out over various performance metrics like traffic type, vehicle speed, and the density of network to evaluate the average latency, cumulative distribution function (CDF), energy efficiency, packet delivery rate, and network throughput.

Figure 17.4 shows the quality of network connectivity between the DNN-IA and existing DNN and ANN models. The range of transmission is varied between 200

TABLE 17.1
Simulation Parameters

Parameters	Value
Total vehicle units	100–300
Channel carrier frequency	5.9 GHz
Frequency	5.9 GHz
Simulation area	1500 m × 1500 m
Maximum transmission	20 mW
Bit rate	18 Mbps
Packet length	Uniform
Path loss coefficient	2
Signal attenuation threshold	–90 dBm
Vehicle velocity	20–50 kmph
Traffic type	CBR
Transmission range	150 m
MAC protocol	802.11 p
CBR rate	4 packets/sec
Beacon interval	0.5 s
Simulation time	1000 s
Data rate of MAC	6 mb
Mobility model	Krauß model

and 500 m and the arrival rate of the vehicle is kept fixed at between 30 and 50 kmph. The simulation result shows that with increasing distance metric, the probability of connection expires and the connection with vehicle gets lost with further increase of distance from a RSU 400 m. However, the placement of RSU at every 350 m helps the VANETs to establish connection with vehicles over the long run. On the other hand, with increasing network traffic (from 100 to 300 vehicles), the connection probability still remains a challenge in establishing connections with vehicles lying within 300 m (which is evident from Figure 17.5). Figure 17.5 shows the results of CDF between the DNN-IA, DNN, and ANN, where the connection probability is tested between 100 and 300 vehicles. In this regard, the network density is labeled as high for 300 vehicles, medium for 200, and low for 100. The result shows that the proposed method establishes longer distance connectivity with vehicles than DNN and ANN models with a low density network. A degradation exists in the network when the network density increases.

Figure 17.6 shows the results of average latency of data transmission between the vehicles by varying the velocity of vehicles. The result of the simulation indicates that with increased velocity, the latency of data transmission increases. By contrast, with increasing network density, the latency increases further. The combination of both increasing velocity and increasing vehicle density contributes to maximum average latency. Such mobility impacts the data transmission due to the failure of

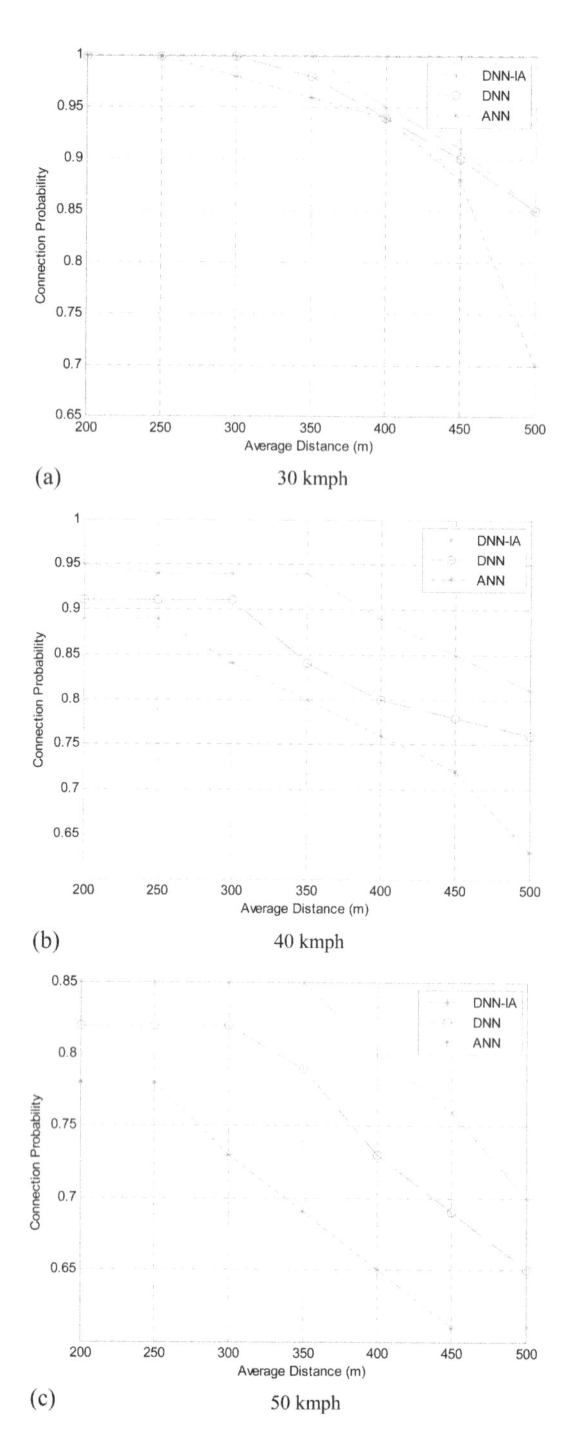

FIGURE 17.4 Network connectivity vs. varying distance and network density.

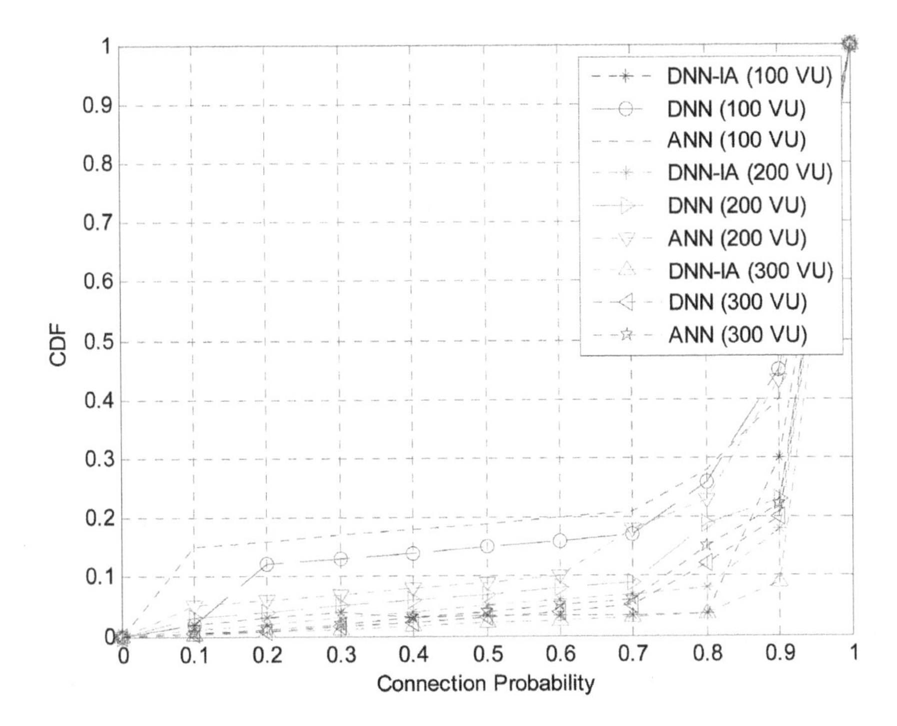

FIGURE 17.5 CDF of successful connection with varying network density (100–300 vehicle units (VU)).

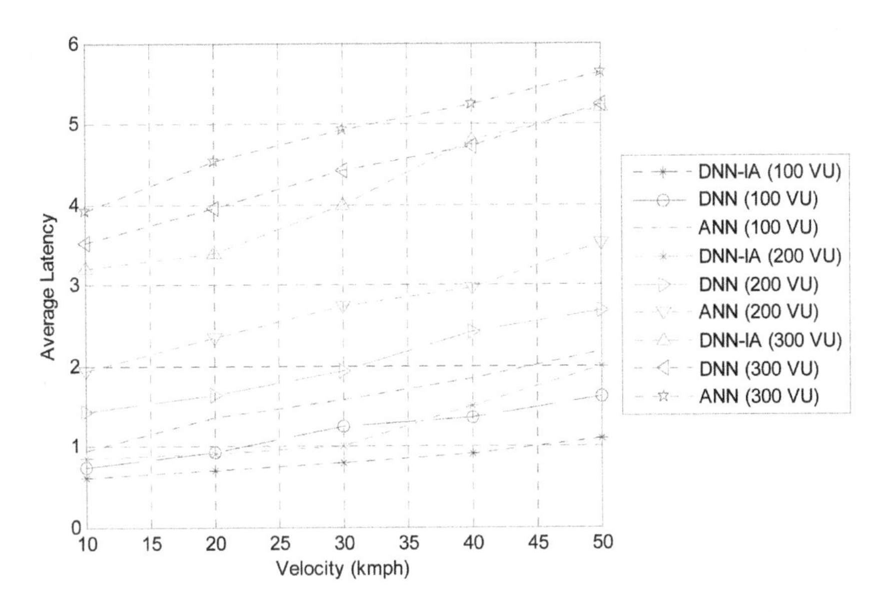

FIGURE 17.6 Impacts on average latency with variable velocity and network density.

establishing a link in forwarding the packets between the vehicles. This directly affects the delivery rate as the average latency is indirectly proportional to the packet delivery ratio (Figure 17.7). The results of simulation, by contrast, show that with minimal vehicle speed and density, the link is established well and hence the average delay is reduced considerably. Overall, the performance of DNN-IA is efficient in terms of minimal average latency rate with reduced failure rate due to the presence of MA than DNN and ANN.

Figure 17.7 shows the results of the packet delivery ratio with varying vehicle speed and vehicle density. With increasing velocity, the packet delivery ratio gets affected due to a break in the link as the forwarders fail in forwarding the packets to the neighboring vehicle. Hence the link breaks and similar conditions exist with increasing network density. The link breakage highly influences the stimulus of packet loss rate and hence the performance is affected highly. On the other hand, with increasing vehicle density the collision of packets between the vehicles increases the packet loss rate and it affects the network overhead leads to reduced functionality. The result of simulation shows that the DNN-IA has a higher packet delivery rate than DNN and ANN with increased functionality.

Figure 17.8 shows the results of throughput with varying data rates. The data rates are varied with respect to varying network density and vehicle velocity. The result of simulation shows that at the road edges, maximum throughput is achieved as the vehicle moves with minimal speed and minimal throughput at straight and curved road segments. Furthermore, the optimal selection of vehicles along the road segments by the DNN-IA has increased the network throughput than other methods.

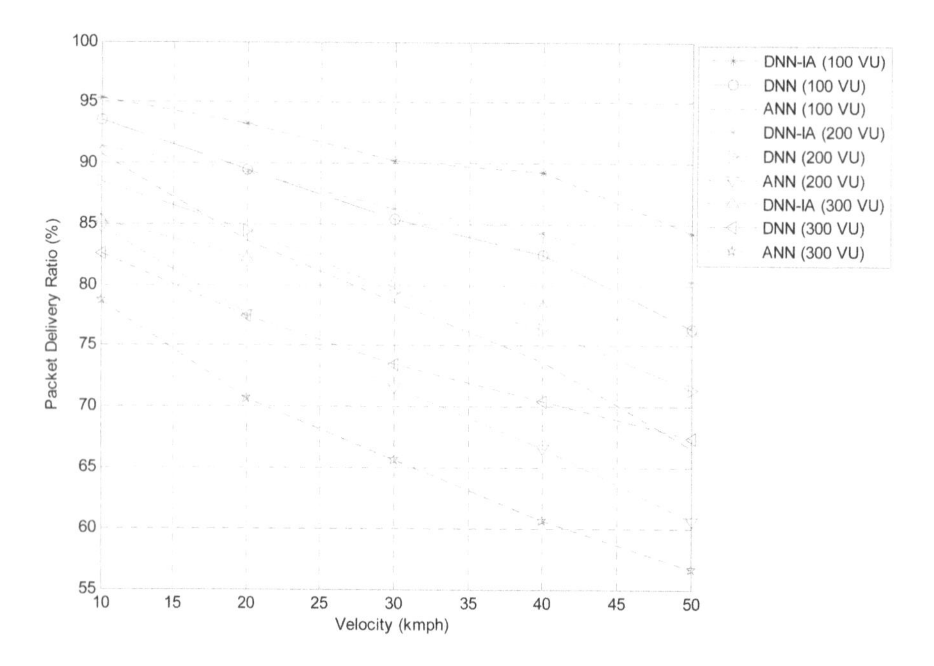

FIGURE 17.7 Impacts on packet delivery rate with variable velocity and network density.

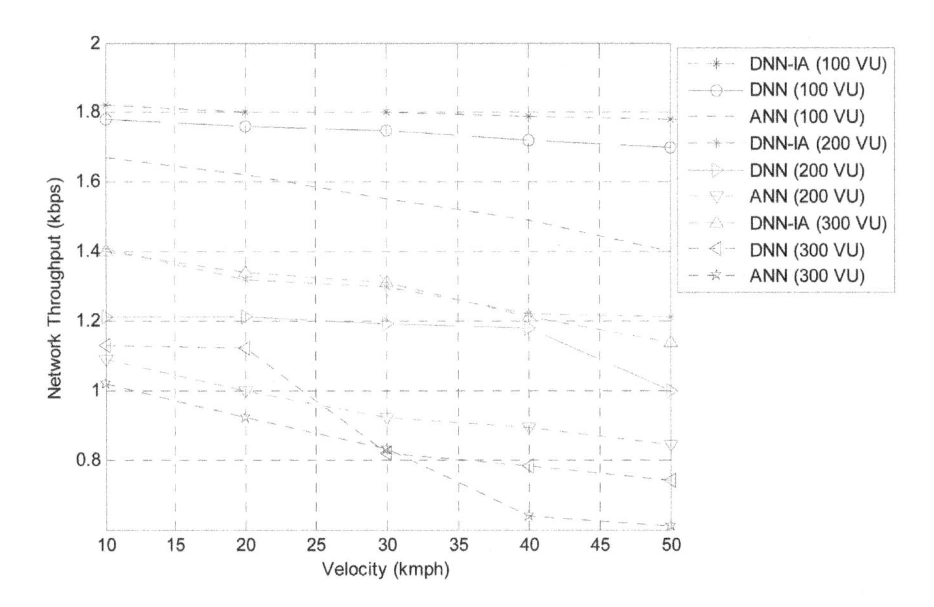

FIGURE 17.8 Impacts on throughput with variable velocity and network density.

The optimal selection reduces the packet loss rate and selection of optimal adjacent vehicle hops effectively maintains the transmission link, thereby the throughput is highly consistent and energy efficiency is high. The throughput degradation is found in DNN and ANN due to the absence of MA in denoting which vehicle can be used to establish the link and, thereby, the energy consumption is avoided strictly on link reliability.

17.5 CONCLUSION

In this chapter, DNN-MA provides effective routing of vehicles under highly congested routes in VANETs. This DNN-MA used in this chapter offers optimal paths to increase the energy efficiency rate. The deep learning-based MA examines the entire network to find the states of each vehicle moving in VANETs. Higher traffic congestion in the network allows the DNN-MA to optimize the routing decision based on the inputs from MA. The DNN processes effectively the routing decision at a faster rate and provides a solution to the network in setting the optimal paths such that the congestion in the network is reduced in a faster instance of time. The utilization of the routing table at RSU for regular updates on the vehicle's state ensures optimal vehicle selection and stable routing decisions. The validation under variable velocity, distance, vehicle density, and data and transmission rate shows that the DNN-MA offers faster routing decisions than existing DNN and ANN models. The results of simulation show that the DNN-MA model has increased connectivity, throughput, packet delivery rate, and end-to-end delay and reduced latency. The result further claims that the connection termination probability is lesser in DNN-MA that supports increased data delivery rate. Thus, the simplified routing decisions using

DNN based on constant monitoring by MA have maintained the traffic density in an optimal way and this ensures efficient delivery of packets to the destination nodes. Finally, the applicability of routing decisions on low mobility vehicle is more efficient than the one with faster vehicle mobility.

REFERENCES

1. Blot, G., Fouchal, H., Rousseaux, F., & Saurel, P. (2016, May). An experimentation of vanets for traffic management. In *2016 IEEE International Conference on Communications (ICC)* (pp. 1–6). IEEE, Kuala Lumpur, Malaysia.
2. Li, S., Liu, Y., & Wang, J. (2019). An efficient broadcast scheme for safety-related services in distributed TDMA-based VANETs. *IEEE Communications Letters*, *23*(8), 1432–1436.
3. Shen, L., Liu, R., Yao, Z., Wu, W., & Yang, H. (2018). Development of dynamic platoon dispersion models for predictive traffic signal control. *IEEE Transactions on Intelligent Transportation Systems*, *20*(2), 431–440.
4. Yao, Z., Jiang, Y., Zhao, B., Luo, X., & Peng, B. (2020). A dynamic optimization method for adaptive signal control in a connected vehicle environment. *Journal of Intelligent Transportation Systems*, *24*(2), 184–200.
5. Srivastava, S., & Sahana, S. K. (2017). Nested hybrid evolutionary model for traffic signal optimization. *Applied Intelligence*, *46*(1), 113–123.
6. Srivastava, S., & Sahana, S. K. (2019). Application of bat algorithm for transport network design problem. *Applied Computational Intelligence and Soft Computing*, *2019*, 1–18.
7. Balta, M., & Özçelik, İ. (2020). A 3-stage fuzzy-decision tree model for traffic signal optimization in urban city via a SDN based VANET architecture. *Future Generation Computer Systems*, *104*, 142–158.
8. Ke, X., Shi, L., Guo, W., & Chen, D. (2018). Multi-dimensional traffic congestion detection based on fusion of visual features and convolutional neural network. *IEEE Transactions on Intelligent Transportation Systems*, *20*(6), 2157–2170.
9. Altiındağ, E., & Baykan, B. (2017). Discover the world's research. *Turk Journal of Neurology*, *23*, 88–89.
10. Nagy, A. M., & Simon, V. (2018). Survey on traffic prediction in smart cities. *Pervasive and Mobile Computing*, *50*, 148–163.
11. Tanwar, S., Vora, J., Tyagi, S., Kumar, N., & Obaidat, M. S. (2018). A systematic review on security issues in vehicular ad hoc network. *Security and Privacy*, *1*(5), e39.
12. Luo, X., Li, D., Yang, Y., & Zhang, S. (2019). Spatiotemporal traffic flow prediction with KNN and LSTM. *Journal of Advanced Transportation*, *2019*, 1–14.
13. Ravi, B., Thangaraj, J., & Petale, S. (2019). Data traffic forwarding for inter-vehicular communication in VANETs using stochastic method. *Wireless Personal Communications*, *106*(3), 1591–1607.
14. Kang, D., Lv, Y., & Chen, Y. Y. (2017, October). Short-term traffic flow prediction with LSTM recurrent neural network. In *2017 IEEE 20th International Conference on Intelligent Transportation Systems (ITSC)* (pp. 1–6). IEEE, Yokohama, Japan.
15. Mackenzie, J., Roddick, J. F., & Zito, R. (2018). An evaluation of HTM and LSTM for short-term arterial traffic flow prediction. *IEEE Transactions on Intelligent Transportation Systems*, *20*(5), 1847–1857.
16. Fu, R., Zhang, Z., & Li, L. (2016, November). Using LSTM and GRU neural network methods for traffic flow prediction. In *2016 31st Youth Academic Annual Conference of Chinese Association of Automation (YAC)* (pp. 324–328). IEEE, Wuhan, China.

17. Ma, X., Tao, Z., Wang, Y., Yu, H., & Wang, Y. (2015). Long short-term memory neural network for traffic speed prediction using remote microwave sensor data. *Transportation Research Part C: Emerging Technologies*, *54*, 187–197.
18. Shao, H., & Soong, B. H. (2016, November). Traffic flow prediction with long short-term memory networks (LSTMs). In *2016 IEEE Region 10 Conference (TENCON)* (pp. 2986–2989). IEEE, Singapore.
19. Ku, I., Lu, Y., Gerla, M., Gomes, R. L., Ongaro, F., & Cerqueira, E. (2014, June). Towards software-defined VANET: Architecture and services. In *2014 13th Annual Mediterranean Ad Hoc Networking Workshop (MED-HOC-NET)* (pp. 103–110). IEEE, Piran, Slovenia.
20. Lu, J., Feng, L., Yang, J., Hassan, M. M., Alelaiwi, A., & Humar, I. (2019). Artificial agent: The fusion of artificial intelligence and a mobile agent for energy-efficient traffic control in wireless sensor networks. *Future Generation Computer Systems*, *95*, 45–51.
21. Venkataram, P., Ghosal, S., & Kumar, B. V. (2002). Neural network based optimal routing algorithm for communication networks. *Neural Networks*, *15*(10), 1289–1298.
22. Reis, J., Rocha, M., Phan, T. K., Griffin, D., Le, F., & Rio, M. (2019, July). Deep neural networks for network routing. In *2019 International Joint Conference on Neural Networks (IJCNN)* (pp. 1–8). IEEE, Budapest, Hungary.
23. Geyer, F., & Carle, G. (2018, August). Learning and generating distributed routing protocols using graph-based deep learning. In *Proceedings of the 2018 Workshop on Big Data Analytics and Machine Learning for Data Communication Networks* (pp. 40–45).
24. Perez-Murueta, P., Gómez-Espinosa, A., Cardenas, C., & Gonzalez-Mendoza, M. (2019). Deep learning system for vehicular re-routing and congestion avoidance. *Applied Sciences*, *9*(13), 2717.
25. McGill, M., & Perona, P. (2017, August). Deciding how to decide: Dynamic routing in artificial neural networks. In *Proceedings of the 34th International Conference on Machine Learning-Volume 70* (pp. 2363–2372). JMLR. org.

18 Data Science and Distributed AI

V. Radhika

CONTENTS

18.1 INTRODUCTION

Research in Distributed Artificial Intelligence (DAI) vows to have wide-extending impacts in intellectual science, for example, mental models, social insight, disseminated frameworks, compositional, and language support. Conveyed AI likewise props up human-PC interaction, that is an assignment distribution, smart interfaces, exchange intelligibility, and discourse acts. Issue interpretation, epistemology, joint idea development, communication thinking, and critical thinking are essential in AI explorations, and the building of AI frameworks are coworking master frameworks, conveyed detecting and information combination, coworking robots, collective structure critical thinking, and so forth. In any case, an investigation into DAI is appealing for basic reasons: to facilitate their activities, smart specialists need to speak to and reason about the information, activities, and plans of different operators.

18.2 INSPIRATION

The advancement in digitalization deeply affects industry, science, and innovation. The prudence of information science and AI calculations establish a basic control point to access, and handling information/data have appeared in Figure 18.1. They are unequivocal for monetary accomplishment for the logical and cultural advancement of people, associations, and even countries [1]. The fourth modern turbulence is utilized to determine the logical bits of knowledge dependent on examining the huge datasets that were created.

Information has become a key factor in creation. It is generally made, and therefore abused, utilizing information science and AI calculations, to deliver new bits of knowledge to all the more likely comprehend (or explain) an issue. It has been distinguished that information and information science. AI innovations are serious differentiators in the information economy: Companies capable of utilizing them become quicker and perform better [2]. Accordingly, claiming information and acing information science just as AI innovations are the key elements for future seriousness. R is the most favored open-source language for investigation and information science. R offers an enormous assortment of measurable (straight and non-linear demonstrating, old-style, factual tests, time-arrangement examination, order, bunching, and so on) and graphical procedures and is exceptionally extensible. It is a lot of valuable in huge information and large science.

Dissemination and parallelism have generally been significant subjects in man-made brainpower. During the fifties, AI and neural systems administration handling strategies are created. The computational structures utilized were layers of "neurons," the procedures they displayed were typically designed acknowledgment assignments, and the principle programming strategy was synaptic learning. There were some hypothetical outcomes in learning hypotheses. Half breed simple and computerized nets were investigated in likelihood state variable (PSV) frameworks Lee [3]. Input in neural systems was examined by Aleksander [4], which summed up the frameworks to systems of RAM components. Be that as it may, the absence of adequately ground-breaking programming or learning strategies, hypothetical

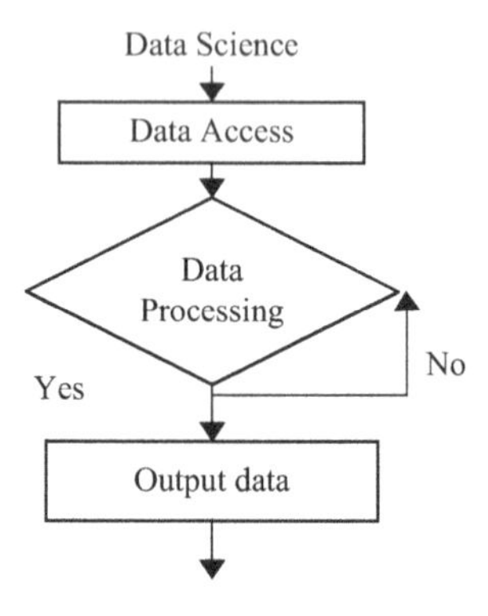

FIGURE 18.1 Flow of data science.

unmanageability, and essentially usage troubles made such work troublesome. It has been taken up another in the current execution partner progressively lucky time, as Parallel Distributed Processing McClelland [5] and Connectionism Feldman [6]. Equal Distribution Processing is a huge grained wise procedure. There are numerous explanations behind dispersing knowledge—these incorporate old-style explanations behind appropriating any program or framework and reasons specific to Distributed AI frameworks. In certain spaces where AI is being applied (e.g., dispersed detecting, clinical conclusion, airport regulation), information or action is characteristically conveyed. The circulation can emerge due to geographic dispersion combined with handling or information transfer speed restrictions, in light of the regular practical conveyance in an issue, or on account of a longing to disseminate control (e.g., for come up short delicate corruption), or for particular information obtained, and so on. In different areas, it might be required to streamline the turn of events and develop wise frameworks by building them as assortments of independent, however associating parts. Ordinary methods of reasoning for appropriating an AI framework incorporate the accompanying:

Versatility: Logical, semantic, worldly, and spatial appropriation permits a DAI framework to give elective viewpoints on rising circumstances and conceivably more noteworthy versatile force.

Cost: A conveyed framework may hold the guarantee of being financially savvy since it could include countless straightforward PC frameworks of low unit cost. On the off chance that correspondence costs are high, nonetheless, incorporated knowledge with disseminated recognition or detecting might be more costly than appropriated insight.

18.3 DISTRIBUTED SENSOR NETWORKS

To get to the information, Distributed Sensor Networks (DSN) are summoned. Advances in microchips, bundle exchanging radio interchanges, sensors, and manmade reasoning strategies for a mechanized critical thinking all consolidate to frame the establishments of an exceptionally robotized, minimal effort, savvy, conveyed sensor arrange. One of the models appears in Figure 18.2. A DSN comprising of cheap smart hubs dispersed over a combat zone is relied upon to be valuable in getting loud insight information, handling it to decrease vulnerability and produce more significant level understandings, and transmitting this disconnected data rapidly to the administrators who require it. The advancement of circulated sensor systems is progressing quickly. Two of the difficult zones are traffic control and mechanical control.

In endeavoring to portray appropriate structures for a DSN, it has been affected by a few general equipment patterns: handling power is expanding at a similarly fast rate that equipment costs are diminishing; correspondence capacities are additionally getting better with less expense, and altogether exceptionally less computational force is required. These are ease and low-power sensors that are probably going to remain respectably untrustworthy. The DSN has request while keeping unlimited by

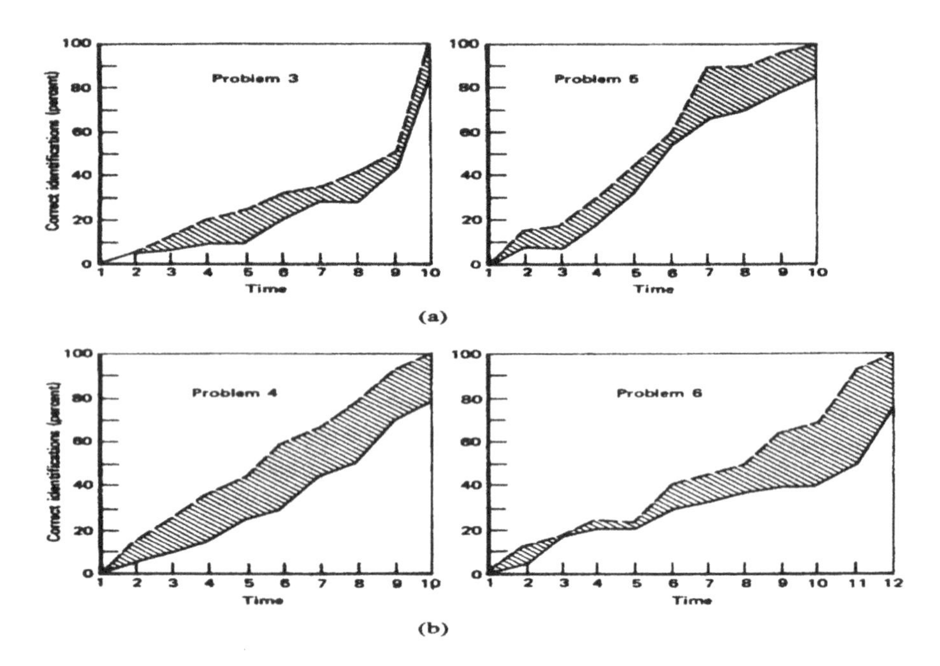

FIGURE 18.2 Overall performance: Problem 3-6 (a) anarchic committee, (b) dynamic hierarchical cone.

detailed current equipment plans and imperatives. Huge scope Situation Assessment (SA) as a rule includes a system of data makers and customers dealing with data that shifts by area, measure of conglomeration, and level of deliberation. Auspicious circumstance appraisals in current military situations require frameworks that can procedure a greater amount of this data quicker than at any other time. New data preparing innovations recommend robotizing a greater SA task Heyes and Wesson [7].

The SA is comprised of two structures: one is a board of trustees structure, encourages correspondence and rearrangement, while another, all the more carefully various leveled, probably cultivates the misgiving of a prevalent worldwide issue point of view. To look at these structures an exploration dependent on a rearranged SA task is performed. This test task was made by changing over many of the components of a front line SA task into progressively recognizable data handling issues. In particular, messages (moving words and expressions) move in a snake-like way through a field meagerly covered with impediments before halting at their last goals. From a DSN point of view, the words speak to moving stages that radiate otherworldly signals. The errand is to distinguish approaching stages as fast as conceivable before they arrive at their goals. Every individual from the association is restricted in the field of view, so collaboration is basic for the association to play out its strategy.

Correspondence necessities can be diminished if a dormant data hypothesis can be displaced with dynamic theories that predict their own advancement after some time. A Process Assembly Network (PAN) empowers hubs to speak to and anticipate the conviction frameworks of its important neighbors. In the interim, it replaces the ordinarily visit information detailing task with a framework for which just the unforeseen should be accounted.

The SA task is reasonable for research center experimentation. The issue wherein disarray, short-lived, and componential data could be totaled to decipher and clarify significant level occasions. Military front line conditions, in the same way as other different situations looked at by associations, are too intricate and poorly characterized to help this kind of lab investigation, so we expected to develop a diminished space that caught the pith of the non-exclusive case. The necessities for this errand include the following.

A combat zone-like situation as appears in Figure 18.2 [2] comprises both of articles moving around exclusively and data, because the issue should bolster common sorts of activities; for example, camouflage and spread, composed assaults, and bluffs;

- numerous sensors, every one of that can see a few, yet not all, of the low-level exercises, because announced information must contain blunders and grant deficiency;
- short-lived information, as in data esteem relies fundamentally upon convenient preparing;
- restricted correspondence among processors with the goal that the correspondence calculation exchange off could be researched;
- a rich information condition, since enough data must be available to permit the framework to defeat the above impediments and accomplish an answer.

A test task configuration showing these attributes and utilized it to consider the hierarchical aspects delineated previously (Figure 18.3).

18.4 ASSOCIATIONS TESTED

Inside this errand condition, two different authoritative structures as a possibility for comparable assignments were tried. The principal association decided to contemplate a "level" non-various leveled association. In computerized reasoning (AI)

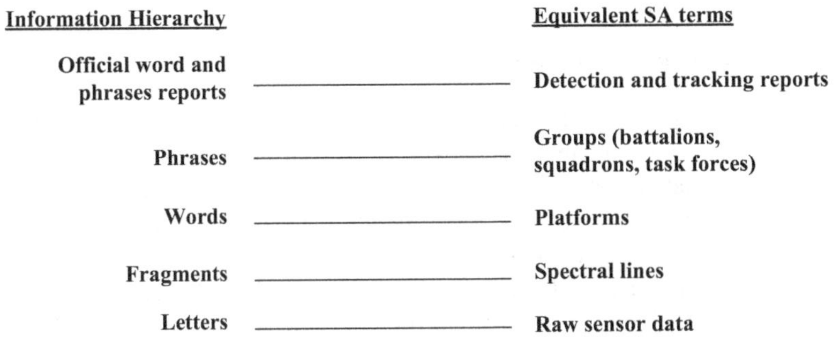

Information Hierarchy	Equivalent SA terms
Official word and phrases reports	Detection and tracking reports
Phrases	Groups (battalions, squadrons, task forces)
Words	Platforms
Fragments	Spectral lines
Letters	Raw sensor data

FIGURE 18.3 Data chain of importance for model DSN.

writing, this kind of structure is ordinarily known as the "participating specialists" worldview. Maybe the most popular AI work of this sort is exemplified by the Hearsay-II discourse understanding framework Erman et al. [8]. By and large, associations of "participating specialists" are made out of authorities with practically zero various leveled structure. Such associations tackle issues by sharing individual points of view, which refine and eventually incorporate nearby translations into a gathered agreement. Sub entrusting, revealing necessities, and asset designation choices are commonly not determined from earlier. The association structures conduct and correspondence designs powerfully in light of the earth and change the examples in an information coordinated way.

In direct complexity to the "coordinating specialists" worldview lies the various leveled "hypothesis Y" or "perceptual cone" associations. The previous name originates from the authoritative hypothesis Lawrence and Lorsch [9], which focuses on huge built up consistent state firms. The last term originates from early AI work in design acknowledgment Uhr [10]. Associations of this class are amassed as strict pecking orders of deliberation levels. At each level, individual components get reports from the levels underneath them, incorporate the reports as indicated by their extraordinary abilities and position in the pecking order, and report upward preoccupied renditions of their outcomes. The most significant level of the system may more than once request its subordinates to modify some past reports as per its worldwide points of view, or it can report the available translation it has framed. As every hub has decisively indicated input/yield (I/O) action and undertaking prerequisites, this sort of association is commonly found in areas requiring schedule, yet mind-boggling, data handling. The "coworking specialists" associations will, in general, be supported in less intricate, however increasingly dubious or quickly evolving situations.

The full SA task display highlights that propose that each hierarchical structure may be appropriate, and tried both. The main structure became known as the anarchic committee (AC) to mirror the absence of obvious legislative structure and the propensity for this association to bring forth many covering boards of trustees to perform explicit errands. The subsequent structure was tried as a dynamic hierarchical cone (DHC). The DHC is a "perceptual cone" association, adjusted to be increasingly receptive to either a spatially uneven or quickly changing information stream.

18.4.1 Human-Based Network Experiments

Contraption, Subjects, and Procedure: A significant objective of the underlying tests, where human players went about as hubs, was to catch data on the action of hubs in a structure appropriate for examination. This drove us from the begin to consider a PC based testbed that could give better instruments to every hub to play out its assignment, and, simultaneously, could record follows and execution information naturally. In the framework that came about, every hub is furnished with a cathode-beam tube (CRT) terminal. Utilizing this terminal, they may send messages to different hubs (with the limitations of the issue), make official reports of puzzle substances to an outside "controller," and show its present perspective on the framework.

Utilizing this testbed and the associations portrayed that they directed a progression of casual analyses. These investigations were not planned with exacting trial controls to allow formal theory testing. Or maybe, they were embraced to create nature with the constraints forced by every one of the authoritative structures and to infer heuristics for use by a PC program that would bolster appropriately controlled analyses. Six separate preliminaries or "issues" were run. Three preliminaries utilized the AC hierarchical structure, and three utilized the DHC structure. An alternate riddle was utilized for each situation.

Before every preliminary, subjects met to get hub assignments (i.e., hub position) and create standard problem-understanding systems and methodology. These measures secured both low-level shows for arranging messages and more elevated level systems (e.g., "To decrease monotonous authority surmises, send formal speculations for outer correspondence to HL who alone would report them"). Hubs likewise traded data about how they moved toward their problem-fathoming errands, for example, techniques for following word or letter movements. These arranging gatherings kept going around 30 minutes. The subjects at that point played out the undertaking. The absolute time for every issue was around 60 minutes. A short questioning followed, in which subjects gave general impressions of the activity, their problem-explaining heuristics, and new thoughts for improving systems later on. They likewise finished a short post-test survey.

Exploratory Analysis and Results: The initial two critical thinking meetings were treated as training meetings to pick up involvement in each authoritative structure. The last four were examined and analyzed, first in quite a while of total problem-fathoming execution (exactness and practicality of theories), and later in increasingly nitty-gritty ways. This examination achieved the finish, and somewhat more delicate theories were uncovered.

Since every issue utilizing an alternate riddle, it cannot evaluate the unwavering quality of watched contrasts between issues. Contrasts could be expected, to some degree, to confuse trouble and hub fluctuation just as to authoritative structure. Perceiving this constraint, it appeared sensible to contrast genuine hub execution and some worldwide presentation rule for that specific issue. In this way, for every issue, it figured benchmark execution by a speculative "worldwide hub." By presumption, a worldwide hub can see the entire riddle framework without blunders and has a rundown of rules or heuristics for defining theories and authority reports. For each progressive cycle (one information update like clockwork), a scorer applied these guidelines to the current view to figuring out which words the worldwide hub would report.

Execution results are introduced in Figure 18.2. Each chart shows the total percentage of the right reports by the worldwide and real hubs at each round. The incubated region between the bends demonstrates the degree to which the real execution misses the mark concerning the worldwide standards. Generally speaking, apparently, AC performs better than DHC. It all the more about accomplishes the worldwide benchmark revealing levels in all cases.

To comprehend the authoritative elements which lead to great execution, it led to further examinations of the critical thinking follows. These examinations concentrated on three general guesses.

1) AC conveyed more productively than DHC, utilizing fewer messages to accomplish equivalent execution.
2) AC advanced quicker correspondence than DHC, with hubs reacting to demand for data all the more rapidly.
3) Higher level, increasingly conceptual data, for example, expression and subject speculations, was shared and seen all the more regularly in the AC structure.

The principal guess was evident. DHC groups utilized about twice the same number of messages as AC groups per issue.

The trouble of moving human ability to machine hubs made to search for basic and usable standards of agreeable conduct that could, in one way or another, guide the general plan process. One of the most significant standards included using models to mimic and foresee other hubs' exercises.

18.4.2 Examinations with Machine Networks

The real inspirations for building a machine system ought to be self-evident. How troublesome it is to move the heuristic look at structures, a component for growing progressively formal methods for putting away and sharing information in a DSN task is required.

The message puzzle task (MPT) space was simply an emblematic substitute for increasingly broad circumstance appraisal issues. The plan in the phrasing of detecting and following items traveling through space (stages).

Program Structure: The essential methodology was to develop a global problem-solver (which could see the entire riddle each round) utilizing the "participating specialists" worldview from man-made reasoning Erman et al. [8]. This worldwide program, stretched out with correspondence abilities, shaped the portion of every hub in a reproduction of a sensor organize. The recreated various processors by utilizing the exemplary methodology of time-cutting among forms, that require no further remark here. In any case, the method of coordinating specialists inside a solitary procedure being less notable requires clarification.

As in Hearsay-II Erman et al. [8] and HASP Nii and Feigenbaum [11], every hub kept a worldwide information structure called a board whereupon data at various degrees deliberation was posted. Alluding back to Figure 18.4, this data comprised crude sensor input information and increasingly dynamic and less certain theories about the particular unearthly lines, stages, and stage bunches that may be derived from that information. Data on the board came from code modules called information sources (KSs), which lived at the hub. Other hubs' data could, in any case, be posted on the writing board by correspondence modules.

The KSs worked by taking as information speculations and information from the board and posting new ones. There can follow their activity by following the change of starting sensor information to a last report. At the end of each round, any information that has not been represented will be utilized by a KS pro to bring forth a progression of conceivable ghastly line blends, which it knows are

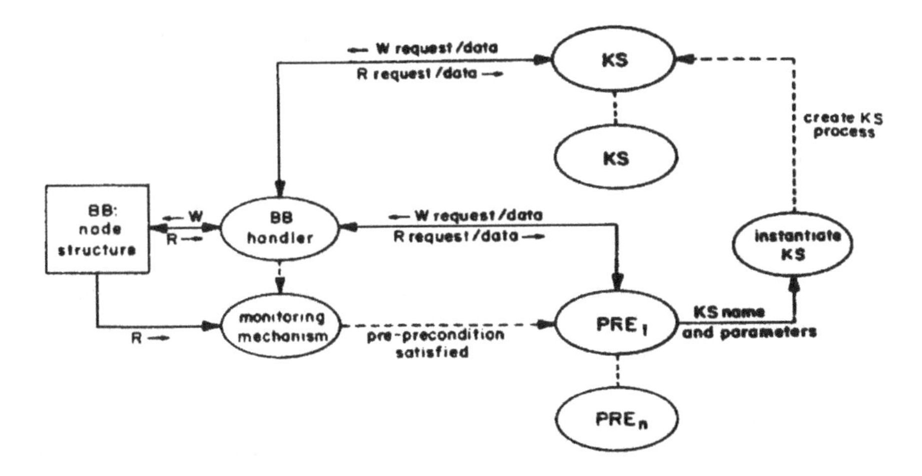

FIGURE 18.4 Simplified HSII system organization.

conceivable clarifications for that new data. At the point when this happens, another KS answerable for changing over ghastly line speculations into stage theories "awakens" and posts its best conjectures about stages that may be creating those lines. As time passes by and new information comes in, the confidences of these different speculations change as one stands apart as the best clarification of the info stream. After a specific point, a theory turns out to be sufficient to impart to different hubs in the framework a KS that "knows" about such standards decides this and passes the speculation to the correspondence module. Afterward, maybe as affirming data from a neighbor shows up, the trust in a speculation turns out to be high to the point that it is formally detailed outside the system. When that occurs, the at first contending speculations have been pruned away by an arrangement of coordinating KS specialists.

Correspondence limitations were loose from those in the human examinations. Every hub could speak with some other hub in the system, just as with the outside world (through sensor information and authority reports). The correspondence medium was great and immediate. Three sorts of data could be shared low-level (LL) sensor reports, moderate level (IL) unearthly line mix theories, and elevated level (HL) stage speculations.

Tests Performed and Results: Three fundamental structures were tried utilizing this system reproduction.

- A worldwide incorporated hub: This structure can likewise be seen as a two-level progression on the off chance that one restricted the activity of all lower-level hubs to the repetition securing and retransmission of crude sensor information. This structure gave a benchmark against which to assess the system structures beneath, since we believed that, given enough handling power, a brought together algorithm would essentially show the ideal critical thinking capacities inside a fixed arrangement of KS modules.

- A uni-level system of non-conveying hubs: Simply lessening the field of perspective on the worldwide hub above and making a system reproduction dependent on it permitted us to see how the execution was influenced by severe decay with no correspondence or coordination.
- A uni-level system of non-conveying hubs: Simply decreasing the field of perspective on the worldwide hub above and making a system re-enactment dependent on it permitted us to see how the execution was influenced by exacting deterioration with no correspondence or coordination.
- A uni-level system of conveying hubs: This structure that the AC relied upon to perform was fundamentally superior to the system of quiet hubs; however, it was poorer than the worldwide hub.

It did not test a DHC-like structure, incompletely due to its horrible showing during the human tests, and mostly because of PC asset and time restrictions made it difficult to include more hubs and the extensively more analyze structures; and required a component for growing increasingly formal techniques for putting away and sharing information in a DSN task.

The message puzzle task is utilized to proceed as our area. Be that as it may, because it was just an emblematic substitute for increasingly broad circumstance appraisal issues, the structure is discussed here in the more broad wording of detecting and following articles traveling through space (stages). The progressive data system on the left-hand side of Figure 18.5 was utilized in building information

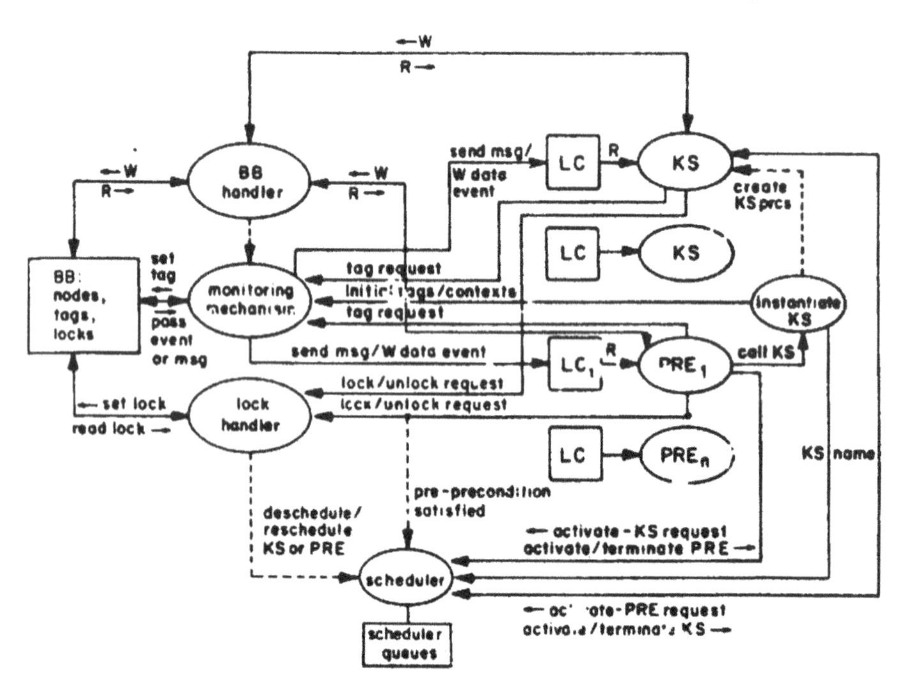

FIGURE 18.5 HSII framework association.

structures in our projects; the proportionate SA terms utilized in this paper are on the right. Note that the data chain of importance looks like different signs understanding disintegrations intently.

The KSs worked by taking as information theories and information from the board and posting new ones there.

The way that our machine organized accomplished human degrees of execution with such decreased correspondence stacking incited us to dissect the components which propel correspondence in a dispersed insight arrangement. Among the reasons hubs converse with each other are the following:

1) to share crude neighborhood data that may have increased worldwide ramifications;
2) to refresh others' information with recently showed up neighborhood data;
3) to coordinate the obtaining, preparing, and transmission of other hubs' data dependent on neighborhood needs;
4) to partition and offer the general errand load in a decent manner;
5) to share handling and detailing aims to forestall repetitive exercises;
6) to share significant level deductions or speculations which may divert or deter the requirement for others' preparing exercises;
7) to anticipate where, what, and when new data will show up to affirm or disconfirm existing speculations.

From these, reasons can be determined by two essential correspondence prerequisites.

- Nodes must share time-shifting data about the outer world they are detecting.
- Nodes must share time-fluctuating data about their interior conditions of preparing.

This basic perception clarifies why correspondence prerequisites in the machine structure were so little. Verifiable in the correspondence handling was the information on what kind of data would be inevitable from different hubs, when it would be incited, and how transmitted data would be utilized. To put it plainly, the hubs in the system verifiably shared models of one another's conduct.

This may not be conceivable, obviously, with progressively complex frameworks. Additionally, this present reality does not oblige us by isolating itself into adjustments that update messages very much organized and advantageous.

18.5 AN ABSTRACT MODEL FOR PROBLEM-SOLVING

In the theoretical, the critical thinking association hidden HSII might be demonstrated as a "creation framework," Newell [12]. A creation framework is a plan for indicating a data preparing framework in which the control structure of the framework is characterized by procedure on a lot of creations of the structure P → A, which work from and on an assortment of information structures. P speaks to an intelligent predecessor, called a precondition, which could is fulfilled by the data encoded inside the progressively current arrangement of information structures. If P is seen as fulfilled by certain information structure, at that point, the related activity An is executed, which apparently will make them change impact upon the database with the end goal that some other (or the equivalent) precondition gets fulfilled. This worldview for sequencing of the activities can be thought of as an information coordinated control structure since the fulfillment of the precondition is needy upon the dynamic condition of the information structure. Creations are executed as long as their precursor preconditions are fulfilled. The procedure ends either when no precondition is seen as fulfilled or when an activity executes a stop activity (in this way, flagging issue arrangement or disappointment, on account of critical thinking frameworks).

The Hearsay-II discourse getting framework (HSII) Lesser et al. [13], Erman, and Lesser [8] as of now a work in progress at Carnegie-Mellon University speaks to a critical thinking association that can adequately abuse a multiprocessor framework. HSII has been planned as an AI framework association reasonable for communicating information-based critical thinking techniques that suitably sorted out topic information might be spoken to as information sources fit for contributing their insight in an equal information coordinated style. An information source (KS) might be depicted as an operator that exemplifies the information on a specific part of a difficult area and helps take care of an issue from that space by performing activities dependent on its information to facilitate the advancement of the general arrangement. The HSII framework association permits these different autonomous and various wellsprings of information to be determined. Their collaborations facilitated to help out each other (maybe non-concurrently and in corresponding) to impact a complex arrangement.

18.5.1 The HSII Organization: A Production System Approach

The HSII framework association, which can be described as an "equal" creation framework, has a robust database that speaks to the dynamic issue arrangement state. This database, which is known as the chalkboard, is a multidimensional information structure, the basic information component of which is known as a hub. For instance, the elements of the HSII discourse understanding framework database are instructive level (e.g., phonetic, surface-phonemic, syllabic, lexical, and phrasal levels), articulation time (discourse time estimated from the earliest starting point of the information expression), and information choices (where numerous hubs are allowed to exist at the same time at a similar level and articulation time). The board is decipherable and writable by any precondition or KS process (where a KS procedure exemplifies a creative activity, to utilize the phrasing of creation frameworks). Preconditions are procedurally arranged and may indicate discretionarily complex tests to be performed on the information structure to choose precondition fulfillment. To abstain from executing these precondition tests superfluously regularly, they, thusly, have pre-preconditions which are basic screens on important crude database occasions (e.g., observing for a change to a given field of a given hub in the database, or a given field of any hub in the database). At whatever point any of these crude occasions happen, those preconditions checking such occasions become schedulable and, when executed, test for full precondition fulfillment. Testing for precondition fulfillment isn't ventured to be a prompt or even a unified activity, and a few such precondition tests may continue simultaneously.

The KS forms speaking to the creative activities are additionally procedurally arranged and may indicate subjectively complex groupings of tasks to be performed upon the information structure. The general impact of some random KS process is normally either to conjecture new information which are to be added to the database or to check (and maybe adjust) information recently positioned in the database. This follows the general theory and test problem-tackling worldview wherein theories speaking to incomplete issue arrangements are created and afterward tried for legitimacy; this cycle proceeds until the confirmation stage ensures the finishing of preparing (and either the issue is explained or disappointment is demonstrated). The execution of a KS procedure is generally transiently disjoint from the fulfillment of its precondition; the execution of some random KS process isn't dared to be inseparable. The simultaneous execution of different KS forms is allowed. Likewise, a preconditioning procedure may conjure various launches of a KS to chip away at the various pieces of the writing board, which autonomously fulfill the precondition's example. In this way, the free information coordinated nature of precondition assessment and KS execution can produce a lot of equal action all through the simultaneous execution of various preconditions, diverse KS's, and different launches of a solitary KS.

The fundamental structure and parts of the HSII association might be delineated, as appeared in the message exchange outline of Figure 18.4. The outline demonstrates a dynamic data stream between the different segments of the problem-tackling framework as strong bolts; ways showing control action have appeared as broken bolts. The significant parts of the outline incorporate an inactive worldwide information structure (the writing board), which contains the present condition of the complex arrangement.

Access to the slate is adroitly incorporated in the chalkboard handler module, [4] whose essential capacity is to acknowledge and respect demands from the dynamic preparing components to peruse and compose portions of the board. The dynamic preparing components which create this information get to demands comprise of KS forms and their related preconditions. Preconditions are enacted by a chalkboard checking system that screens the different compose activities of precondition becomes schedulable [6]. If, upon further assessment of the board, the precondition gets itself "fulfilled," the precondition may demand a procedure launch of its related KS to be set up, passing the subtleties of how the precondition was fulfilled as parameters to this launch of the KS'. Once started up, the KS procedure can react to the writing board information condition which was distinguished by its precondition, potentially mentioning further changes be made to the chalkboard, maybe in this way activating further preconditions to react to the most recent adjustments. This specific portrayal of the HSII association, while excessively rearranged, shows the information-driven nature of the KS initiations and collaborations. An increasingly complete message exchange outline for HSII will be introduced in an ensuing area.

18.5.2 Hearsay-II Multiprocessing Mechanisms

Multiprocessing conditions, systems must be given to help the individual restricted executions of the different dynamic and prepared procedures and to shield the procedures from meddling with each other, either legitimately or in a roundabout way. Then again, instruments should likewise be given so the different dynamic procedures may speak to accomplish the ideal procedure participation. Since the constituent KSs are thought to be freely evolved and are not to assume the express presence of different KSs, correspondence among these KS's must fundamentally be backhanded. The longing for a particular KS structure emerges from the way that generally a wide range of individuals are engaged with the usage of the arrangement of KSs, and, for motivations behind experimentation and KS execution investigation, the framework ought to have the option to be reconfigured effectively utilizing elective subsets of KSs. Correspondence among KSs takes two essential structures: database observing for gathering relevant information occasion data for sometime later (nearby settings and precondition enactment), and database checking for distinguishing the event of information occasions which damage earlier information suspicions (labels and messages). The accompanying passages examine these types of database checking, and their relationship to the information get to synchronization instruments required in a multiprocessing framework association.

18.5.3 Nearby Context

The worldwide database (the board) is proposed to contain just powerfully current data. Since preconditions (being information coordinated) are to be tried for fulfillment upon the event of significant database changes (which are authentic information occasions), and since neither precondition testing nor activity execution (nor the consecutive mix of the two) is thought to be an indissoluble activity, confined databases must be accommodated each procedure unit (precondition or activity) that requirements to recollect applicable verifiable information occasions. These limited databases, called neighborhood settings in HSII, which record the progressions to the writing board since the preconditioning procedure was last executed or since the KS procedure was made, give customized working situations to the different precondition and KS forms. A nearby setting jelly just those information events4 and state changes applicable to its proprietor. The creation time of the nearby setting (i.e., the time from which it starts gathering information occasions)

18.4.4 Data Integrity

One approach to moving toward information trustworthiness is to ensure the legitimacy of information suspicions by prohibiting interceding forms the capacity to alter (or maybe even to look at) basic information. The HSII framework gives two types of locking natives, hub and area locking, which can be utilized to ensure selective access to the ideal information. Hub locking ensures restrictive access to an expressly determined hub on the board. In contrast, district locking ensures privileged access to an assortment of hubs that are indicated certainly dependent on many hub qualities. In the present usage of HSII, the area attributes are indicated by a specific data level and timeframe of a hub. On the off chance that the writing board is considered as a two-dimensional structure with directions of data level and time, at that point, district locking licenses the locking of a subjective rectangular region in the slate. Locale locking has the extra property of forestalling the production of any new hub that would be put in the slate zone indicated by the area by other than the procedure which had mentioned the district lock. Extra bolting adaptability is acquainted by permitting forms with demand read-just access to hubs or districts (called a hub or area looking at); this diminishes conceivable conflict by allowing various perusers of an offered hub to exist together while barring any scholars of that hub until all perusers are done. The framework likewise gives a "super lock," which permits a discretionary gathering of hubs and locales to be bolted simultaneously. A predefined direct requesting system for non-preemptive information get to designation is applied by the "super lock" crude to the ideal hub and district bolts to keep away from the chance of database halt.

18.5.5 Contextual Analysis

The current areas will talk about the different tests used to describe the multiprocessing execution of the HSII association in the discourse getting task.

18.5.6 HSII Multiprocessor Performance Analysis through Simulation

To pick up knowledge into the different effectiveness issues, including multiprocessing problem-tackling associations, a reproduction model was consolidated inside the uniprocessor form of the HSII discourse getting framework. The HSII problem-fathoming association was not displayed and reenacted. Yet, rather the actual HSII usage (which is a multiprocessing association in any event, when executing on a uniprocessor) was altered to allow the recreation of a firmly coupled multiprocessor equipment condition.

There were four essential targets of the recreation tests: (1) to quantify the product overheads engaged with the plan and execution of a confounded, information coordinated multiprocedure (or) control structure, (2) to decide if they truly exist a lot of equal movement in the discourse getting task, (3) to see how the different types of between process correspondence and impedance, particularly that from information get to synchronization in the chalkboard, influence the measure of viable parallelism acknowledged, and (4) to pick up understanding into the structure of a suitable booking calculation for a multiprocess critical thinking structure. Unquestionably, any outcomes introduced will mirror the point by point efficiencies and wasteful aspects of the specific framework usage being estimated, however ideally, the association of HSII is adequately broad that the different explanations will have more extensive quantitative relevance for those considering comparative multiprocess control structures. To give a premise to the conversation in the ensuing segments of this part, Figure 18.6, delineating the different segments of the HSII authoritative structure, is advertised. The outline is an increasing point by point variant of the message exchange model introduced beforehand. The new segments of this chart are principally a consequence of tending to

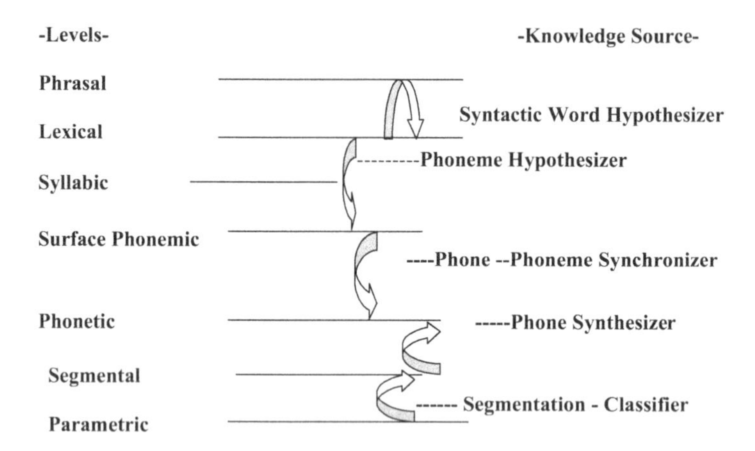

FIGURE 18.6 Streamlined HSII, KS, and data-level arrangement.

multiprocessing contemplations. As in the prior, progressively streamlined authoritative chart, the powerfully current condition of the complex arrangement is contained in the slate. The writing board contains information hubs and records information checking data (labels), and information gets to synchronization data (locks). Access to the writing board is theoretically concentrated in three modules. As in the past, the slate handler module acknowledges peruse and composes information to get to demands from the dynamic preparing components (the KS forms and their precondition forms). A lock handler arranges hub and locale lock demands from the KS procedures and preconditions, with the capacity to obstruct the advancement of the mentioning procedure until the synchronization solicitation might be fulfilled. A checking component is answerable for tolerating information labeling demands from the KS procedure and preconditions and sending messages to the labeling forms at whatever point a labeled information field is altered. The checking system likewise must disseminate information occasions to the different neighborhood settings of the KS procedures and preconditions, just as to enact precondition forms at whatever point adequate information occasions important to those preconditions have happened in the writing board.

Every dynamic preparing component is a nearby information structure, the neighborhood setting, which records information occasions that have happened on the writing board and are important to that specific procedure. The nearby settings might be perused by their related procedures to discover which information hubs have been adjusted as of late and what the past estimations of specific information fields were. The nearby settings are naturally kept up by the writing board observing system.

18.5.7 The HSII Speech Understanding System: The Simulation Configuration

The arrangement of the HSII discourse getting framework, whereupon the accompanying reproduction results were based, comprises eight conventional KSs that work on a board containing six data levels (Figure 18.5).

The KSs utilized in the reproduction was as per the following: the section classifier, the telephone synthesizer (comprising of two KSs), the phoneme estimate, the telephone phoneme synchronizer (comprising of three KSs), and the rating approach module (see Figure 18.6). These KSs are initiated by about six for all time started up precondition forms that are consistently checking the slate database for occasions and information designs pertinent to their related KSs. Because of the over the top expense of the recreation exertion (and because of the constrained phases of improvement of some accessible KSs), this design speaks to just a subset of a progressively complete framework containing around 15 KSs. Lesser et al. [13] contain an increasingly nitty-gritty depiction of the board and the different KSs for the complete HSII discourse getting framework.

18.6 HIERARCHICAL DISTRIBUTION OF WORK

The director delegates employments to these units, seeing them as natives [9] in the authoritative arrangement. Every unit at that point deciphers the control directions and develops them to control the natives inside the unit.

The epitome of the two components and data is essential to the best possible organizing of an association. It is required by limited reasonability. The following are attributes of units that fulfill the requirements forced by limited reasonability.

1) View the various activities (programs) contained in a unit as a solitary activity (deliberation).
2) Control units as though they were crude activities (arranging in deliberation spaces).
3) Delegate authority. Orders to a unit are explained by and inside the unit.
4) Reduce the data stream. The unit can sum up data inside a unit.
5) Hide detail. Data and control not required by different units are covered up inside the unit.

Deciphering the above imperatives in the structure of appropriated frameworks requires that:

1) the results of the procedure must be completely characterized;
2) the communication between forms must be negligible (close to decomposability);
3) the impacts of a procedure upon different procedures must be comprehended;
4) away from power must be perceived;
5) away from of data stream must be perceived. The first between process limitation is a base necessity. It must be recognized what a procedure creates before it tends to be utilized. The second lessens the multifaceted control nature of the association, which may decrease its viability. The third is vital with the goal that one procedure's activity won't sabotage another procedure. The last two focuses guarantee appropriate control, and the data whereupon to base that control.

The design of Hearsay-II complies with a large number of these necessities. A KS is a unit that can be seen as a solitary activity. In Hearsay-II, improvement reaction outlines are utilized to portray the impact of a KS as a solitary activity to the scheduler. A KS conceals its calculation and transitory outcomes from different KSs and confines its correspondence to a couple of theories on the board. The board structure system characterizes how KSs impart and how they interface (basically, posting a theory may make another KS respond).

On the other hand, straight speedups can be acquired if calculations and data stockpiling are painstakingly broke down and appropriately organized. Thusly, more noteworthy consideration regarding task decay in program associations and critical thinking must be paid.

Following are some authoritative bases that decrease the multifaceted nature of coordination.

1) Slack Resources: One part of multifaceted coordination nature is the coordination of coupled undertakings. Assignments are coupled when the contribution of one relies upon the yield of another. Errands are firmly coupled when state changes in a single undertaking promptly influence the condition of another assignment. To decrease the snugness of the coupling, slack assets are presented. Support inventories are embedded between coupled errands so that on the off chance that one undertaking has something turn out badly with it, different assignments are not promptly influenced.

   Two understandings of slack in dispersed frameworks are:

   i) the replication of undertakings (forms, modules) on interchange processors in the event of a processor disappointment and
   ii) the substitution of technique calls by message lines; requests and messages to an errand are put in a line to lessen the synchronization (tight coupling) of assignments.

Slack has likewise been utilized broadly in PC equipment. The duplication of functional units has reached out to the dependability of equipment. Space vehicles, for instance, copy basic units. Slack shows up in various manners in Hearsay-II. KSs can be copied on numerous processors with the goal that a similar undertaking can be conveyed in equal. Besides, the creation and position of different theories on the writing board can be seen as making a cushion stock. A KS has numerous speculations to take a shot and can keep preparing without monitoring what another KS is doing or isn't doing.

2) Function Versus Product Division: The coupling of assignments can likewise be decreased by legitimate deterioration. Association hypothesis recognizes two sorts of association dividing.
   - The first is an item or independent division. This division necessitates that units be based on the item that will be created by the association.
   - The second sort is a practical division. A useful division situates the units to the capacities important to deliver the items (e.g., buying,

advertising, materials, and so on.). Contingent upon attributes of the issue being illuminated by the association (e.g., delivering a plane), one division diminishes intricacy while different builds unpredictability.

A significant proportion of unpredictability is the measure of coordination. Any division of an issue except a more significant association inside a unit than between units (communication area). A framework that displays cooperation territory is almost decomposable (Simon [14]). When the association area does not exist anymore, the coordination of units turns out to be excessively mind-boggling. The public square has various attributes that make it especially reasonable to advance complex frameworks in a multiprocessor situation. These include: the intricacy of equal preparing can be covered up by building "reusable" custom conditions that manage a client in portraying, investigating, and running an application without engaging in equal programming; calculations can be communicated in various dialects; the structure of a framework can be altered while the framework is running; KSs are initiated by designs processed on the information produced by different KSs; KSs are depicted in a way that permits Agora to coordinate the accessible engineering and its assets with the necessities of the calculation; Custom models can undoubtedly be incorporated with segments running on universally useful frameworks.

A marketplace that bolsters the development of huge, computationally costly, and inexactly organized frameworks, such as information-based frameworks for discourse and vision understanding. Public square can be redone to help the programming model that is increasingly appropriate for a given application. The marketplace has been structured unequivocally to help different dialects and exceptionally equal calculations. Frameworks worked with Agora can be executed on various universally useful and custom multiprocessor structures.

18.7 AGORA

Agora is Greek for a marketplace or gathering place where individuals examine open issues (similar to a commercial center). In this way, the AI Agora is an activity from the Department of Information Science and Media Studies. It establishes a talk and conversation arrangement about Artificial Intelligence in principle and as applied science. Computer-based intelligence Agora presents worldwide visitors just as neighborhood speakers.

Online discussions are untidy. Voices lose all sense of direction in the group. Thoughts get covered in the commotion. That is why we made Agora engage people to take care of the issues together with each noteworthy discussion in turn. Public square auto-mysteriously finds noteworthy bits of knowledge utilizing AI and common language handling. Improve your work process with these bits of knowledge, and you're on the way to amazingness. Public square's crucial to enable people to take care of issues together – each noteworthy discussion in turn. Public square uses AI + characteristic language handling to find noteworthy experiences in huge scope discussions.

Highlights of Agora:

- Prioritize cleverly. Immediately anticipate earnestness, beat probability, and goals way dependent on any approaching client message.
- Detect inclining issues. Keep steady over the most revealed issues, recommendations, and new patterns continuously.
- Suggested reaction. Spare time and divert generally posed inquiries with a recommended reaction.

There are 13 main alternatives to Agora:

1. Vox neural

Vox is the conclusive general intrigue news hotspot for the 21st century. Vox's crucial basis: Explain the news. It treats genuine points honestly, openly controlling crowds through complex subjects running from governmental issues, open strategy, and world undertakings to mainstream society, science, business, food, sports, and everything else that issues Vox is one of the quickest developing news locales (comScore) and part of Vox Media. Vox Media is reclassifying the cutting edge media organization by engaging the most astute imaginative voices with the innovation to make and disseminate premium substance and associates with a crowd of people who seeks us for keen and educational substance encounters.

- Manage approaching calls, outcoming calls, and SMS.
- Provide support for web talk, WhatsApp, Facebook Messenger, Telegram, and Line.
- Handle cooperations of Facebook, Instagram, Twitter, and email.
- Retrieve continuous reports for each channel.
- Monitor all parts of each cooperation.
- Classify and spare all interchanges.

Prescient DIALER

- Horizontal and vertical dialing techniques.
- Capable of up to 10 concurrent calls for every operator.
- Define set up parameters for ring limit times and dialing endeavors.
- Customize email and SMS layouts.
- Detect replying mail.
- Create database channels.
- Configure boycotts.
- Enable Smart Dialing (Big Data examination).

Discourse ANALYTICS

- Analyze clients' voices naturally.
- Generate continuously advanced agreements.
- Capture discourse examination for quality affirmation.
- Qualify explicit catchphrase settings.

- Support voice acknowledgment for IVR.
- Create shrewd labels when catchphrases are spoken.

3. Reports and analytics
   1. Generate reports and documentation of numerous classes of data, including:
      - Agent abilities
      - Campaigns
      - Speech examination
      - Gamification
      - Interaction and talks
      - Routing and that's only the tip of the iceberg
   2. Fare over 60 administration reports.
   3. Coordinate with business intelligence systems.
   4. Facilitate extra API's from wolkvox.

4. RapidMiner

RapidMiner is an information science programming stage created by a similar name that gives an incorporated domain to information readiness, AI, profound learning, content mining, and prescient investigation.

Highlights of RapidMiner

- Data Prep Seamlessly incorporated and improved for building ML models
- Machine Learning Design models utilizing a visual work process architect or mechanized displaying
- Model Operations Deploy and oversee models and transform them into prescriptive activities.

5. Kapiche

Item details:

Kapiche is set to assist associations with comprehension and relate to their clients at scale utilizing our exceptional programming. Kapiche utilizes one of a kind Topic Modeling system that permits clients to construct an away from vital subjects in unstructured information. While contender programming depends on word references or thesauri to do this, Kapiche programming is based on a solo AI calculation that accomplishes a similar result in substantially less time and without the inclinations of human-assembled word references. Delivering a comprehension of nuanced human language as utilized by ordinary individuals.

Highlights of Kapiche:

- Take long periods of work off your plate. Let Kapiche's common language handling calculation prepare so you can invest more energy being the information legend you realize you are.

- Automatically spot developing patterns, before it gets monstrous Kapiche's AI calculation naturally recognizes emanant subjects each time new information is included, empowering you to tell the pertinent offices and follow up on the issues, before they become a more concerning issue for your association.
- Easily see how to improve CX measurements. Impact scores permit you to recognize the client gives that need fixing the most rapidly. Comprehend the specific effect singular issues are having on your NPS, CSAT, or CES scores and perceive how improving these issues would drive up your general scores.
- Track issues and patterns after some time. Measure the effect of your business choices on consumer loyalty and perceive how input changes as you address client issues or present new items and administrations.
- Deeper bits of knowledge initially Instantly perceive how various topics or client portions are performing, and organize as needs are. Dig as profound as you need to and cut the information any way you need. You'll have more power over your information than any time in recent memory.
- Happy, pitiful, and everything in the middle of Kapiche distinguishes any slant in your client criticism at the in general and per-reaction level, giving you another measurement to understanding your clients.
- Sharing is minding with adjustable dashboards. Insights are just helpful if the correct individuals think about them. Utilize adjustable dashboards to share bits of knowledge over the whole association or just with explicit areas, groups, or representatives to actualize the required fixes. Make, modify, and share dashboards loaded up with the profound, significant bits of knowledge from your investigation.
- Supercharge your experiences with reconciliations. Prevent information storehouses and enable your business to settle on laser-centered business choices with Kapiche mixes. Kapiche permits you to interface different information hotspots for more profound client examination, including segment, conduct, spend, and CX information from CRMs, review apparatuses, and distributed storage suppliers.

6. MonkeyLearn

MonkeyLearn is an AI stage that permits you to investigate content with Machine Learning to robotize business work processes and spare long periods of manual information preparation. Arrange and concentrate noteworthy information from crude writings like messages, talks, website pages, records, tweets, and then some! You can characterize writings with custom classes or labels like feeling or point and concentrate specific information like associations or watchwords. MonkeyLearn's motor can be effectively incorporated using direct combinations (no coding required) or API and SDKs.

Highlights of MonkeyLearn are given below.

For support teams:

- Automatically course and organize tickets. MonkeyLearn will consequently label your tickets dependent on theme, issues, expectation, supposition, or need. You can utilize the labels inside your help programming in blend with directing and prioritization rules.
- Elimination of manual and redundant preparing. Automaticlabeling will empower your group to concentrate on shutting more tickets as opposed to manual information handling.
- Obtain revealing and bits of knowledge. Consistent and programmed labeling will enable your group to find new bits of knowledge from your help tickets.

For product and CX teams:

- Analyze feedback at scale. MonkeyLearn helps via consequently labeling your input in minutes not weeks.
- Customized, centralized, and accurate. MonkeyLearn encourages you to believe your information by keeping reliable labeling standards.
- Customized reporting. Build custom reports and dashboards by interfacing your preferred AI apparatuses.

For developers:

- Customized text investigations. Obtain high exactness results via preparing your custom content classifier or extractor.
- Quick integration tests. Integrate your models in minutes.
- Beautiful API, SDKs, and Docs. Built for engineers by designers.

7. Find TEXT

A shared book investigation stage for AI.

We give many multi-lingual, content mining, information science, human explanation, and AI highlights. Discover Text offers a scope of easy to cutting edge cloud-based programming devices engaging clients to rapidly and precisely assess a lot of content information. Our clients sort unstructured free content basic in statistical surveying, just as related metadata, additionally found in client criticism stages, CRMs, talks, email, enormous scope HR or different studies, open remark to government organizations, Twitter, RSS channels, and different types of content information.

- Highlights of Discover Text.
- Collect, clean, and investigate content information.
- Humans and machines group content.
- Discovery of instruments that work.

- Boolean characterized search, n-grams, word mists, and custom point word references.

8. Advize.ai

Advize consolidates AI-controlled profound learning classification, point displaying, opinion investigation, and pattern identification to help shopper organizations comprehend their clients, and spotlight on high effect enhancements that expansion development and client devotion.

Highlights of Advize.ai:

- Categorization. Deep learning arrangement precisely containers client remarks. For exceptionally precise outcomes, we manufacture custom models that spread the client venture and each key part of a given business.
- Sentiment analysis. Advize utilizes propelled point demonstrating and semantic job naming to ascribe notion to how the client portrays their encounters precisely.
- Prioritization. Advize consolidates client criticism, fulfillment scores, feeling, and reaches to assist you with settling on information-driven prioritization choices.
- Segmentation. Use propelled division to see how every client portion sees your business.
- Anomaly recognition. Uncover patterns with a continuous, information-driven methodology that recognizes oddities in criticism and alarms your group.
- Powerful announcing. Monitor advance and uncover the purposes for changes. Improve KPIs like NPS by seeing how each issue influences consumer loyalty.

9. CITIBEATS

Citibeats is the AI Advance Analytics Platform extraordinary in recognizing social patterns and concerns. It finds, orders, and integrates individuals' feelings from enormous amounts of information. It offers subjective data and estimation experiences that give all the more significance to your information. Citibeats answers for Smart Cities, Mobility, Tourism, Finance, and Insurance are depended on by customers around the world. Our innovation has been created with MIT and Singularity University and granted by the United Nations and the European Union. Citibeats is a Social Coin item, intended to improve social orders dependent on the requirements of the individuals.

Highlights of CITIBEATS are given below.

For subject matter experts:

- Made simple. Steer your task without the requirement for coding or information science.

- Adaptable to your setting. Create your own model for your neighborhood setting and difficulties.
- Scalable effect. Extend your task across geologies and 50+ dialects.

For data experts:

- Custom models, quick. Create custom models in minutes and see them develop with the discussion after some time.
- All in one spot. Connect to information sources, train, and investigate one stage.
- Enrich your investigations. Use our API to correspond your numerical information with content information.

10. GAVAGAI

Gavagai is the best content investigation programming. In any language, Gavagai Explorer assists with picking up bits of knowledge from the entirety of your client criticism information. An adventurer is simple, quick, and works for 46 dialects. It's an AI-driven investigation for everybody!

Highlights of Gavagai:

- Access to audit information from online sources. Now you can scratch and download survey information from more than 50 diverse online sources; information that you can examine in Explorer. At no additional charge.
- Customizable notions. Your space may require customization of assessment estimations. 'Delicate' tissue paper is acceptable; 'delicate' teeth are terrible. Peruse progressively about estimation customization here.
- Save and re-use models. A model of investigation can be put something aside for later use, for instance, on the off chance that you need to screen drifts in the responses to comparable inquiries after some time.
- Discover key points in the writings. Gavagai Explorer lets you find key subjects; consequently there is no compelling reason to predefine them. You can alter the points that the framework proposes and even include new ones. Also, you can even make point definitions change naturally when new information is added to the undertaking: read increasingly about the auto-included terms that are highlighted.
- A more extensive scope of estimations. Gavagai Explorer characterizes Positivity and Negativity in the writings, as well as Love, Fear, Hate, Violence, Desire, and Skepticism, for each language.
- Get the greater part of metadata. Meta-information and NPS can influence the arrangement and qualities of points. Gavagai Explorer permits you to apply different channels to find and model these associations.
- Cross-language investigation. Start examination in one language, spare the model, and it is consequently meant to extend to different dialects. This

makes working with numerous dialects simple and furthermore permits you to control cross-venture consistency.
- Easy mix utilizing APIs. Gavagai Explorer usefulness is anything but difficult to incorporate into any creation pipeline utilizing the gave RESTful APIs.
- Custom concepts. You can characterize your own custom ideas like obscenity or purchasing inclination and measure them after some time or across socioeconomics.
- Automatic multiword articulations. "San Francisco", "Feliznavidad", "God morgon"; a few successions of words simply have a place together. We call them N-grams and Explorer thinks about them naturally so you don't need to monitor them yourself. In any language.

10. Agara

Agara is the eventual fate of business-to-buyer correspondence. Agara utilizes propelled Voice AI to process client service brings progressively. Agara can process calls completely independently or gives live proposals to a human specialist to determine the call rapidly and effectively. Agara evacuates all grating focused on help calls. Endeavors can improve their help experience drastically with zero hold times, fundamentally improved first call goals rate, and shorter call spans. This improved experience goes ahead top of over half cost reserve funds and operational efficiencies.

Highlights of Agara:

- Automatic speech recognition. Agara naturally creates and stores an excellent call transcript for each bring continuously.
- Advanced NLU. Agara's exclusive NLU motor empowers a profound investigation of the deciphered voice and content stream and orders them into legitimate goals.
- Deep learning. The profound learning model prepared with 10 million+ client care explicit information focuses, at that point utilizes this contribution to locate the correct goals to a client's issue.
- Text-to-speech. Our human-like text-to-speech framework at that point answers to the client with the means to investigate and shuts the consider like how a human operator would do.
- Analytics. Agara likewise gives a rich dashboard that gives all of you the information going from key help measurements to operator execution details.
- Flexible deployment. Agara fits in non-intrusively with your CRM and cloud communication administrations with insignificant advancement necessities, so you can convey business sway immediately.

11. MeaningCloud

MeaningCloud is the least demanding, generally amazing, and most reasonable approach to extricate the importance of a wide range of unstructured substance:

social discussion, articles, records, web content, and so on. We give content examination items to separate the most precise bits of knowledge from any media content in numerous dialects. What's more, we do it SaaS and On-prem. We work for various enterprises (pharma, account, media, retail, accommodation, telco, and so on) creating customized and industry-arranged arrangements. Our situations incorporate – Insight extraction in showcase knowledge – Analysis of the voice of the client, worker, resident, or patient. (Client or client experience investigation all in all.) – Intelligent archive computerization – Smart ticketing (for helpdesks) – Semantic distributing – Social crisis the executives

Highlights of MeaningCloud:

- Powerful. It consolidates the most cutting edge innovations to give complex functionalities: include level notion examination, online life language handling.
- Customizable. It gives realistic interfaces to permit the client to alter effectively the framework utilizing his/her own word references and models.
- Easy to utilize and incorporate. Use it from the include for Excel, coordinate it without coding through our modules, or create over our SDKs and web administrations. An open stage, simple to learn, and use.
- Without duties. Without programming to introduce or framework to convey. All the unwavering quality and adaptability of arrangements in the cloud, and the chance of testing it for nothing.
- Multiple dialects. The main instrument you have to break down substance in Spanish, English, French, Portuguese, or Italian.
- Affordable. Pay just for what you use, with no actuation expenses, least time duty, and with the most liberal free arrangement of the market. On the off chance that you don't care for it, you can quit utilizing it, much the same as that.

12. Proxem

Proxem is a French pioneer in the semantic examination of Big Data and Digital Transformation. We offer SaaS programming answers for break down huge volumes of printed information with the goal that our customers can separate applicable data progressively from the web, online life, chatbot, messages, fulfillment overviews, and that's only the tip of the iceberg. We total all your data onto a multi-source, multi-lingual, and multi-subject programming arrangement. Proxem Software offers the best in easy to understand content mining apparatuses, by permitting clients to break down the content without composing any code whatsoever.

Highlights of Proxem:

- Proxem Insight: significant understanding from each client and representative criticism in one stage
- Proxem Dialog: Customizable Virtual Assistants and Chatbots

- Proxem Knowledge: data extraction and information the executives through AI and Semantic Analysis

13. Rosette

Rosette, from Basis Technology, is the content examination foundation of decision for driving firms in a wide scope of verticals, including undertaking search, business knowledge, information mining, e-Discovery, monetary consistency, and national security. A similar superior, thoroughly tried NLP suite utilized via Airbnb, Attivio, Microsoft, Oracle, Pinterest, Salesforce, Stumble Upon, Yelp, and innumerable others is accessible to new businesses.

Highlights of Rosette:

- Chat translation
- Base linguistics
- Categorization
- Entity extraction
- Entity linking
- Language identification
- Name matching
- Name translation
- Relationship extraction
- Semantic similarity
- Sentiment analysis
- Topic extraction

Some applications in data science and distributed AI involve image and video.

- A Quantum registering framework is adjusted to numerous applications, for example, Quantum Image Processing (QIMP) Quantum Image portrayal, Image division, and so on. This QIR model contains every one of the three pictures that structure the pioneer QIRs and the later ones. The entirety of the QIRs in this gathering use the shading model or visual contrast to encode the substance of a picture (Yan et al. [1]).
- Super-goals of a picture can be performed by various levels per outline.
- A picture thick coordinating reliant on locale developing method can be performed with a versatile window, to diminish the multifaceted nature in calculation and sparing of time in short order.
- The scope of the picture (from 2D to 3D) can be disentangled by utilizing Quadric mistake matrix (Sang and Wang [15]).

18.8 EXPLORATORY OUTCOMES FOR IMAGE PROCESSING

These days the Novel Image Processing turns out to be so famous and has appeal in space science, medication, remote detecting, machine vision, programmed change

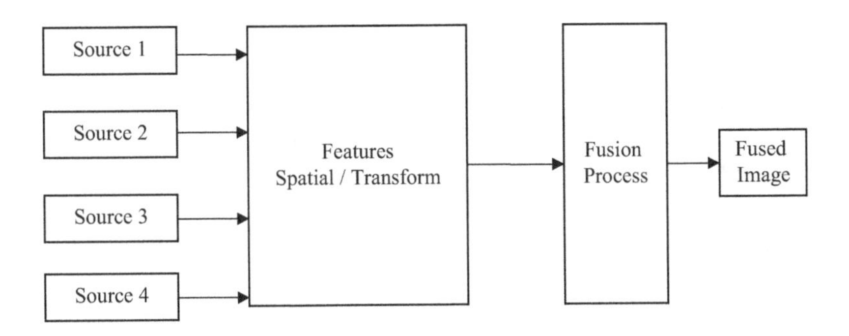

FIGURE 18.7 Block diagram of the novel image fusion process.

identification, biometrics, mechanical autonomy, minute vision, and so on. There are a lot of handling procedures that existed, among them; Image combination is interesting. The image fusion calculation requires the whole striking data contained in the info images. The calculation should not present any antiques or irregularities, which can occupy image processing steps. It must be dependable, strong, open-minded to commotion, and blunder in enrolments.

The Image fusion score is a measure for blocking antiques. In the event that the score is less, at that point the blocking antiques are high, and the other way around. The general fusion process is given in Figure 18.7. The required feature information is acquired from a number of source images and is transformed into frequency domain. The information can be compressed and efficiently used in the fusion process.

18.9 SUMMARY AND CONCLUSIONS

This chapter introduced a review of data science that incorporates information access and information preparation methods. The association for an information-based AI problem-understanding framework which is intended to exploit any distinctness of the handling or information parts accessible inside that association. KSs are expected to be to a great extent free and fit for offbeat execution as KS forms. By and large framework control is appropriated and basically information coordinated, depends on occasions happening in a universally shared writing board database. The intercommunication (and association) of the different KS forms is limited by making the slate database the essential methods for correspondence, in this manner displaying an indirection regarding correspondence like the aberrant information coordinated type of procedure control. Such a problem-settling association is accepted to be especially manageable to execution in the equipment condition of a system of firmly coupled offbeat processors that share a typical memory. The HSH, which has been created utilizing the procedures for framework association portrayed here, has given a setting to assessing the multiprocessing parts of this framework design.

REFERENCES

1. F. Yan, A.M. Iliyasu, and S.E. Venegas-Andraca. A survey of quantum picture portrayal. *Quantum Information Procedure*, 15, 2016, pp. 1–35.
2. A.H. Bond, and L. Gasser. *Readings in Distributed Artificial Intelligence*. Text Book, Morgan Kaufmann Publishers, Inc., San Mateo, CA, 1988.
3. B.J. Lee, L.O. Gilstrap Jr., B.F. Snyder, and M.J. Pedelty. *Hypothesis of Probability State Variable Systems Volume 1: Summary and Conclusions*. Specialized Report ASD-TDR-63-664, Vol. No. 1, AF Avionics Laboratory, Wright-Patterson Air Force Base, Ohio, 1963.
4. I. Aleksander. Activity oriented learning networks. *Kybernetes*, 4, 1975, pp. 39–44.
5. J.L. McClelland, D.E Rumelhart, and the PDP Research Group. *Equal Distributed Processing: Explorations in the Microstructure of Cognition (2 Volumes)*. MIT Press, Cambridge, 1987.
6. J.A. Feldman, and D.H. Ballard. Connectionist models and their properties. *Intellectual Science*, 6(3), 1982, pp. 205–254.
7. F. Hayes-Roth, and R. Wesson. *Conveyed Insight for Circumstance Evaluation*. Rep. N-1447-ARPA, Rand, Corporation, Santa Monica, CA, January 1980.
8. L.D. Erman, F. Hayes-Roth, V.R. Lesser, and D.R. Reddy. *The Hearsay-II Discourse Getting Framework: Integrating Information to Determine Vulnerability*. Tech Rep. CMU-CS-79-156, Carnegie-Mellon University, Pittsburgh, PA, in press.
9. P.R. Lawrence, and J.W. Lorsch. *Organization and Environment*. Harvard University, Boston, MA, 1967.
10. L. Uhr. *Pattern Recognition, Learning and Thought*. Prentice-Hall, Englewood Cliffs, NJ, 1973.
11. H.P. Nii, and EA. Feigenbaum. Rule-based comprehension of signs. In *Pattern-Directed Inference System*, D.A. Waterman and F. Hayes-Roth, Eds. Academic, New York, 1978, pp. 483–501.
12. A. Newell. Creation frameworks: models of control structures. In *Visual Information Processing*, W. C. Pursue, Ed. Academic Press, New York, 1973, pp. 463–526.
13. V.R. Lesser, and D.D.Corkill. Practically exact, helpful distributive frameworks. *IEEE Transactions on Man, Systems and Cybernetics*, SMC-11, 1981, pp. 81–96.
14. H. A. Simon. The engineering of multifaceted nature. *Proceedings of the American Philosophical Society*, 106, 1962, pp. 467–487. (Likewise in Simon 1968.)
15. S.C. Park, and G.N. Wang. 3D range image simplification. In *International Conference on Digital Image Processing*, IEEE, 2009, pp. 270–274. doi:10.1109/ICDIP.2009.

Index